AF872355

# Sustainable Development Materials Based on Electrochemical Energy Storage

This book provides a comprehensive review of recent advancements in the electrochemical energy storage and conversion of two-dimensional (2D) nanomaterials and their composites, including their synthesis and associated challenges. It covers the precise design and construction of local atomic and electronic structures at the interface of catalysts, the interaction mechanism, and the regulation of catalyst solid surface interface properties under environmental and external field influences.

This book:

- Highlights the importance of designing and developing efficient electrocatalysts to accelerate the oxygen reduction reaction.
- Discusses recent breakthroughs in 2D-based oxygen evolution reaction electrocatalysts for designing 2D-based composite oxygen evolution reaction catalysts.
- Covers the challenges and opportunities faced by 2D-based nanomaterials in electrochemical carbon dioxide conversion.
- Provides a comprehensive review of recent advancements in the electrochemical energy storage and conversion of 2D nanomaterials and their composites, including their synthesis and associated challenges.
- Explains the aspects of supercapacitors, batteries, electrocatalytic reactions, green hydrogen production, and the conversion of carbon dioxide into solar fuels.

It is written primarily for senior undergraduates, graduate students, and academic researchers in chemical engineering, materials science, and chemistry.

## Artificial Intelligence for Sustainability

*S. Velliangiri and P. Karthikeyan*

Artificial Intelligence is used to process and analyze massive amounts of data and is used in a variety of industries, including finance, renewable energy, environmental health, and transportation. Its applications can process high-dimensional data in extensive data sets and analyze them with goals for predictions, recognition, and interaction challenges for more sustainable environmental protection in climate change, energy transformation, and pollution. This series discusses intelligent computing design and implements methodological and algorithmic long-term solutions by addressing the sustainability problems. It showcases sustainable technologies including the internet of things (IoT), cloud computing, blockchain, and machine learning. Aimed at graduate students, academic researchers, and professionals, the proposed series will focus on key topics including drone data analytics in aerial computing, deep learning techniques for Industry 4.0, autonomous robots in UAV, and applications of computer vision and natural language processing.

**Applications of Blockchain and Artificial Intelligence in Finance and Governance**
*A M Viswa Bharathy, Dac-Nhuong Le, and P. Karthikeyan*

**Sustainable Farming through Machine Learning: Enhancing Productivity and Efficiency**
*Suneeta Satpathy, Bijay Paikray, Ming Yang, and Arun Balakrishnan*

**Artificial Intelligence and Sustainable Networks: Innovations Beyond 5G**
*Edited by R. Krishnamoorthy, Kazuaki Tanaka, and Mohamed S. Soliman*

**Artificial Intelligence for Sustainable Supply Chain Management**
*Edited by N. Nasurudeen Ahamed, P. Karthikeyan, and Polinpapilinho F. Katina*

**Sustainable Development Materials Based on Electrochemical Energy Storage**
*Edited by Karthik Kannan, Yong-Song Chen, and Yuan-Yao Li*

# Sustainable Development Materials Based on Electrochemical Energy Storage

Edited by

Karthik Kannan

Yong-Song Chen

Yuan-Yao Li

CRC Press
Taylor & Francis Group
Boca Raton London New York

CRC Press is an imprint of the
Taylor & Francis Group, an **informa** business

Designed cover image: Shutterstock

First edition published 2027
by CRC Press
2385 NW Executive Center Drive, Suite 320, Boca Raton FL 33431

and by CRC Press
4 Park Square, Milton Park, Abingdon, Oxon, OX14 4RN

*CRC Press is an imprint of Taylor & Francis Group, LLC*

© 2027 selection and editorial matter, Karthik Kannan, Yong-Song Chen and Yuan-Yao Li; individual chapters, the contributors

Reasonable efforts have been made to publish reliable data and information, but the author and publisher cannot assume responsibility for the validity of all materials or the consequences of their use. The authors and publishers have attempted to trace the copyright holders of all material reproduced in this publication and apologize to copyright holders if permission to publish in this form has not been obtained. If any copyright material has not been acknowledged please write and let us know so we may rectify in any future reprint.

Except as permitted under U.S. Copyright Law, no part of this book may be reprinted, reproduced, transmitted, or utilized in any form by any electronic, mechanical, or other means, now known or hereafter invented, including photocopying, microfilming, and recording, or in any information storage or retrieval system, without written permission from the publishers.

For permission to photocopy or use material electronically from this work, access www.copyright.com or contact the Copyright Clearance Center, Inc. (CCC), 222 Rosewood Drive, Danvers, MA 01923, 978-750-8400. For works that are not available on CCC please contact mpkbookspermissions@tandf.co.uk

For Product Safety Concerns and Information please contact our EU representative GPSR@taylorandfrancis.com. Taylor & Francis Verlag GmbH, Kaufingerstraße 24, 80331 München, Germany.

*Trademark notice*: Product or corporate names may be trademarks or registered trademarks and are used only for identification and explanation without intent to infringe.

ISBN: 978-1-032-95411-0 (hbk)
ISBN: 978-1-032-95413-4 (pbk)
ISBN: 978-1-003-58474-2 (ebk)

DOI: 10.1201/9781003584742

Typeset in Sabon
by SPi Technologies India Pvt Ltd (Straive)

This book is dedicated to the cherished memory of my beloved friend, Dr. S. Karthi, who passed away on 7 August 2020. His friendship, integrity, and encouragement were a constant source of strength and inspiration. May this work serve as a lasting tribute to his memory and the profound impact he had on my life.

Dr. Karthik Kannan

# Contents

# About the editors

**Karthik Kannan** is an assistant research fellow (Global Research Assistant Professor) in the Department of Mechanical Engineering at National Chung Cheng University, Taiwan. He also serves as an honorary research fellow at Shinawatra University, Thailand, and as a distinguished adjunct professor at Karpagam Academy of Higher Education, India, and Saveetha Institute of Technical and Medical Sciences, India. He is a recipient of the Taiwan Employment Gold Card (Science and Technology) approved by the National Science and Technology Council and was awarded the prestigious Brain Pool Fellowship from the National Research Foundation of Korea. His international research experience spans Israel, Australia, and Qatar. Dr. Kannan recently secured competitive research funding of NTD 1.15 million from the National Science and Technology Council for his work on green hydrogen production via advanced electrocatalysis. He has published over 190 peer-reviewed articles with more than 9,500 citations (h-index: 64) and serves on the editorial boards of several international journals. He serves on the editorial boards of several international journals and has edited one Royal Society of Chemistry (RSC)-sponsored book and three IGI Global-sponsored books in the fields of sustainable energy, electrochemical materials, and advanced technologies. In recognition of his outstanding contributions to scholarly publishing, he was named among the MDPI Top 1000 Reviewers for 2025, selected from a global community of more than 210,000 reviewers for his excellence and consistency in peer review. He has also been recognized among Stanford University's World's Top 2% Scientists for six consecutive years (2020–2025).

**Yong-Song Chen** is a professor and chair in the Department of Mechanical Engineering at National Chung Cheng University, Taiwan. He earned his Ph.D. in Mechanical Engineering from the University of Michigan, Ann Arbor, USA, where his research focused on modeling water transport in proton exchange membrane fuel cells. He also holds M.S. and B.S. degrees from National Taiwan University, Taiwan, with specializations in materials science and mechanical engineering. His research interests include flow batteries and fuel cells, particularly for applications in energy storage systems, unmanned aerial vehicles, and unmanned underwater vehicles. He has extensive experience in electrochemical energy systems and advanced materials, with a strong focus on improving performance and efficiency. Prior to his current position, he served as a Specialist in the Materials and Electro-Optics Research Division at Chung-Shan Institute of Science and Technology, Taiwan. At National Chung Cheng University, he has held academic positions as Assistant Professor, Associate Professor, and currently Professor. Prof. Chen has received several recognitions, including the Excellent Teaching Award (2021), Excellent Mentor Awards (2017, 2020), and Young Scholar Awards (2017), reflecting his contributions to both research and education.

**Yuan-Yao Li** is a distinguished professor at the National Chung Cheng University, Taiwan. He earned his Ph.D. from the University of Bath, UK, in 1998 and later conducted postdoctoral research at the University of Tokyo from 1998 to 2001. In 2018–2019, he served as a visiting scholar at Stanford University. His research focuses on carbon nanomaterials, electrocatalysts, metal-air batteries, metal-ion batteries, and fuel cells.

# Contributors

**Kaushlendra Agrahari**
Graduate Institute of Photonics
National Changhua University of Education
Changhua, Taiwan

**Akram AlFantazi**
Department of Chemical Engineering, Khalifa University
Abu Dhabi, United Arab Emirates

**Mariappan Anandkumar**
High-Entropy Materials Research Laboratory
South Ural State University
Chelyabinsk, Russia

**R. R. Awasthi**
Faculty of Engineering and Technology
Khwaja Moinuddin Chishti Language University
Lucknow, India

**Anandha Babu**
Department of Physics
Bannari Amman Institute of Technology
Sathyamangalam, India

**Vanshika Bhardwaj**
Department of Physics
Chaudhary Charan Singh University
Meerut, India

**G. Bhavani**
Department of Physics
PSG Institute of Technology and Applied Research
Coimbatore, India

**Yong-Song Chen**
Department of Mechanical Engineering and Advanced Institute of Manufacturing with High-tech Innovations
National Chung Cheng University
Chiayi County, Taiwan

**Jyoti Prakash Das**
Nanomaterials & System Lab
Major of Mechatronics Engineering, Faculty of Applied Energy System
Jeju National University
Jeju, Republic of Korea

**Deepa**
Department of Chemistry
Prathyusha Engineering College
Tiruvallur, India

**Nitika Devi**
Electrochemical Power Sources Division
CSIR-Central Electrochemical Research Institute
Karaikudi, India

**Durvesh Gautam**
Department of Physics
Chaudhary Charan Singh University
Meerut, India

**Krishnamoorthy Gurushankar**
Department of Physics,
School of Computational and Physical Sciences,
Kristu Jayanti (Deemed to be University), Bengaluru, India.

**Kim Sang Jae**
Nanomaterials & System Lab, Major of Mechatronics Engineering, Faculty of Applied Energy System
Jeju National University
Jeju, Republic of Korea

**G. Kaladevi**
Department of Chemistry, Faculty of Engineering and Technology
SRM Institute of Science and Technology
Chennai, India

**Karthik Kannan**
Department of Physics, Centre for Energy and Environment
Karpagam Academy of Higher Education
Coimbatore, India

and

Department of Mechanical Engineering and Advanced Institute of Manufacturing with High-tech Innovations
National Chung Cheng University
Chiayi County, Taiwan

**P. K. Kannan**
Department of Physics
PSG Institute of Technology and Applied Research
Coimbatore, India

**K. Karuppasamy**
Department of Chemical Engineering
Khalifa University
Abu Dhabi, United Arab Emirates

**Manikandan Krishnamurthy**
Department of Chemistry, Faculty of Engineering and Technology
SRM Institute of Science and Technology
Chennai, India

**Vinotha Krishnasamy**
Research Institute of Sciences and Engineering
University of Sharjah
Sharjah, United Arab Emirates

**Aman Kumar**
Department of Physics, K.V. Subharti College of Science
Swami Vivekanand Subharti University
Meerut, India

**Dev Kumar**
Graduate Institute of Photonics
National Changhua University of Education
Changhua, Taiwan

**Vipin Kumar**
Department of Physics
CCS University Campus Meerut
Meerut, India

**Kamana K. Mishra**
Department of Physics
Indian Institute of Technology (Banaras Hindu University)
Varanasi, India

**Lakshmana Reddy Nagappagari**
Normandie Université
ENSICAEN, UNICAEN, CNRS, LaboratoireCatalyse et Spectrochimie
Caen, France

**Valarmathi Narayanan**
Department of Chemistry, School of Advanced Science (SAS)
Vellore Institute of Technology (VIT)
Vellore, India

**Vivek Kumar Nautiyal**
Department of Physics
Chaudhary Charan Singh University
Meerut, India

**Hemalatha Parangusan**
PG and Research Department of Physics
Mannar Thirumalai Naicker College
Madurai, India

**Sundara Venkatesh Perumalsamy**
Nanomaterials Laboratory, Department of Physics
Sri. S. Ramasamy Naidu Memorial College
Sattur, India

**Mahadevi Pichandi**
Department of Science and Humanities
Velammal Institute of Technology
Chennai, India

**Deepalekshmi Ponnamma**
Centre for Advanced Materials
Qatar University
Doha, Qatar

**E. Praveen**
Department of Basic Engineering (Physics Branch), 126
Government Polytechnic College
Kottur, India

**Swati Rani**
Department of Physics
Chaudhary Charan Singh University
Meerut, India

**V. Navakoteswara Rao**
Nanomaterials & System Lab, Major of Mechatronics Engineering, Faculty of Applied Energy System
Jeju National University
Jeju, Republic of Korea

**A. Shalini**
Department of Chemistry, Faculty of Engineering and Technology
SRM Institute of Science and Technology
Chennai, India

**Uma Sharma**
Department of Physics
Indian Institute of Technology (Banaras Hindu University)
Varanasi, India

**Ekta Singh**
Department of Physics and Material Science and Engineering
Jaypee Institute of Information Technology
Noida, India

**Prabhakar Singh**
Department of Physics
Indian Institute of Technology (Banaras Hindu University)
Varanasi, India

**Pankaj Kumar Tripathi**
Department of Physics
Sharda University
Greater Noida, India

**Teerthraj Verma**
Department of Radiotherapy
King George Medical University
Lucknow, India

**Yu-Wu Wang**
Graduate Institute of Photonics
National Changhua University of Education
Changhua, Taiwan

**Anil K. Yadav**
Department of Physics
Chaudhary Charan Singh University
Meerut, India

# Acknowledgments

The editor, **Dr. Karthik Kannan**, gratefully acknowledges the financial support provided by the **National Science and Technology Council (NSTC), Taiwan**, under Project No. **NSTC 115-2222-E-194-002**, which contributed to the completion of this book. This work was financially/partially supported by the Advanced Institute of Manufacturing with High-tech Innovations (AIM-HI) from The Featured Areas Research Center Program within the framework of the Higher Education Sprout Project, by the Ministry of Education (MOE), Taiwan. The editor also sincerely thanks all contributing authors, reviewers, and the editorial staff at CRC Press/Taylor & Francis for their valuable support throughout the publication process.

Chapter 1

# A fundamental approach towards sustainable development of electrochemical energy storage and conversion technologies

*Hemalatha Parangusan, K. Karuppasamy, Deepalekshmi Ponnamma, Sundara Venkatesh Perumalsamy, Anandha Babu, and Akram AlFantazi*

## 1.1 INTRODUCTION

Energy is essential to life and economic development, impacting billions of people worldwide. In today's world, electricity is particularly essential for sustaining transportation, industry, telecommunications, agriculture, and general economic growth (Tari & Patil, 2017; Kannan & Tari, 2025). The International Energy Agency predicts that by 2050, the world's energy consumption is expected to have doubled. Both industrialized and emerging nations are experiencing an increase in demand, which emphasizes the need for renewable energy sources to ensure constant supplies of electricity (Mohamed et al., 2024a). Devices that convert and store energy electrochemically, as well as the reactions between their individual electrodes, are significant environmental issues worldwide. To achieve sustainable energy development and reduce our dependence on fossil fuels, there are several topics of interest, including electrochemical capacitors, rechargeable batteries, electrolyzers, and fuel cells (FCs) (Bello et al., 2023; Zhao et al., 2021; Tari et al., 2024; Gurushankar et al., 2024).

In addition, among other objectives, the Sustainable Development Goals (SDGs) of the United Nations have included climate action and the source of more inexpensive energy for everyone (Siddiqui et al. 2022; Tari et al., 2024; Yalçın et al., 2022). Therefore, the sectors of energy research, especially research and industry, have placed a strong emphasis on combining these objectives (Iqbal et al., 2019). In this context, various effective and innovative materials have been investigated for use in photovoltaic (PV) and electrochemical applications. The basic components of many electrochemical systems are electrode materials, electrolytes, and membranes (Azeem et al., 2024). The performance of the materials in these systems is greatly influenced by a wide range of properties. One of the most important characteristics of materials is their electrical conductivity, which determines how easily electrons can move during an electrochemical reaction. Additionally, specific surface area is important since it expands the number of active sites

DOI: 10.1201/9781003584742-1 

that are available for electrochemical reactions (Abdel Maksoud et al., 2021). For electrochemical devices to remain in use, electrode materials need to be chemically stable. For instance, metal oxides and nitrides have a long lifespan and can withstand a variety of operating conditions without losing their original shape or functioning (Aissani et al., 2022). Redox activity is essential for energy storage devices because it permits ongoing reversible oxidation and reduction reactions. High-performance energy systems require rapid electrochemical reactions, which are made possible by catalysts and materials with different crystal structures. Additionally, problems like cost-effectiveness and environmental impact have encouraged researchers to investigate sustainable, eco-friendly alternatives. As a result, electrode materials have a variety of characteristics that together influence the performance, effectiveness, and environmental impact of electrochemical devices. Finding new materials and learning more about their characteristics are necessary for advancements in energy storage, conversion, and sensing technologies (Iro et al., 2016; Zhu et al., 2023).

This chapter thoroughly examines the ways in which electrochemical energy conversion and storage technologies have the potential to transform the energy landscape. In order to describe these two cutting-edge technologies' types, uses, and reliability as electrochemical energy conversion and storage solutions, we first give a basic environmental and technical background. We have discussed new developments and strategic moves that might promote the advancement of these advanced technologies. This chapter aims to present new ideas on electrochemical energy conversion and storage technologies that might guide investments and policies to accelerate the shift to a cleaner and greener future.

### 1.1.1 Energy storage devices

Devices for storing excess energy produced by renewable or non-renewable sources are particularly important. Furthermore, it is their responsibility to supply this energy in a way that minimizes losses and costs while being dependable, safe, and efficient (Gabbar et al., 2021). As a result, several types of energy storage devices, each with unique properties and natures, contribute to the stability and security of power systems. The four primary categories of energy storage systems are mechanical, electrochemical, electrical, and chemical (Emad et al., 2021). In this context, supercapacitors (SCs) are regarded as mechanical and electrical kinds of energy storage devices, whereas batteries are regarded as electrochemical devices. For a more dependable functioning, energy storage devices must be combined due to technological difficulties in power systems (Dashtdar et al., 2022). Because of this, integrating SCs with batteries will increase their power and energy density when they are charged or discharged.

The SC and capacitor are electrostatic systems. A dielectric layer separates the electric capacitor's two metal plates, known as electrodes. The metallic electrodes' surface stores the energy. This kind has a very short energy storage lifespan (Al Shaqsi et al., 2020). The dielectric material, plate size, and spacing between plates all influence the capacitor's energy capacity (Guney & Tepe, 2017). Because of their long life cycle and ability to recharge instantly, the capacitors are appropriate for power applications on a small scale. In the case of large-scale applications, the cost will be high (Babu et al., 2020). In order to overcome the aforementioned issues with capacitors, the Japanese business Nippon Electric business patented SCs, also known as ultracapacitors, in 1975. SCs are superior to regular capacitors in terms of output power, charge density, life likely, times life cycle, and self-discharge. SCs are therefore appropriate for applications that need immediate high-output power, such as permanent magnet synchronous generators and hybrid vehicle engine turning applications. Capacitors and SCs work on the same principles. But in SCs, the electrolytic physical barrier composed of activated carbon (AC), which serves as a dielectric and leads to the SC's high energy density and high specific area, promotes ionic conduction.

Electrochemistry converts chemical energy into electrical energy and vice versa (Guney & Tepe, 2017). Due to their low maintenance requirements, high efficiency (70–80%), superior electrical energy storage, and ease of building, electrochemical conversion and energy storage systems (ECESS) are regarded as a significant candidate in energy storage applications. However, main problems include expensive maintenance costs in large-scale facilities, as well as the fact that their durability depends on the depth of discharge and relatively short cycling intervals (K.P. Kampouris, et al., 2020). The ECESS are flow battery, sodium-sulfur, nickel, lead-acid, and lithium batteries. The lead-acid battery is the most often used rechargeable battery. They consist of a positive electrode made of lead dioxide and a negative electrode made of lead dioxide immersed in a sulfuric acid electrolyte mitigated solution. Lead-acid batteries have very short response times, low regular self-discharge rates, relatively high efficiencies, and cost-effective investment requirements. They can also be stabilized for large-scale applications (Dunn et al., 2011). The lithium-ion battery's low weight, excellent performance, high cycle efficiency, and quick response time make it ideal for portable devices like laptops and cell phones. Graphitic carbon serves as the anode, and lithium metal oxide as the cathode in a Li-ion battery. Current research on Li-ion batteries focuses on employing nanoscale materials to enhance innovative electrolyte solutions and electrode materials, increasing the battery's specific energy and power capability. In addition to having relatively high efficiency and low maintenance requirements, nickel-cadmium batteries are up to ten times more expensive than Li-ion batteries. For renewable energy systems, NiCd batteries can be used to store huge amounts of energy. The technology employed in the manufacturing of NiCd batteries determines their storage efficiency. Compared to other battery types, the sodium-sulfur battery can have a higher rated capacity. So, it is designed for applications requiring high power energy storage (X. Luo et al., 2015). The main advantages of this storage are its high energy densities, high energy efficiency, lack of self-discharge, cheap maintenance and material costs, non-toxic materials, and recyclable materials. There are many articles discussing challenges and limitations associated with batteries. Batteries offer the benefits of lowering costs, avoiding voltage fluctuations, optimizing PV utilization, and reducing losses in a variety of systems, whether they are grid-connected or isolated. There are other optimization strategies that can be applied to get these advantages. One of the first energy storage systems with numerous uses is the mechanical energy storage system (MES). It is easily transformed into and out of other forms of energy and may be stored for extended periods of time. Flywheel energy storage frameworks, which store kinetic energy, air compressed energy storage systems, which store potential energy, and pumped hydro storage are the three forms of MESs that are proposed. Devices that store thermal energy, store warm or cold for later use. These systems are made up of a control system, an energy transfer medium, and a heat storage tank. By using these technologies, energy consumption is decreased and the short-term nature of renewable energy sources is resolved (F. Nadeem et al., 2019). The chemical bonds that bind molecules and atoms together store chemical energy. One type of chemical energy storage system is FCs. FCs generate electricity from fuel input from an external source. This fuel can be either methanol, hydrogen, or it can be indirectly produced by turning hydrocarbon gases, natural gas, ammonia, or ethanol into hydrogen. Currently, the majority of FCs are still in their development stage and are only utilized in a few specific fields, such as aerospace and large-scale grid backup generation. A reservoir is used to store the hydrogen that is generated; an electrolyzer converts electrical energy into hydrogen; and a conversion system, such as FCs, converts chemical energy into electrical form. Water electrolysis is used to separate water into its essential components (oxygen and hydrogen) using excess power from renewable energy sources, and the resulting water is then kept in tanks (S. Verma, et al., 2021). Hydrogen technology is an ideal candidate when energy storage capacity is more important than efficiency because it is inexpensive compared to other energy storage technologies, but its efficiency is not very high (Yu et al., 2022).

### 1.1.2 Energy conversion devices

There is less environmental dependence on solar energy, which is clean, renewable, and abundant. PV technology is expected to account for around one-third of all new power generation capacity globally by 2030, according to the "Global Renewable Energy Market Outlook." (Si et al., 2024). A number of new technologies related to sustainable and environmentally friendly solar energy conversion are extensively investigated. Another sustainable method to address the growing demand for energy is solar water splitting, which offers an environmentally friendly way to produce chemical fuels like hydrogen. Any conventional water-splitting photocatalytic/photoelectrochemical system that has been documented so far usually needs a liquid electrolyte (Walter et al., 2010). Since continuous photoelectrochemical reactions require extra water input, which limits their large-scale application, it has proven impossible to implement a zero-water input system for effective water splitting. Additionally, solar water splitting usually requires two electrodes. One for the evolution of $H_2$ (hydrogen) and another for that of $O_2$ (oxygen) evolution. A hybrid single-nanoparticle-based photocatalyst was used by Yang et al. (2019) to demonstrate the simultaneous evolution of $H_2$ and $O_2$ without the need for manual water input or external bias. In addition to solar water splitting, harvesting mechanical energy connected to water is another sustainable and inexpensive method to develop. The sustainable energy sector has seen tremendous developments and as a result of the recent increase in wearable electronics and the growing necessity to abandon traditional energy production methods like burning fossil fuels. Since Zhong Lin Wang and his colleagues developed the Triboelectric nanogenerator (TENG) in 2012, mechanical energy, which has been suggested to be the most abundant of the many energy sources (Fan et al., 2012). Machine learning, Internet of Things, and human-machine interaction all require cleaner energy producing methods and renewable energy sources. TENG converts potentially wasted mechanical energy into electrical power by using the triboelectric effect and electrostatic induction (Niu & Wang, 2015).

Hybrid energy harvesters, which scavenge several green energy sources like solar, water, and wind energy, have been developed recently by numerous researchers with a synergistic coupling effect. It is usually expected that hybrid energy harvesters will become more productive, more stable, and more versatile. As previously stated, one of the advanced technologies is solar energy harvesting. However, solar cells always function properly in a location with full sun. At night or on cloudy, rainy days, solar cell panel production is always significantly reduced or diminished (Zheng et al., 2015). Additionally, a hybrid energy harvester was developed by some researchers with basic structure and various modes based on TENGs. In order to simultaneously scavenge vibrational energy and solar energy, Pu et al. integrated fabric TENG with fiber-shaped DSSC to achieve textile-based energy harvesting. Shao et al. developed a multifunctional power unit by combining solar cells, electromagnetic generators, and TENGs to collect both solar and water motion energy (Pu et al., 2016).

### 1.1.3 Materials for storing and converting energy

Energy conversion and storage device operation depend largely on energy materials. Due to their restricted accessible active regions and moderate diffusion kinetics, these materials frequently experience challenges. The mechanical and electrical properties of new energy materials are essential for improving the performance of energy devices. Furthermore, these materials' overall behavior is greatly influenced by the interaction between their bulk and surface properties. Due to the technological revolution of the 1980s, rechargeable batteries with higher capacities, smaller sizes, and lighter weights were required for the specified uses (Development of the Lithium-Ion Battery and Recent Technological Trends, 2014). Energy storage materials are a wide range of compounds that are designed to store energy in several forms, such as mechanical, electrical, or chemical energy. The electrodes provide the primary contribution to energy storage performance. Batteries, SCs, and traditional capacitors, commonly referred to as ultracapacitors, are the most common energy storage technologies utilized in a variety of applications. The main difference between a battery and a SC is their capacity; batteries can store more energy per unit weight, commonly a few hundred Wh $Kg^{-1}$, because they have an increased density of energy. As seen in Figure 1.1, SCs provide a higher power density (they can release more power within a short period of time, typically tens of thousands of W $Kg^{-1}$) (Winter & Brodd, 2004). SCs greatly increase their capacitance by utilizing porous carbon electrodes and electrolytes to create electric double layers or facilitate quick redox processes, in contrast to traditional capacitors that store energy through a dielectric between metal plates. SCs can achieve significantly larger energy storage, fast charge-discharge speeds, and remarkably long cycle lives, often surpassing one million cycles, mainly due to this structural improvement. Although they function at lower voltages per cell than conventional capacitors, their power density and dependability make them perfect for uses like advanced energy storage systems, backup power, and regenerative braking that call for rapid energy bursts and high durability.

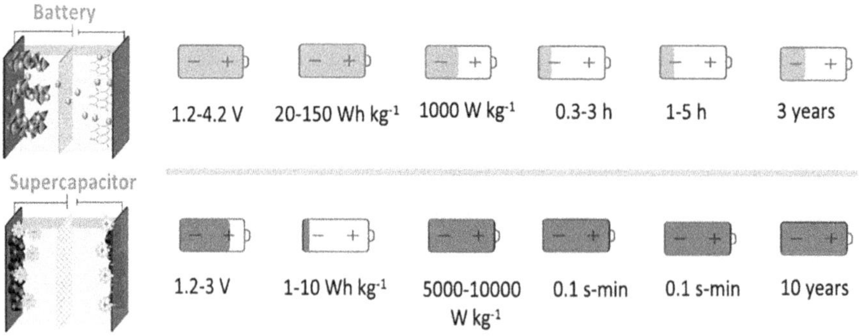


*Figure 1.1* A comparison between batteries and supercapacitors in terms of performance.

**(Copyrights @ 2023, The American Chemical Society-open access unrestricted use, Flores-Diaz et al., 2023.)**

When it comes to material characterization, the active material can be used as the working electrode in an electrochemical cell with a counter and reference electrode, providing analysis of both SC and batteries materials in the traditional three-electrode configuration. The most common active component in commercial SCs is AC, which has capacitance values between 100 and 400 F $g^{-1}$ (Fischer et al., 2021). Higher capacitance is offered by a variety of substitute materials for AC, including carbon nanotubes, graphene (G), (reduced) graphene oxides, and chalcogenides like $RuO_2$, $MnO_2$, $MoS_2$, and transition metal oxides. However, these materials' cost, simplicity of synthesis, and scale-up present major challenges to their application (Wang et al., 2012). In general, carbon-based materials have a large effective surface area, which leads to high power and capacitance.

Metal oxides have received a lot of attention as electrode materials for electrochemical energy storage devices due to their high theoretical capacitance, changeable oxidation states, and strong pseudocapacitive behavior. Materials including manganese oxide ($MnO_2$), nickel oxide (NiO), cobalt oxide ($Co_3O_4$), and ruthenium oxide ($RuO_2$) have redox activity that allows for faradaic charge storage, resulting in higher energy densities than electrostatic double-layer capacitors. Their structural plasticity enables the creation of nanostructures (nanowires, nanosheets, and hollow spheres) that improve ion diffusion and surface accessibility (Zang et al., 2008). Although some metal oxides have low intrinsic conductivity and structural instability during cycling, combining them with conductive materials (such as carbon nanotubes or graphene) has been proven to dramatically improve performance. Thus, metal oxides play an important role in SC technologies by bridging the gap between high energy storage and long cycle life.

Metal nitrides are developing as extremely attractive electrode materials due to their strong electrical conductivity, chemical stability, and great mechanical characteristics. Titanium nitride (TiN), vanadium nitride (VN), and molybdenum nitride ($Mo_2N$) have metallic-like conductivity, allowing for fast electron transport and effective charge storage, particularly in high-power applications. Unlike many metal oxides, nitrides frequently exhibit improved structural stability with repeated electrochemical cycling, resulting in longer device life. Furthermore, their durable surfaces are corrosion-resistant and allow for steady contact with electrolytes. Nanostructured metal nitrides' innate toughness and large surface area make them ideal for high-rate charge-discharge processes. Their exceptional conductivity and endurance make them ideal materials for next-generation high-power SCs and hybrid capacitive systems. Conducting polymers can also exhibit pseudocapacitive behavior, including poly (3, 4-ethylenedioxythiophene), poly (pyrrole), and poly (aniline) (Huang et al., 2021). Among the many advantages of conducting polymers are their quick charge-discharge kinetics, great chemical stability, and high electrical conductivity when doped. The system's supercapacitance is limited by the shrinking and swelling of polymers during extended discharge/charge cycles, which makes the commercialization of polymer-based SCs challenging. SCs with solid or gel electrolytes are typically designed to improve the devices' mechanical strength, flexibility, and cycling stability (Yang et al., 2022). The membrane is an essential part of a high-performance SC because it keeps the electrodes apart, preventing the stored energies on each electrode from recombining and causing short circuits. Membranes need to be thermally stable, nonconductive, and provide little barrier to ion diffusion in the electrolyte. One of the essential components in enhancing LIB performance is the synthesis of high-quality anode material. As an electrode in the LIB, nanomaterial with a large surface area, strong chemical stability, and conductivity is needed (Figure 1.2).

SCs can be categorized as hybrid capacitors, redox SCs, double-layer electrochemical capacitors, or pseudocapacitors based on the electrode materials. SCs and battery materials are combined in hybrid capacitors. Chen et al. (2016) developed a bimetallic zeolitic imidazolate of hierarchical $Co_3ZnC$/CNT incorporated carbon concave-polyhedrons doped with nitrogen.

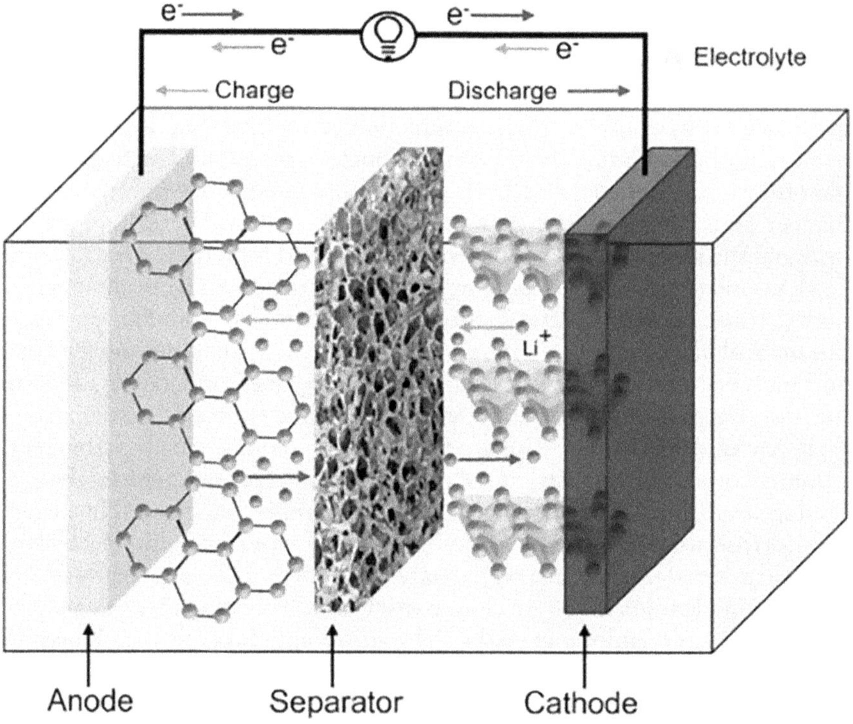


*Figure 1.2* Illustration of a typical lithium-ion battery.

**(Copyrights @ 2025, Elsevier, Bello et al., 2023.)**

The anode's long-term stability and quick electron transport are improved by the synergistic interaction between the metals and the structural components. In the LIB energy storage, the electrode has demonstrated approximately 800 $mAhg^{-1}$ reverse capacity at 0.5 $Ag^{-1}$, and 408 mAh $g^{-1}$ rate capacity at 5.0 A $g^{-1}$. $Ti_3C_2Tx$ MXene was found to have a high capacity for storing ammonium ions and to be an effective electrode material in further investigations. $Ti_3C_2Tx$ MXene has been successfully prepared by Feng et al. (2025) by employing zirconium hexafluoride to selectively decrease the aluminum layer from the $Ti_3AlC_2$ MAX phase. They achieved remarkable energy and power densities by using $Ti_3C_2Tx$ MXene as a negative electrode and AC as a positive electrode. Furthermore, after 10,000 cycles, the capacitor retained 86% of its capacitance.

Layer-structured rare earth nanoparticles, among many other materials, are expected to play a bigger role in energy storage and conversion devices due to their special qualities, such as high reversibility and reactivity (Yan et al., 2022). Rare earth element-based layer structures or composites increase battery performance, SC capacitance and charge transfer, and serve as efficient catalysts, which makes them perfect for energy storage and conversion devices. Abid et al. (2023) reported that $EuZrSe_3$ in a metal-organic framework (MOF) obtained a power density of 658.8 W/Kg and energy density of 97 Wh/Kg because of its larger surface area. In a new method, Mondal et al. (2024) found that a 39% pseudocapacitive contribution allowed Gd-impregnated $\alpha$-$MnO_2$ to achieve 92.4% charge retention and a specific capacitance of 798 F/g at 1 A/g after 5000 cycles.

Vilvanatha Prabu et al. (2025) investigated energy conversion and storage applications using a polymer polyaniline (PANI)/$AgBiSe_2$ polymer nanocomposite. The composite enhances energy storage and conversion efficiency by combining a thermoelectric material $AgBiSe_2$ with a conductive organic PANI. They investigated the electrochemical properties of the developed PANI/$AgBiSe_2$ electrode using cyclic voltammetry (CV). With a constant potential window between −0.70 and 0.50 V and scan speeds ranging from 10 to 50 mV/s, the size of the CV curves increased in direct proportion to the scan rate, indicating significant capacitive properties (Figure 1.3a). The $CsPbI_2Br_1$ perovskite hetero-system was developed by Jaffri et al. (2024) for sustainable energy storage and conversion capacity. Electrochemical Impedance Spectroscopy (EIS) electrochemical methods were used to evaluate charge storage capacity. As shown in the Nyquist curve in Figure 1.3b, the excellent electrochemical output is achieved possible by a lower resistance (Ippili et al., 2023). They found that the prepared electrode is an effective and sustainable material that may replace the expensive components that are typically used. The excellent electrical properties of the perovskite host and the outstanding stability characteristics of the lanthanide dopants are responsible for this increased activity (Kitajou et al., 2012).

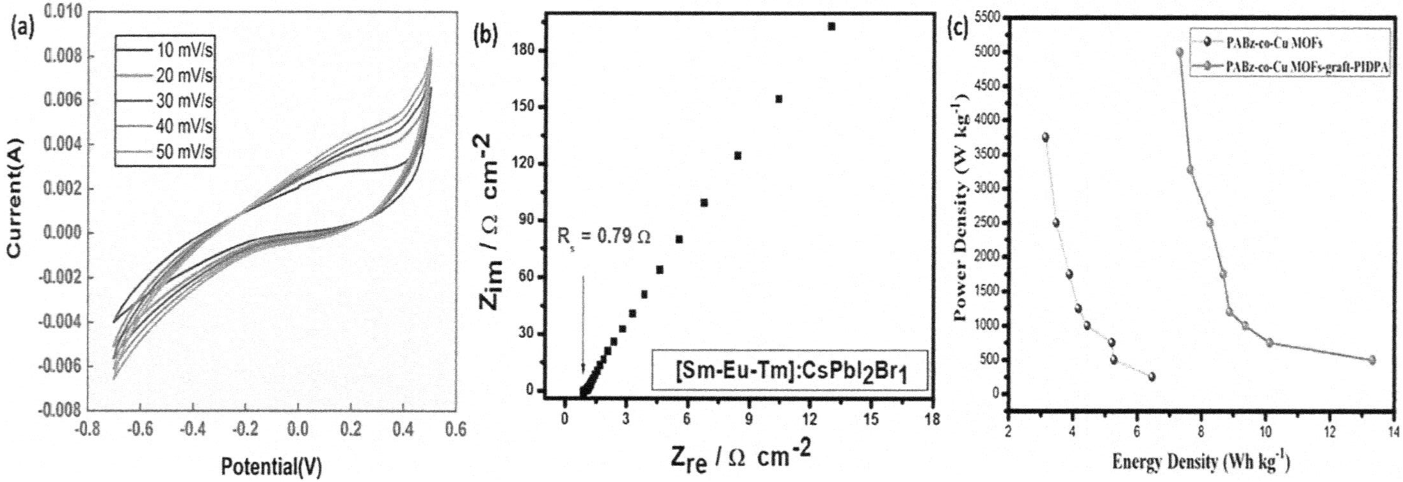


*Figure 1.3* (a) Cyclic voltammetry curves of PANI/$AgBiSe_2$. (Copyrights @ 2025, Elsevier, Vilvanatha Prabu et al., 2025.) (b) EIS impedance spectrum-derived Nyquist plot. (Copyrights @ 2024, Elsevier, Jaffri et al., 2024.) (c) Ragone plots of Pure and MOFs-graft PIDPA polymer. (Copyrights @ 2024, Elsevier, Murugan et al., 2024).

When assessing the real-world use of SCs, energy and power density numbers are typically essential factors. The results of the calculations for the maximum power and energy density of pure and cross-linked poly(imidazole-co-diphenylamine) (PIDPA) are shown in Figure 1.3c. According to the Figure 1.3c, the pure polymer produces values of 6.45 Wh $Kg^{-1}$ and 250 W $Kg^{-1}$, respectively, whereas the maximum energy and power densities are much higher at 13.33 Wh/$Kg^{-1}$ and 499.99 W $Kg^{-1}$. This study finally shows that the cross-linked polymer exhibits significant promise as a material for supercapacitive electrodes (Jaffri et al., 2024; Murugan et al., 2024; Vilvanatha Prabu et al., 2025).

Materials for energy conversion are essential for lowering energy use, increasing energy efficiency, and reducing the impact on the environment. We can promote a more sustainable and greener future by developing and using such materials, which will result in a healthier world and improved quality of life for future generations. Because of their high conductivity, advantageous optical characteristics, and adjustable band gaps, MXenes have become significant prospects for light-absorbing materials and transparent conductive electrodes in the field of solar cells (Yin et al., 2021). Recently, Wei et al. (2025) utilized perovskite solar cells with the 2D $Ti_3CNT_x$ MXene as a novel electron transport layer (ETL). They found that in order to enhance the perovskite layer, the ETL materials needed to have high photoelectric properties and appropriate energy levels.

The stability function of the selected lanthanide dopants in transforming the perovskite host is reflected in the exceptional performance of the developed solar cell functioning in completely ambient environments. As a result, the most effective photo-active rare earth doped perovskite absorber in these manufactured solar cells is capable of absorbing solar radiation and converting it into sustainable electrical energy (Figure 1.4). For example, using $LaNa_{0.06}Co_{0.06}Mn_{0.88}O_3$ and $Na_{0.56}Co_{0.17}Mn_{0.8}La_{0.03}O_2$, Xia et al. (2021) attempted to develop a conversion system with an higher efficiency and strength. The compounds' stability and durability were enhanced by the rare earth ions present in these materials, indicating their function as stabilizing agents. Reports indicate that lanthanide dopants contribute to improved photo-stability in addition to mechanical stability (He et al., 2021).

The development of clean and sustainable energy solutions depends largely on hydrogen production (Naik Shreyanka et al., 2022). Hydrogen, a high-energy and adaptable fuel, has enormous potential to power a variety of industries, transportation, and electricity production with little impact on the environment (Theerthagiri et al., 2020). More hydrogen can be developed effectively using cutting-edge techniques like electrolysis, steam methane reforming, biomass gasification, and others, providing an opportunity for lowering greenhouse gas emissions and promoting the shift to a more environmentally friendly energy future (Yu et al., 2022). Recent research is

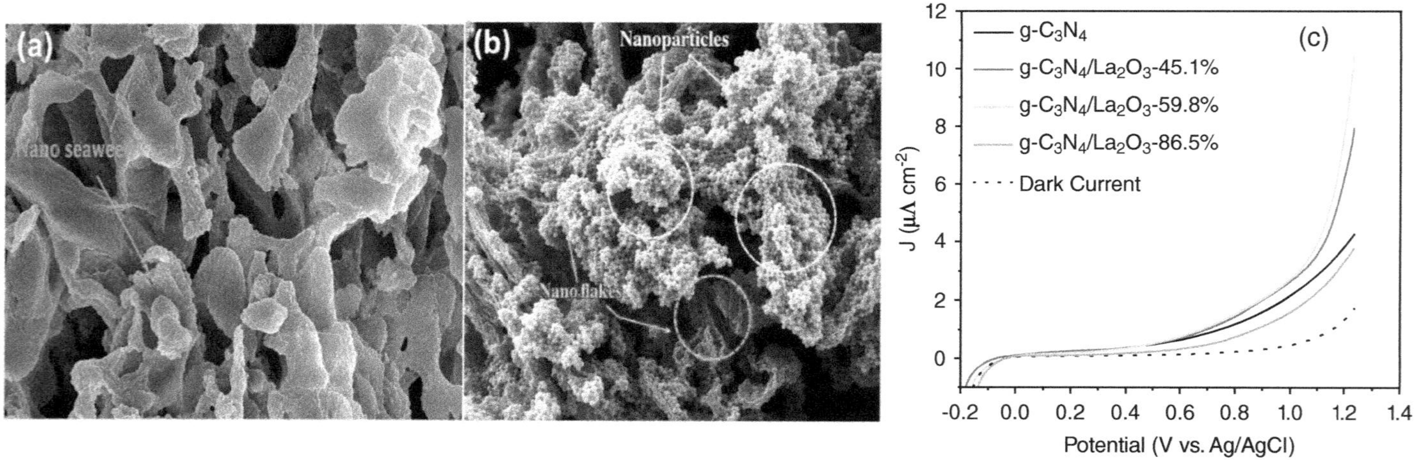


*Figure 1.4* (a, b) g-C3N4's composite morphology with nanoparticles and nanoflakes decorating it and (c) composite photoanode's LSV photocurrent response in 0.5 M $Na_2SO_4$.

**(Copyrights @ 2024, Elsevier, Mohamed et al., 2024a.)**

focused on the electrochemical conversion of small molecules, like methanol, water, formic acid, and ethanol, etc., that can produce electricity in FCs. Energy conversion and generation were significantly improved by electrocatalysts. However, developing large-scale hydrogen production from water splitting presents financial challenges. In terms of energy, atom, and conversion efficiency, combined electrolysis is highly valuable (Liu & Li, 2021; Sbei et al., 2021; Tackett et al., 2017). To date, the reactions have been produced using expensive metals like ruthenium oxide and iridium oxide as prospective electrocatalysts; however, the reactions' high cost and rare metal content have prevented their commercialization (Li et al., 2022). Therefore, Earth-abundant transition metal (TM)-based nanomaterials and their composites are used in hydrogen evolution reaction (HER) and oxygen evolution reaction (OER). Because of their very similar electronic structures, transition metal chalcogenides (TMCs) improve electron transfer between the carbide material and the metal, increasing the metal electrocatalyst's electrical conductivity, chemical stability, and intrinsic activity for selective electrocatalytic reactions (Gebre, 2022). Using the methanolic dispersion method, researchers developed a g-$C_3N_4$/$La_2O_3$ composite with a composition of roughly 59.8%, producing a photostable photoanode. This composite significantly improved surface area, light absorption, and catalytic sites. The improved performance resulted in an approximately threefold enhancement of photocurrent density, reaching about 10.16 μA $cm^{-2}$, in contrast to the control g-$C_3N_4$, which showed 4.56 μA $cm^{-2}$ at 1.23 V (Figure 1.4) (Mohamed et al., 2024b).

Nanogenerators, such as piezoelectric nanogenerators and TENGs, are a type of new energy-harvesting technology that is suggested to scavenge mechanical energy from the environment and convert it in a distributed method into electricity (Figure 1.5). (Wang & Song, 2006). With the benefits of a high output voltage, lightweight, small size, and easy fabrication, TENGs are made using the triboelectric effect and electrostatic inductive coupling, whereas piezoelectric nanogenerators (PENGs) are made using the piezoelectric effect and polarization to generate currents (Sriphan et al., 2020). The versatile hybrid nanogenerator can capture thermal and mechanical energy from the surrounding environment. A key component of any TENG fabrication process is material selection, and options will differ substantially based on the TENG's intended purpose, the amount of stress and strain it will be subjected to, and the material's function within the TENG device.

A piezoelectric-triboelectric nanogenerator (PTCNG) with two electrodes was developed by Qian et al. (Qian & Kang, 2018). The multifunctional layer (PDMS/ZnO NFs/3DGr) possesses both piezoelectric and tribological properties. A current loop will form between the lower and upper electrodes as long as the external force is exerted. After releasing the pressure, all

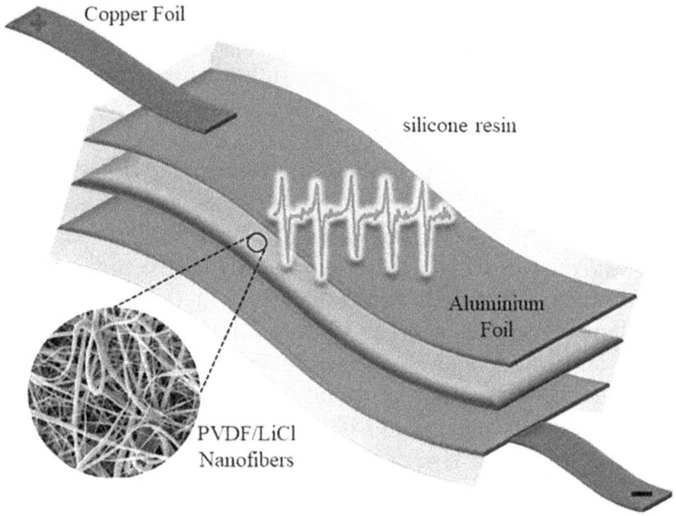


*Figure 1.5* Schematic structure of the nanogenerator device for energy harvesting applications.

**(Copyrights @ 2023, The American Chemical Society-open access unrestricted use.)**

charges will revert to their initial condition and achieve electrostatic stability. A three-electrode PTCNG was designed by Jung et al. (2015). Gold coating is deposited on both sides of the polyvinylidene difluoride (PVDF) (piezoelectric layer). The other electrode is made of aluminum friction layer. With an output power density of 4.44 $mWcm^{-2}$, a current density of 12 $\mu Acm^{-2}$, and an open-circuit output voltage of 370 V, the device exhibits an excellent output enhancement. The TENG component operates in the contact-separation mode due to its three-layer electrode design. Suo et al. (2016) developed a composite structure by combining $BaTiO_3$ and PDMS (Figure 1.6). The device demonstrated exceptional output performance. The incorporation of barium titanium oxide (BTO) was the primary factor influencing the piezoelectric output, and the fabrication of composite films with this material increased the polydimethylsiloxane's (PDMS's) charge density and dielectric constant. The advantages of PTCNG made with BTO material include its simplicity, affordability, and superior performance. A comparative overview of electrode materials for energy storage and conversion technologies is shown in Table 1.1.

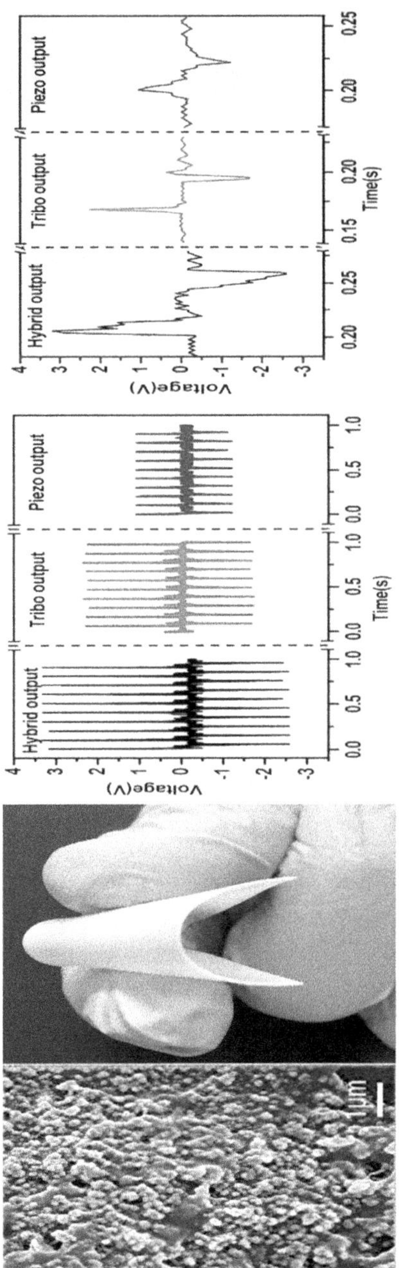


*Figure 1.6* (a) A cross-sectional SEM image of the BTO/PDMS composite film and a photograph of the final product (c) Output voltage in the three test modes (piezoelectric, triboelectric, and hybrid).

**(Copyrights @ 2016, American Chemical Society, Suo et al., 2016.)**

*Table 1.1* Comparative overview of electrode materials for energy storage and conversion technologies

| *Electrode material* | *Chemical composition/ structure* | *Application in energy storage* | *Application in energy conversion* | *Key properties* | *Sustainability aspects* | *Electrode material* |
|---|---|---|---|---|---|---|
| **Carbon-based Materials** | Graphene, CNTs, AC | SCs, LIBs, Sodium-ion batteries | Supports in FCs, electrocatalysts | High surface area, conductivity, stability | Abundant, renewable sources (biomass), recyclable | **Carbon-based Materials** |
| **Transition Metal Oxides** | $MnO_2$, NiO, $Co_3O_4$, $Fe_3O_4$ | Pseudocapacitors, LIBs, hybrid capacitors | Electrocatalysis in OER/ORR/HER | High pseudocapacitance, redox-active centers | Earth-abundant (Mn, Fe), moderate toxicity | **Transition Metal Oxides** |
| **Conductive Polymers** | Polyaniline, Polypyrrole | SCs, hybrid batteries | Electrocatalysts, sensing electrodes | Tunable conductivity, lightweight, flexibility | Synthesized from non-toxic monomers, biodegradable options | **Conductive Polymers** |
| **Metal Sulfides** | $MoS_2$, NiS, CoS, $FeS_2$ | Anodes in LIBs, SIBs, SCs | HER and OER electrocatalysts | Layered structure, high theoretical capacity, catalytic activity | Low-cost synthesis, some metals abundant (Fe, Ni) | **Metal Sulfides** |
| **Metal Phosphides** | $Ni_2P$, CoP, FeP | Anodes in batteries, hybrid capacitors | HER/OER catalysts | High conductivity, robust catalytic sites | Can be synthesized via green routes | **Metal Phosphides** |
| **Metal Nitrides/ Carbides** | TiN, VN, MXenes, $Mo_2C$ | SCs, high-rate batteries | HER, OER, ORR | High electrical conductivity, mechanical stability | Some MXenes can be derived from layered precursors sustainably | **Metal Nitrides/ Carbides** |

| | | | | | | |
|---|---|---|---|---|---|---|
| **Perovskite Oxides** | $LaNiO_3$, $SrCoO_3$, $Ba_{0.5}Sr_{0.5}Co_{0.8}Fe_{0.2}O_3$ | Not widely used in storage (emerging) | OER/ORR catalysts in FCs, metal–air cells | Flexible composition, high catalytic activity | Element substitution offers eco-friendly design | **Perovskite Oxides** |
| **Chalcogenides** | CoTe, $MoSe_2$, $WS_2$ | Battery anodes, SCs | HER, OER catalysts | Tunable electronic structure, layered morphology | Some chalcogenides (e.g., Te-based) require toxicity control | **Chalcogenides** |
| **Ni-based Alloys** | Ni-Fe, Ni-Co | Battery electrodes | OER in alkaline media | High stability and activity in alkaline conditions | Nickel is relatively abundant and recyclable | **Ni-based Alloys** |
| **Bio-derived Materials** | AC from coconut shell, banana peel | SCs | Enzyme electrodes in biofuel cells | Porous morphology, low cost | Renewable, low carbon footprint | **Bio-derived Materials** |

## 1.2 CONCLUSION

The latest advancements in energy technologies are summarized in this chapter. The development of green energy harvesters is becoming increasingly significant as the need for renewable energy around the world increases. A brief overview of the materials with different functional groups that can be produced is also provided. Additionally, the latest developments in energy storage and conversion technologies, including solar cells, batteries, FCs, SCs, and catalysis, are thoroughly examined. Developing hybridized generators, like PTCNG units, has emerged as a promising area of study for environmental and energy-harvesting systems. The hybrid device can be used for a variety of energy harvesting methods, including vibrational energy, wind flow, thermal and solar energy, and mechanical hand clapping. Because of its exceptional flexibility, biocompatibility, and low weight, the hybrid PENG-TENG can meet the needs of health and ambient signal monitoring as a wearable electronic device. It is essential to develop a high-performance nanogenerator for particular applications and promote it in innovative ways.

The future of electrochemical energy storage and conversion technologies lies in the development of low-cost, high-performance, and environmentally friendly materials capable of meeting the world's growing demand for clean energy. As the world shifts to low-carbon energy systems, there is an urgent need to develop next-generation batteries, SCs, and FCs with higher energy and power densities, longer cycle life, and faster charging capabilities. Nanomaterials, hybrid designs, and solid-state electrolytes are predicted to provide significant advances in performance and safety. Additionally, the development of intelligent, flexible energy storage technologies can be greatly accelerated by incorporating AI and machine learning into materials discovery and system optimization. In order to create sustainable, scalable, and effective energy technologies, interdisciplinary approaches such as materials science, electrochemistry, and systems engineering must come together.

Electrochemical energy systems have the potential to be essential in facilitating a robust and sustainable energy future. These technologies will be essential for stabilizing renewable energy grids, enabling decentralized energy systems, and promoting the creation of green hydrogen through water splitting, in addition to powering electric cars and portable gadgets. To guarantee long-term environmental sustainability, emphasis must also be made on circular economy principles, such as the utilization of plentiful, non-toxic elements and the recyclability of materials. These technologies will have a greater worldwide influence if policies and international cooperation encourage innovation, standardization, and responsible implementation. In the end, promoting the sustainable development of electrochemical energy storage and conversion systems will require a basic and comprehensive strategy that tackles both scientific obstacles and socioeconomic ramifications.

## REFERENCES

Abdel Maksoud, M. I. A., Fahim, R. A., Shalan, A. E., Abd Elkodous, M., Olojede, S. O., Osman, A. I., Farrell, C., Al-Muhtaseb, A. H., Awed, A. S., Ashour, A. H., & Rooney, D. W. (2021). Advanced materials and technologies for supercapacitors used in energy conversion and storage: A review. *Environmental Chemistry Letters*, *19*(1), 375–439. https://doi.org/10.1007/s10311-020-01075-w

Abid, A. G., Gouadria, S., Manzoor, S., Katubi, K. M. S., Jabbour, K., Abdullah, M., Un Nisa, M., Aman, S., Al-Buriahi, M. S., & Ashiq, M. N. (2023). Uniformly dispersed flowery $EuZrSe_3$ derived from the europium-based metal–organic framework for energy storage devices. *Fuel*, *336*, 127066. https://doi.org/10.1016/j.fuel.2022.127066

Aissani, L., Alhussein, A., Zia, A., Mamba, G., & Rtimi, S. (2022). Magnetron sputtering of transition metal nitride thin films for environmental remediation. *Coatings*, *12*(11), 1746. https://doi.org/10.3390/coatings12111746

Al Shaqsi, A. Z., Sopian, K., & Al-Hinai, A. (2020). Review of energy storage services, applications, limitations, and benefits. *Energy Reports*, *6*, 288–306. https://doi.org/10.1016/j.egyr.2020.07.028

Azeem, W., Shahzad, M. K., Wong, Y. H., & Tahir, M. B. (2024). Ab-initio calculations for the study of the hydrogen storage properties of $CsXH_3$ (X= Co, Zn) perovskite-type hydrides. *International Journal of Hydrogen Energy*, *50*, 305–313. https://doi.org/10.1016/j.ijhydene.2023.07.072

Babu, T. S., Vasudevan, K. R., Ramachandaramurthy, V. K., Sani, S. B., Chemud, S., & Lajim, R. M. (2020). A comprehensive review of hybrid energy storage systems: Converter topologies, control strategies and future prospects. *IEEE Access*, *8*, 148702–148721. https://doi.org/10.1109/ACCESS.2020.3015919

Bello, I. T., Guan, D., Yu, N., Li, Z., Song, Y., Chen, X., Zhao, S., He, Q., Shao, Z., & Ni, M. (2023). Revolutionizing material design for protonic ceramic fuel cells: Bridging the limitations of conventional experimental screening and machine learning methods. *Chemical Engineering Journal*, *477*, 147098. https://doi.org/10.1016/j.cej.2023.147098

Chen, T., Cheng, B., Chen, R., Hu, Y., Lv, H., Zhu, G., Wang, Y., Ma, L., Liang, J., Tie, Z., Jin, Z., & Liu, J. (2016). Hierarchical ternary carbide nanoparticle/carbon nanotube-inserted N-doped carbon concave-polyhedrons for efficient lithium and sodium storage. *ACS Applied Materials & Interfaces*, *8*(40), 26834–26841. https://doi.org/10.1021/acsami.6b08911

Dashtdar, M., Flah, A., Hosseinimoghadam, S. M. S., & El-Fergany, A. (2022). Frequency control of the islanded microgrid including energy storage using soft computing. *Scientific Reports*, *12*(1), 20409. https://doi.org/10.1038/s41598-022-24758-6

Development of the Lithium-Ion Battery and Recent Technological Trends. (2014). In A. Yoshino, *Lithium-Ion Batteries* (pp. 1–20). Elsevier. https://doi.org/10.1016/b978-0-444-59513-3.00001-7

Dunn, B., Kamath, H., & Tarascon, J. M. (2011). Electrical energy storage for the grid: A battery of choices, Science. *Science*, *334*, 928–935. https://doi.org/10.1126/science.1212741

Emad, D., El-Hameed, M. A., & El-Fergany, A. A. (2021). Optimal techno-economic design of hybrid PV/wind system comprising battery energy storage: Case study for a remote area. *Energy Conversion and Management*, *249*, 114847. https://doi.org/10.1016/j.enconman.2021.114847

Fan, F.-R., Tian, Z.-Q., & Lin Wang, Z. (2012). Flexible triboelectric generator. *Nano Energy*, *1*(2), 328–334. https://doi.org/10.1016/j.nanoen.2012.01.004

Feng, K., Xiao, B., Wang, Y., Yang, H., Xiong, L., Zhao, M., Ma, J., Li, B., Guan, Y., & Ding, Y. (2025). Hexafluorozirconic acid-etched MXene for ammonium-ion capacitors. *Journal of Alloys and Compounds*, *1010*, 177962. https://doi.org/10.1016/j.jallcom.2024.177962

Fischer, J., Thümmler, K., Fischer, S., Gonzalez Martinez, I. G., Oswald, S., & Mikhailova, D. (2021). Activated carbon derived from cellulose and cellulose acetate microspheres as electrode materials for symmetric supercapacitors in aqueous electrolytes. *Energy & Fuels*, *35*(15), 12653–12665. https://doi.org/10.1021/acs.energyfuels.1c01449

Flores-Diaz, N., De Rossi, F., Das, A., Deepa, M., Brunetti, F., & Freitag, M. (2023). Progress of photocapacitors. *Chemical Reviews*, *123*, 9327–9355. https://doi.org/10.1021/acs.chemrev.2c00773

Gabbar, H., Othman, A., & Abdussami, M. (2021). Review of battery management systems (BMS) development and industrial standards. *Technologies*, *9*(2), 28. https://doi.org/10.3390/technologies9020028

Gebre, S. H. (2022). Synthesis and potential applications of trimetallic nanostructures. *New Journal of Chemistry*, *46*(12), 5438–5459. https://doi.org/10.1039/D1NJ06074K

Guney, M. S., & Tepe, Y. (2017). Classification and assessment of energy storage systems. *Renewable and Sustainable Energy Reviews*, *75*, 1187–1197. https://doi.org/10.1016/j.rser.2016.11.102

Gurushankar, K., Karuppaiya, C., Tari, V., & Kannan, K. (2024). Exploring MXenes and MBenes in metal sulfur and selenium battery technologies. In K. Kannan & V. Tari (Eds.), *Novel energy storage and conversion technologies for two-dimensional MXenes and MBenes*. IGI Global. https://doi.org/10.4018/979-8-3693-9546-2.ch005

He, P., Xu, H., Lan, L., Deng, C., Wu, Y., Lin, Y., Chen, S., Ding, C., Li, X., Xu, M., & Peng, J. (2021). The effect of charge transfer transition on the photostability of lanthanide-doped indium oxide thin-film transistors. *Communications Materials*, *2*(1), 86. https://doi.org/10.1038/s43246-021-00193-4

Huang, Z., Ji, Z., Feng, Y., Wang, P., & Huang, Y. (2021). Flexible and stretchable polyaniline supercapacitor with a high rate capability. *Polymer International*, *70*(4), 437–442. https://doi.org/10.1002/pi.5982

Ippili, S., Kim, J. H., Jella, V., Behera, S., Vuong, V.-H., Jung, J.-S., Cho, Y., Ahn, J., Kim, I.-D., Chang, Y. H., Kim, H.-S., & Yoon, S.-G. (2023). Halide double perovskite-based efficient mechanical energy harvester and storage devices for self-charging power unit. *Nano Energy*, *107*, 108148. https://doi.org/10.1016/j.nanoen.2022.108148

Iqbal, T., Ijaz, M., Javaid, M., Rafique, M., Riaz, K. N., Tahir, M. B., Nabi, G., Abrar, M., & Afsheen, S. (2019). An optimal Au grating structure for light absorption in amorphous silicon thin film solar cell. *Plasmonics*, *14*(1), 147–154. https://doi.org/10.1007/s11468-018-0787-2

Iro, Z. S., Subramani, C., & Dash, S. S. (2016). A brief review on electrode materials for supercapacitor. *International Journal of Electrochemical Science*, *11*(12), 10628–10643. https://doi.org/10.20964/2016.12.50

Jaffri, S. B., Ahmad, K. S., Abrahams, I., Almanqur, L., Alderhami, S. A., & Alharbi, Y. T. (2024). [Sm-Eu-Tm]:$CsPbI_2Br_1$ perovskite hetero-system: Probing rare earth driven stabilization for sustainable energy conversion, generation, and storage capacity. *Ceramics International*, *50*(5), 7645–7654. https://doi.org/10.1016/j.ceramint.2023.12.087

Jung, W.-S., Kang, M.-G., Moon, H. G., Baek, S.-H., Yoon, S.-J., Wang, Z.-L., Kim, S.-W., & Kang, C.-Y. (2015). High output piezo/triboelectric hybrid generator. *Scientific Reports*, *5*(1), 9309. https://doi.org/10.1038/srep09309

Kampouris, K. P., Drosou, V., Karytsas, C., & Karagiorgas, M. (2020). Energy storage systems review and case study in the residential sector. *IOP Conference Series: Earth and Environmental Science*, *410*, 1.

Kannan, K., & Tari, V. (Eds.). (2025). *Novel energy storage and conversion technologies for two-dimensional MXenes and MBenes*. https://www.amazon.com/dp/9798369395462

Kitajou, A., Komatsu, H., Chihara, K., Gocheva, I. D., Okada, S., & Yamaki, J. (2012). Novel synthesis and electrochemical properties of perovskite-type $NaFeF_3$ for a sodium-ion battery. *Journal of Power Sources*, *198*, 389–392. https://doi.org/10.1016/j.jpowsour.2011.09.064

Li, Y., Wu, Y., Hao, H., Yuan, M., Lv, Z., Xu, L., & Wei, B. (2022). In situ unraveling surface reconstruction of $Ni_5P4$@FeP nanosheet array for superior alkaline oxygen evolution reaction. *Applied Catalysis B: Environmental*, *305*, 121033. https://doi.org/10.1016/j.apcatb.2021.121033

Liu, H., & Li, W. (2021). Recent advances in paired electrolysis of biomass-derived compounds toward cogeneration of value-added chemicals and fuels. *Current Opinion in Electrochemistry*, *30*, 100795. https://doi.org/10.1016/j.coelec.2021.100795

Luo, X., Wang, J., Dooner, M., & Clarke, J. (2015). Overview of current development in electrical energy storage technologies and the application potential in power system operation. *Applied Energy*, *137*, 511–536. https://doi.org/10.1016/j.apenergy.2014.09.081

Mohamed, N. A., Ismail, A. F., & Kiong, T. S. (2024a). G-$C_3N_4$/$La_2O_3$ nanocomposite as a photo-electrocatalyst in solar water splitting. *Surfaces and Interfaces*, *44*, 103639. https://doi.org/10.1016/j.surfin.2023.103639

Mohamed, N. A., Kiong, T. S., & Ismail, A. F. (2024b). Pioneering sustainable energy solutions with rare-earth nanomaterials: Exploring pathways for energy conversion and storage. *International Journal of Hydrogen Energy*, *93*, 607–649. https://doi.org/10.1016/j.ijhydene.2024.10.299

Mondal, D., Kundu, M., Paul, B. K., Bhattacharya, D., Sarkar, S., Sau, S., Senapati, D., Mandal, T. K., & Das, S. (2024). Rare earth ion-doped $\alpha$-$MnO_2$ nanorods for an asymmetric supercapacitor. *ACS Applied Nano Materials*, *7*(5), 4913–4926. https://doi.org/10.1021/acsanm.3c05666

Murugan, E., Munusamy, K., & Vallipparambil Babu, A. (2024). Development of aryl ether-free cross-linked polymer membranes for sustainable electrochemical energy conversion and storage applications. *Chemical Engineering Journal*, *501*, 157473. https://doi.org/10.1016/j.cej.2024.157473

Nadeem, F., Hussain, S. M. S., Tiwari, P. K., Goswami, A. K., Ustun, T. S. (2019). Comparative review of energy storage systems, their roles, and impacts on future power systems. *IEEE Access*, *7*, 4555–4585. https://doi.org/10.1109/ACCESS.2018.2888497

Naik Shreyanka, S., Theerthagiri, J., Lee, S. J., Yu, Y., & Choi, M. Y. (2022). Multiscale design of 3D metal–organic frameworks (M–BTC, M: Cu, Co, Ni) via PLAL enabling bifunctional electrocatalysts for robust overall water splitting. *Chemical Engineering Journal*, *446*, 137045. https://doi.org/10.1016/j.cej.2022.137045

Niu, S., & Wang, Z. L. (2015). Theoretical systems of triboelectric nanogenerators. *Nano Energy*, *14*, 161–192. https://doi.org/10.1016/j.nanoen.2014.11.034

Pu, X., Song, W., Liu, M., Sun, C., Du, C., Jiang, C., Huang, X., Zou, D., Hu, W., & Wang, Z. L. (2016). Wearable power-textiles by integrating fabric triboelectric nanogenerators and fiber-shaped dye-sensitized solar cells. *Advanced Energy Materials*, *6*(20). https://doi.org/10.1002/aenm.201601048

Qian, Y., & Kang, D. J. (2018). Poly(dimethylsiloxane)/ZnO nanoflakes/three-dimensional graphene heterostructures for high-performance flexible energy harvesters with simultaneous piezoelectric and triboelectric generation. *ACS Applied Materials & Interfaces*, *10*(38), 32281–32288. https://doi.org/10.1021/acsami.8b05636

Sbei, N., Hardwick, T., & Ahmed, N. (2021). Green chemistry: Electrochemical organic transformations via paired electrolysis. *ACS Sustainable Chemistry & Engineering*, *9*(18), 6148–6169. https://doi.org/10.1021/acssuschemeng.1c00665

Si, H., Zhao, X., Zhang, Z., Liao, Q., & Zhang, Y. (2024). Low-temperature electron- transporting materials for perovskite solar cells: Fundamentals, progress, and outlook. *Coordination Chemistry Reviews*, *500*, 215502.

Siddiqui, N., Faiyaz, T., & Tari, V. (2022). Ganga and Yamuna: Through the lens of National Green Tribunal. *Nature, Environment and Pollution Technology*, *21*(4), 1657–1664. https://doi.org/10.46488/NEPT.2022.v21i04.018

Sriphan, S., Charoonsuk, T., Maluangnont, T., Pakawanit, P., Rojviriya, C., & Vittayakorn, N. (2020). Multifunctional nanomaterials modification of cellulose paper for efficient triboelectric nanogenerators. *Advanced Materials Technologies*, *5*(5), 2000001. https://doi.org/10.1002/admt.202000001

Suo, G., Yu, Y., Zhang, Z., Wang, S., Zhao, P., Li, J., & Wang, X. (2016). Piezoelectric and triboelectric dual effects in echanical-energy harvesting using $BaTiO_3$ / polydimethylsiloxane composite film. *ACS Applied Materials & Interfaces*, *8*(50), 34335–34341. https://doi.org/10.1021/acsami.6b11108

Tackett, B. M., Sheng, W., & Chen, J. G. (2017). Opportunities and challenges in utilizing metal-modified transition metal carbides as low-cost electrocatalysts. *Joule*, *1*(2), 253–263. https://doi.org/10.1016/j.joule.2017.07.002

Tari, V., Siddiqui, N., Rathi, D., Siddiqui, N. N., & Wahyuni, D. K. (2024). Need for an evolved groundwater justice in rural area Uttar Pradesh, India. *Nature, Environment and Pollution Technology*, *23*(1), 125–137. https://doi.org/10.46488/NEPT.2024.v23i01.009

Tari, V. S., & Patil, P. Y. (2017). Challenge of contamination of pesticide for Alphonso in Ratnagiri district, Maharashtra, India. *Research Journal of Chemistry and Environment*, *21*(12), 57–67.

Theerthagiri, J., Lee, S. J., Murthy, A. P., Madhavan, J., & Choi, M. Y. (2020). Fundamental aspects and recent advances in transition metal nitrides as electrocatalysts for hydrogen evolution reaction: A review. *Current Opinion in Solid State and Materials Science*, *24*(1), 100805. https://doi.org/10.1016/j.cossms.2020.100805

Verma, S., et al., (2021). A comprehensive review on energy storage in hybrid electric vehicle. *Journal of Traffic and Transportation Engineering*, *8*, 621–637. https://doi.org/10.1016/j.jtte.2021.09.001

Vilvanatha Prabu, A., Vijayaraghavan, G. V., Suriakarthick, R., Priscilla, J., Shyju, T. S., & Mani, J. (2025). Electrical, electrochemical and thermoelectric properties of PANI/$AgBiSe_2$ multi-functional polymeric composite material for energy storage and conversion applications. *Journal of Alloys and Compounds*, *1021*, 179534. https://doi.org/10.1016/j.jallcom.2025.179534

Walter, M. G., Warren, E. L., McKone, J. R., Boettcher, S. W., Mi, Q., Santori, E. A., & Lewis, N. S. (2010). Solar water splitting cells. *Chemical Reviews*, *110*(11), 6446–6473. https://doi.org/10.1021/cr1002326

Wang, G., Zhang, L., & Zhang, J. (2012). A review of electrode materials for electrochemical supercapacitors. *Chemical Society Reviews*, *41*(2), 797–828. https://doi.org/10.1039/c1cs15060j

Wang, Z. L., & Song, J. (2006). Piezoelectric nanogenerators based on zinc oxide nanowire arrays. *Science*, *312*(5771), 242–246. https://doi.org/10.1126/science.1124005

Wei, X., Sun, Y., Zhang, Y., Yu, B., & Yu, H. (2025). A X-bit optimized 2D solid solution $Ti_3CNTx$ MXene as the electron transport layer toward high-performance perovskite solar cells. *Nano Energy*, *133*, 110513. https://doi.org/10.1016/j.nanoen.2024.110513

Winter, M., & Brodd, R. J. (2004). What are batteries, fuel cells, and supercapacitors? *Chemical Reviews*, *104*(10), 4245–4270. https://doi.org/10.1021/cr020730k

Xia, F., Tie, D., Wang, J., Song, H., Wen, W., Ye, X., Wu, J., Hou, Y., Lu, X., & Zhao, Y. (2021). Ultrahigh rate and durable sodium-ion storage at a wide potential window via lanthanide doping and perovskite surface decoration on layered manganese oxides. *Energy Storage Materials*, *42*, 209–218. https://doi.org/10.1016/j.ensm.2021.07.020

Yalçın, M., Turgut, N., Gökbulut, C., Mermer, S., Sofuoğlu, S. C., Tari, V., & Turgut, C. (2022). Removal of pesticide residues from apple and tomato cuticle. *Environmental Science and Pollution Research*. https://doi.org/10.1007/s11356-022-23269-1

Yan, F., Bai, H., Ge, G., Lin, J., Zhu, K., Li, G., Qian, J., Shen, B., Zhai, J., & Liu, Z. (2022). Boosting energy storage performance of lead-free ceramics via layered structure optimization strategy. *Small*, *18*(34), 2202575. https://doi.org/10.1002/smll.202202575

Yang, L., Ravi, S. K., Nandakumar, D. K., Alzakia, F. I., Lu, W., Zhang, Y., Yang, J., Zhang, Q., Zhang, X., & Tan, S. C. (2019). A hybrid artificial photocatalysis system splits atmospheric water for simultaneous dehumidification and power generation. *Advanced Materials*, *31*(51). https://doi.org/10.1002/adma.201902963

Yang, Y., Zhu, T., Shen, L., Liu, Y., Zhang, D., Zheng, B., Gong, K., Zheng, J., & Gong, X. (2022). Recent progress in the all-solid-state flexible supercapacitors. *SmartMat*, *3*(3), 349–383. https://doi.org/10.1002/smm2.1103

Yin, L., Li, Y., Yao, X., Wang, Y., Jia, L., Liu, Q., Li, J., Li, Y., & He, D. (2021). MXenes for solar cells. *Nano-Micro Letters*, *13*(1), 78. https://doi.org/10.1007/s40820-021-00604-8

Yu, Y., Lee, S. J., Theerthagiri, J., Fonseca, S., Pinto, L. M. C., Maia, G., & Choi, M. Y. (2022). Reconciling of experimental and theoretical insights on the electroactive behavior of C/Ni nanoparticles with AuPt alloys for hydrogen evolution efficiency and Non-enzymatic sensor. *Chemical Engineering Journal*, *435*, 134790. https://doi.org/10.1016/j.cej.2022.134790

Zang, J., Bao, S.-J., Li, C. M., Bian, H., Cui, X., Bao, Q., Sun, C. Q., Guo, J., & Lian, K. (2008). Well-aligned cone-shaped nanostructure of polypyrrole/$RuO_2$ and Its electrochemical supercapacitor. *The Journal of Physical Chemistry C*, *112*(38), 14843–14847. https://doi.org/10.1021/jp8049558

Zhao, S., Zuo, Y., Liu, T., Zhai, S., Dai, Y., Guo, Z., Wang, Y., He, Q., Xia, L., Zhi, C., Bae, J., Wang, K., & Ni, M. (2021). Multi-functional hydrogels for flexible zinc-based batteries working under extreme conditions. *Advanced Energy Materials*, *11*(34), 2101749. https://doi.org/10.1002/aenm.202101749

Zheng, L., Cheng, G., Chen, J., Lin, L., Wang, J., Liu, Y., Li, H., & Wang, Z. L. (2015). A hybridized power panel to simultaneously generate electricity from sunlight, raindrops, and wind around the clock. *Advanced Energy Materials*, *5*(21). https://doi.org/10.1002/aenm.201501152

Zhu, Y., Zhu, R., Guan, P., Li, M., Wan, T., Hu, L., Zhang, S., Liu, C., Su, D., Liu, Y., Liu, D., Li, Q., Yu, J., & Chu, D. (2023). Designing MXene-wrapped AgCl@ Carbon core shell cathode for robust quasi-solid-state Ag-Zn battery with ultralong cycle life. *Energy Storage Materials*, *60*, 102836. https://doi.org/10.1016/j.ensm.2023.102836

Chapter 2

# Synthesis and characterization of two-dimensional hybrid composites

*Krishnamoorthy Gurushankar, Karthik Kannan, and Mariappan Anandkumar*

## 2.1 INTRODUCTION

Nanocomposite (NC) materials are multiphase materials with nanoscale dimensions (usually less than 100 nanometers) in at least one phase, offering improved properties and a variety of applications (Gupta et al., 2024; Tari et al., 2025). NC materials, formed by combining non-metal, metal, and polymeric materials through specific processes, offer benefits like maintaining key properties, mitigating flaws, and exhibiting novel characteristics (Vinyas et al., 2019). NC materials are achieved by combining two or more individual building components, providing plausible unique properties due to the effectiveness of small size, large surfaces, and interphase interface interactions (Henrique et al., 2009; Gurushankar et al., 2024; Kannan and Tari, 2025). The usual possibilities of their additions are used smoothly to improve the biological properties of many drugs, biomaterials, catalysts, and some high-quality materials (Chinnaiah et al., 2024a, 2024b, 2024c). NC offers innovative solutions to the challenges of the healthcare, pharmaceuticals, food packaging, electronics, and energy supply sectors (Tari et al., 2022). The main advantages of these materials are their high surface-to-volume ratio. This leads to small filler size, leading to excellent mechanical properties (high plasticity without sacrificial stability, scratch resistance) and improved optical properties.

## 2.2 CLASSIFICATION OF NC

NC materials can be categorized by the presence or absence of polymeric material in the composite. The NC in which the composition doesn't contain any polymers or polymer-derived materials are called non-polymer-based NC. These non-polymer-based NC are also termed inorganic NC. It can be further divided into three categories: metal-based NC (examples: polymer NCs with metal nanoparticles, metal-ceramic composites, and metal-matrix

DOI: 10.1201/9781003584742-2

NCs), ceramic-based NC (examples: ceramic-matrix NCs reinforced with graphene (GP) or carbon nanotubes (CNTs), and NCs featuring nanoscale SiC, C, or SiOC reinforcements), and ceramic-ceramic-based NC (examples: Alumina/Yttria ($Al_2O_3/Y_2O_3$), Zirconia/Alumina ($ZrO_2/Al_2O_3$), Calcium Sulfate/Biomimetic Apatite, Hydroxyapatite/Titania (HAP/$TiO_2$), $SiO_2$/Fe, $SiO_2$/Ni, and $PbTiO_3/PbZrO_3$) (Khandoker et al., 2011).

## 2.3 SIGNIFICANCE OF MXENE, GRAPHENE, AND METAL OXIDES

MXenes, GP, and metal oxides (MOs) are highly significant materials known for their distinctive properties and impending applications in various technologies, especially in the areas of energy storage and sensing. MXenes and GP are particularly valued for their exceptional conductivity, while MOs are known for enhancing gas adsorption and semiconducting characteristics (Varadarajan et al., 2024). MXenes demonstrate great promise in energy storage applications especially batteries and supercapacitors because of high surface area, and excellent electrical conductivity. Additionally, MXenes are utilized in gas sensors due to their high sensitivity and selectivity, offering benefits in various detection applications and also exhibit potential in diverse fields including biomedical applications, water purification, and as transparent conducting electrodes (Tari et al., 2024). The widely recognized MXene $Ti_3C_2Tx$ (titanium carbide) is noted for its excellent biocompatibility and high conductivity, making it suitable for various applications.

GP's application extends to batteries and supercapacitors, contributing to improved energy density and power output in these energy storage devices. Moreover, GP-based sensors are well-known for their high sensitivity and selectivity, making them appropriate for detecting a wide range of gases and chemicals. GP also finds applications in electronics, composites, and biomedical devices, showcasing its versatility in various technological domains. Additionally, graphene oxide (GO) which is a derivative of GP with oxygen functional groups, holds potential for diverse applications, including water purification and gas adsorption. MOs are widely utilized in gas sensors due to their semiconducting properties and their ability to adsorb gases effectively, making them valuable for detection purposes. Furthermore, MO serves as electrode materials in batteries and supercapacitors, contributing to enhanced performance in energy storage devices. MO also finds applications in catalysis, photocatalysis, and corrosion protection, demonstrating their versatility in various industrial and technological processes. For instance, tin oxide ($SnO_2$) and zinc oxide (ZnO) are frequently used in gas sensors and other diverse applications due to their unique properties and reliability.

## 2.4 CHARACTERIZATION OF 2D NANOCOMPOSITE MATERIALS

### 2.4.1 Characterization of MXenes

The structural and physical properties of MXenes are investigated by the most commonly used characterization techniques X-ray Diffraction (XRD) and scanning electron microscopy (SEM). XRD gives information on whether MXene was formed (Shekhirev et al., 2021). In case the precursor is transformed into a few-layered MXene, characteristic XRD patterns of the 2D material can be seen. Meanwhile, the hydrophilic nature of MXenes, and the consequent wide space between the layers, makes them capable of accommodating intercalating agents and modifying the XRD patterns (Mu et al., 2019).

SEM is a technique that gives us images and information about the progress of structure formation of 2D materials. SEM is also often used with energy dispersive spectroscopy (EDS) to provide more accurate data about the molecule both in qualitative and quantitative manner (Shekhirev et al., 2021). In addition, other analyses similar to pair distribution function (PDF) analysis (Shi et al., 2014), X-ray absorption spectroscopy (XAS), and atomic force microscopy (AFM) methods are also utilized for the characterization of MXenes. From the PDF studies, we can get confirmation of surface groups on the MXenes (Ghidiu et al., 2014), the effects of etching process on the 2D materials (Wang et al., 2016), and the location of each atom in the MXene molecule (Sun et al., 2020). Meanwhile, XAS is used to explore the electronic structure of these 2D materials at the level of the subatomic particle (Shekhirev et al., 2021). The adsorption and catalytic reaction studies on the MXene surface were also enabled through this method (Wang et al., 2020).

### 2.4.2 Characterization of graphene

UV-vis spectroscopy, SEM, TEM, XRD, and Raman spectroscopy (RS) are used most commonly for the characterization of GP family.

The quality and layers of GP and its derivatives are analyzed and studied using monochromatic laser-based RS. This consequence is the production of atomic or frequently molecular vibrations, which modify the molecules' laser energy. Depending on the molecule, this then causes various peaks to appear (Bin et al., 2018; Xu et al., 2018). UV–Vis spectroscopy is commonly used to examine the confirmation of GP molecule formation with variety of form such as GO, decreased GO or pristine GP and a wide variety of layers within the molecule. The absorbance effects can be acquired owing to the electron transfer on the carbon pi bonds (pi to pi* transitions) (Liu et al., 2014). The range of layers or the width of the molecule is analyzed through ultraviolet transmittance. While the range of layers in the GP molecule will increase it will affect transmittance and will be reduced in the materials (Li et al., 2018a). Transmission electron microscopy (TEM) is a powerful technique used to study the structural quality and layer number of GP. The transmittance of electron beam of molecules has been studied by the obtained results. Any impurities, unevenness, or folding inside the GP molecules are studied by SEM studies. XRD studies are used to study unit cell dimensions of the molecule.

### 2.4.3 Characterization of 2D oxides

The structure and stability of 2D oxides are observed with the help of AFM and TEM (Radha et al., 2021). Vibrational spectra, the phonon modes, and the crystalline nature of the 2D oxides are examined with the assistance of RS (Sucharitakul et al., 2017).

## 2.5 SYNTHESIS OF 2D NC MATERIALS

### 2.5.1 Synthesis of MXene

Typically, MXenes are derived from either inorganic ternary carbides or nitrides, commonly known as MAX phases (Figure 2.1). MXenes can possess varying quantities of M and X layers within their two-dimensional structure, denoted by n, which can range from 1 to 4. This results in n + 1 M layers interspersed with n X layers. The Tx term includes elements from groups 16 and 17 of the periodic table, along with hydroxyl and imido groups (T = O, OH, NH, F, Cl, Br, S, Se, Te). The terminations may be either singular or a combination, contingent upon the synthesis method employed (Ronchi et al., 2019). Various top-down and bottom-up synthesis techniques have been developed for MXenes, including the urea glass method, chemical vapor deposition (CVD), molten salt etching, hydrothermal fabrication, and electrochemical preparation (Zhan et al., 2020). Among these, CVD and wet etching methods have been extensively utilized for MXene fabrication. Importantly, high-quality MXenes featuring terminations have been produced through different wet etching techniques, resulting in MXenes that exhibit predominantly hydrophilic characteristics (Vacik et al., 2021). In the top-down synthesis approach, the precursor originates from the three-dimensional structure of the MAX phase. This method employs chemical etchants such as hydrofluoric acid, and the reaction typically occurs at room temperature or lower. The resulting morphology can yield large, unpredictable MXene sheets, with lateral dimensions reaching several hundred nanometers and a thickness ranging from 10 to 20 nanometers. Additionally, the surface properties of the synthesized MXenes should be functionalized with hydroxyl (–OH) and oxide (–O) groups post-synthesis (Ma et al., 2015).

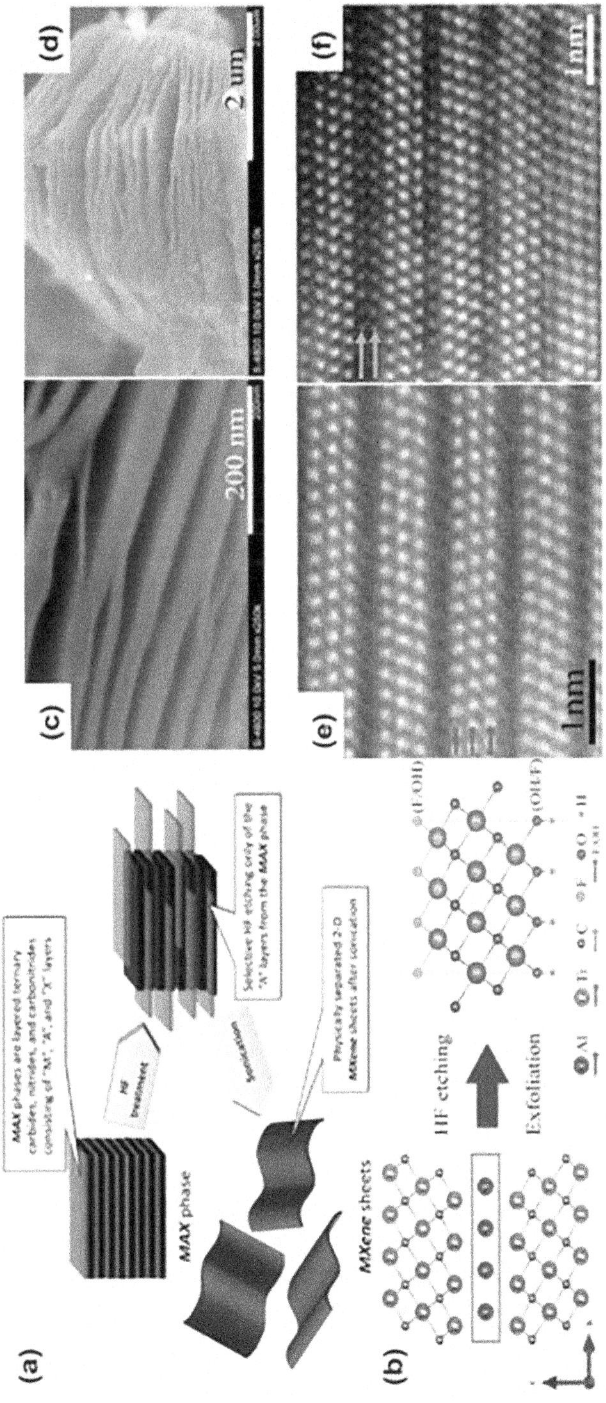


*Figure 2.1* The synthesis process of MXene produced from MAX phases. (b) Structural changes during $Ti_3C_2X$ synthesis. **(Reproduced from Zhan et al., 2020.)**

#### *2.5.1.1 Bottom-up approach*

In this method, the synthesis of MXene involves the growth of a crystalline MXene film from a precursor, initiated at the atomic level. This can be achieved through techniques such as CVD, self-templating, or plasma-enhanced deposition (Xu et al., 2015). During the process, a carbon source is maintained through gas flow, requiring temperatures around 1000°C. The resulting morphology yields a defect-free, highly crystalline multilayer film with lateral dimensions ranging from 10 to 100 μm. However, the synthesis of MXenes often leads to a deficiency of functional groups on their surfaces. A widely used technique for etching aluminum atoms (Al) from the precursor MAX phase is the application of aqueous hydrofluoric acid (HF), which leaves surface terminations, for instance, oxygen (–O), hydroxyl (–OH), or fluorine (–F) groups on the resultant MXenes. This method was initially demonstrated by Naguib and his team (Naguib et al., 2012). To accomplish solid-liquid separation, the precursors are submerged in etching solution, usually HF or acid-fluorides at the proper concentration, and then centrifuged or filtered. After that, the substance is cleaned with deionized (DI) water until its pH falls between 4 and 6. The solution was treated with shearing forces or sonication to produce a single-layered stack. Various functional groups (i.e., Tx) terminate the stack of 'A' layers that is produced by this process in the multilayered MXene. According to various reports (Urbankowski et al., 2016; Sun et al., (2017); Li et al., 2018b; Salim et al., 2019; Liu et al., 2019; Zhan et al., 2020), the A-depleted multilayered MXene undergoes a delamination phase in which various cations, such as $Li^+$, are injected to improve the interlayer spacing.

### 2.5.2 Synthesis of GP

CVD, mechanical exfoliation from graphite, and heating to reduce GO are some of the ways used to create GP (Mishra et al., 2016).

The top-down approach to GP fabrication involves techniques, for example, mechanical exfoliation and liquid exfoliation. Mechanical exfoliation is also known as the 'Scotch-tape' method, is an attractive technique for producing GP due to its simplicity and efficiency. By using the direct exfoliation method, highly pure graphite can be separated into few layers of GP using a scotch tape, resulting in thin and high-quality GP with high electrical conductivity. However, this method has a low production yield and often leads to GP products being contaminated with organic impurities.

Liquid-phase chemical exfoliation is one of another technique employed in the top-down strategy. High-yield GP manufacturing is possible using this technique. It entails turning powdered graphite into premium oxidized graphite and graphite flakes. With specific solvents, GP may be distributed at a concentration of 0.01mgmt$^{-1}$. The flakes can be deposited by spray coating, drop casting, or vacuum filtration to accomplish the dispersion.

Although 1–10 GP flakes may ultimately stack together, thicker flakes are removed by centrifuging the fluid at 500–2000 rpm. Due to the drawbacks of both strategies, low production quality and excessive impurity, an alternative strategy is being considered. The bottom-up method uses small carbon atom molecules to create GP using CVD or epitaxial growth on a SiC substrate. To obtain the required thickness, the epitaxial technique regulates a number of substrate catalysts and process variables. Large area coverage, excellent quality, and the possibility of mass producing high-purity GP are benefits of CVD techniques (Mbayachi et al., 2021).

#### 2.5.2.1 *Epitaxial growth on SiC substrates*

High-temperature, ultra-high vacuum, or atmospheric layer graphitization of silicon carbide wafers results in epitaxial GP. Controlling growth parameters to regulate the number of GP layers on the SiC (0001) substrate has been investigated in a review paper (Telychko et al., 2015). It has been shown that altering the development parameters can regulate the quantity of GP layers. There has been evidence of a correlation between the annealing temperature and the number of GP layers, showing that the latter increases with temperature. Additionally, it has been proposed that longer sintering times are more efficient at generating a stable GP layer at lower temperatures than just raising the temperature. But according to a different publication, the bulk SiC substrate has drawbacks with various factors such as size, cost, and micromachining challenges. By heating to the proper temperature, silicon and carbon sublimation take place during the epitaxial development process of GP on a silicon carbide wafer. GP is created when silicon and carbon stay on the substrate's surface after the sublimation process. This process describes the synthesis of GP on a SiC wafer. Recent developments have made it possible to create GP structures with the required size, purity, and crystallinity using a variety of techniques, including mechanical exfoliation from graphite, CVD, and heating to reduce GO. Some physical and chemical approaches have been used for the synthesis of MXene/GP composite materials as well as mechanical mixing, self-assembly, hydrothermal techniques, heat treatment, and reagent reduction treatments (Sun et al., 2018; Jun et al., 2019; Verger et al., 2019).

### 2.5.3 Synthesis of 2D metal oxide (MO) nanomaterials

The production of 2D MO is broadly categorized into top-down exfoliation, bottom-up synthesis, and liquid metal strategy, and has advanced significantly in recent years. The primary crystal precursors of MOs (usually layered oxides) (MOXs) are placed into the element layers to accomplish the top-down exfoliation. The objective of bottom-up synthesis is to acquire the necessary metal and oxygen atoms or molecules and, under

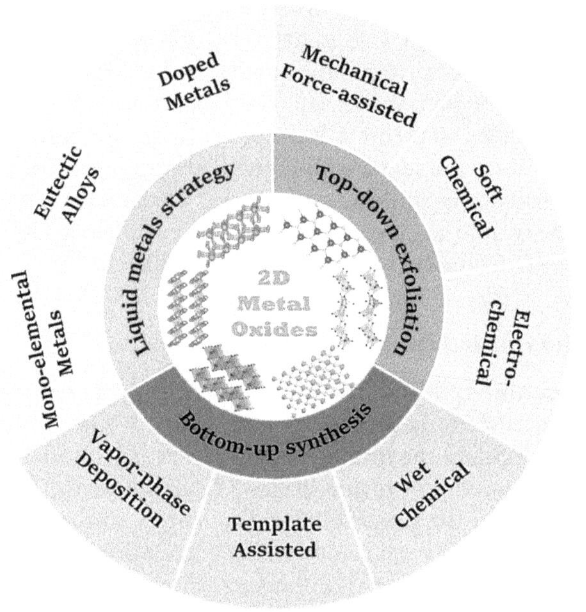


*Figure 2.2* The synthesis methods of 2D metal oxides.

**(Reproduced from Xie et al., 2022.)**

certain circumstances, restrict their growth in a two-dimensional orientation (Figure 2.2). The liquid metal approach uses a self-limiting reaction to create ultrathin 2D MOXs on the surface of liquid metals or alloys.

In general, 2D MOXs are gaining a lot of attention due to their fascinating qualities, and significant advancements in synthetic techniques have been made. Hexagonal layered MOXs, 2D large-area non-layered MOXs, ultrathin amorphous MOXs, and 2D rare-earth oxides are some of the new highlights, despite the fact that several articles have already covered some aspects of the synthesis techniques for 2D MOXs. Significant progress has also been made in creating the 2D ultrathin morphology via its associated synthetic techniques, including the orientation-driven crystalline cleavage, gas-assisted exfoliation, metal-gas interfacial limited reaction, and liquid metal-derived fabrication.

## 2.6 APPLICATION OF 2D NANOCOMPOSITE MATERIALS

In recent times, 2D hybrid NC materials have evolved as a potential class of nanomaterials. Due to their unique crystal structure and their combination with other two-dimensional structure, 2D hybrid NC materials tend to exhibit unique set of properties and functional applications. As an outcome, the electronic and optical properties of 2D hybrid NC materials can be altered by means of optimizing the particle size, and composition. Therefore, the prepared 2D hybrid NC materials are used in various fields such as energy storage, photocatalyst, electrocatalyst and electrode materials, etc. Next, we will discuss the applications of some 2D hybrid NC materials.

One type of electrical energy storage device that may release energy with a high density over a short period of time is a supercapacitor or ultracapacitor. These devices have high power density and specific capacitance. The difference between the supercapacitors and conventional capacitors is the higher specific surface area exhibiting exceptional specific capacitance. In addition, the important feature of supercapacitors includes rapid charging with better safety operation, better cyclic stability, and high reliability with less maintenance.

Transition metal molybdates have been found to be one of the possible electrode materials for supercapacitor due to their enhanced electrochemical activity possessing higher rate capability. However, pure oxides cannot be used due to limited electric conductivity. To improve this, preparing composites with conductive GP like materials are added to address this issue. The $CoMoO_4$/r-GO composite electrode for supercapacitors was primed via a one-step hydrothermal method (Prabhu et al., 2021).

The fabricated $CoMoO_4$/r-GO composite electrode exhibited a specific capacity of 425 $Cg^{-1}$ at current density 1 $Ag^{-1}$. Besides, the electrode's retained 99.2% of specific capacity during 10,000 charge-discharge cycles at current density 10 $Ag^{-1}$. The supercapacitor fabricated using the $CoMoO_4$/r-GO composite electrode displayed high energy density of 63.5 Wh $kg^{-1}$ at a power density of 500 W $kg^{-1}$.

Another carbon-based material, graphitic carbon nitride (g-$C_3N_4$) is a multilayer semiconductor and can be used as a composite nanomaterial with several oxides (Zahra et al., 2024). The addition of g-$C_3N_4$ to MOs develops the electrochemical performance due to synergistic interaction between g-$C_3N_4$ and MOs and displays better stability with prolonged cycle life during repeated charge/discharge cycles. The $BiFeO_3$@g-$C_3N_4$ NC electrodes prepared using a hydrothermal route established a specific capacitance of 1164 F/g while pure $BiFeO_3$ has a value of 557 F/g. This increased specific capacitance was attributed to the material's structural stability, increased surface area, and a nitrogen-rich structure of g-C3N4, which is responsible for improved electrochemical activity facilitating better ion transport. In terms of electrochemical stability, the prepared $BiFeO_3$@g-$C_3N_4$ NC electrode material was stable after 5000 cycles after 50 h.

Other transition metal-based oxides such as CuO are used as electrode materials because of their low cost and environmental friendliness. However, several drawbacks have been observed such as low electrical conductivity, poor cyclic stability, and poor chemical stability in alkaline environments. To improve this, g-$C_3N_4$, carbon-based NCs was used. The fabricated

$CuO/g-C_3N_4$ electrode material displayed high specific capacitance value 384 F/g at the current density 2 A/g with 98.02% capacitance retention after testing over 5,000 cycles. The addition of $g-C_3N_4$ as a carbon NC will enhance the electrical conductivity of the final electrode material.

In addition to oxides, metal phosphides are a promising option for energy storage applications as supercapacitor electrode materials (Zhong et al., 2025). These metal phosphides have narrower band gaps and greater conductivity. The kinetics of the redox processes will be improved as a result of increased electron transport in the electrode materials. However, the intrinsic conductivity, rate performance, and structural stability of these metal phosphides are all lacking. As a NC, rGO can be used to address these issues. As an electrode material for supercapacitor applications, rGO nanosheet-anchored $Co_2P$ nanoparticles with phosphorus vacancies (V-Co2P/rGO//PPrGO) were used. The fabricated electrode achieves a decent capacitance 357 F $g^{-1}$ at 1 A $g^{-1}$ with an impressive energy density 77 Wh $kg^{-1}$ at 631.1 W $kg^{-1}$. The electrode material retained 89.7% of its initial capacitance until 8000 cycles.

To enhance electrochemical characteristics specific capacitance, energy density, and power performance (various characteristics), rGO can also be mixed with MOs like $TiO_2$ ($TiO_2$-rGO) (Jadoon & Pham, 2024). The fabricated electrodes show evidence of a higher specific capacitance value 152 F $g^{-1}$ for the $TiO_2$-rRGO composites at a comparatively faster scan rate 25 $mVs^{-1}$. The synergistic interaction between rGO and $TiO_2$ enhanced effective electron transport, which is the cause of the increased specific capacitance. In addition, the analysis revealed that the incorporation of RGO reduced the $TiO_2$ bandgap from 3.87 eV to 3.02 eV and improved the specific capacitance, enhancing the $TiO_2$-RGO electrode's supercapacitive performance.

Secondly, we will discuss the use of 2D hybrid NC materials in photocatalyst applications. Photocatalyst is decomposes the adverse substances when illuminated with a light source such as ultraviolet light, visible light, or sunlight. Water and air treatment, hydrocarbon generation, hydrocarbon production, the transformation of certain molecules, photocatalytic water splitting, and self-cleaning coatings are further uses for photocatalyst materials. An $rGO/TiO_2$ NC (GTO/NC) is produced by hydrothermally anchoring $TiO_2$ nanoparticles to the rGO sheets (Parveen et al., 2025). Methyl orange was broken down using the provided photocatalyst. The NC's first order rate constant of 0.0193 $min^{-1}$ indicated improved dye degradation. Furthermore, at a current density 1.0 $Ag^{-1}$, it was employed as an electrode material outlining with a specific capacitance of 320 $Fg^{-1}$.

The use of composite materials, especially carbon-based nanomaterials, hinds/delays recombination of electron-hole pairs, improving the photocatalytic activity. A new type of 2D/1D CN/PANI NC which was decorated on activated carbon was successfully synthesized (Saraee et al., 2025). The synthesized NC was used to degrade the Congo red (CR) dye. The photocatalyst can effectively degrade 91% of CR dye within 100 min. During the photocatalytic degradation, the free radicals, superoxide radicals, and positive holes are involved to effectively degrade CR dye. The interaction between CN nanosheets and PANI NC efficiently separated the electron-hole pair from recombination and improved the CR dye degradation.

The carbon-based materials with different forms can be utilized to improve the photocatalytic activity (Sasikumar et al., 2025). For example, various forms of $gC_3N_4$ including nanoribbons, nanofibers, nanorods, and nanosheets have been extensively researched as a photocatalytic material. In addition, forming heterojunction can improve efficient charge transfer pathway during the photocatalytic reaction. To explain this, a novel Z-scheme heterostructure $NiV_2O_6@gC_3N_4$ (NVO@gCN) was prepared to degrade various antibiotics, such as Ciprofloxacin (CIP) and Norfloxacin (NF). The use of $NiV_2O_6@gC_3N_4$ NC degraded 93.90% and 91.16% for CIP and NF within 120 min and the role of $\bullet O_2^-$ and •OH radicals was found to play a significant role in the degradation.

Photo-electrocatalyst is another application which addresses environmental challenges such as water purification and hydrogen evolution reaction (HER) (Rostami et al., 2025). The most popular photo-electrocatalyst materials include $TiO_2$, $Bi_2O_3$, ZnO, and $ZnFe_2O_4$, which were used to promote clean water and renewable energy solutions. The major drawbacks in these nanomaterials like $TiO_2$ lie in the wide bandgap of the material, which will be active only in the UV regions of the solar spectrum. In addition, high recombination rates limit their availability for use as a photocatalytic material. Therefore, for the electrocatalytic HER, composites using carbon nitride are added to the MOs. As a result, the composite nanoparticles possess excellent photoactivity with improved electron migration and photoactivity in the visible light region. Here, a photocatalyst was designed using $TiO_2@Ti_3C_2Cl_2$ MXene fragments attached to $C_3N_5$ for the HER reaction. In a 1 M KOH solution, the HER performance of $C_3N_5$-$TiO_2@Ti_3C_2Cl_2$/Cu NC was improved by the addition of Cu, which aids in increased conduction of the NC.

The carbon-based composite made of rGO, NiO, and $Fe_3O_4$ was employed as an electrode material for alcohol electrooxidation as well as supercapacitors. Excellent electrocatalytic activity in alcohol oxidation was the result of the synergistic interaction between two MO semiconductors (NiO and $Fe_3O_4$). Chronoamperometric measurement showed that the catalyst stability for the methanol oxidation reaction was 98.7% over 20 hours. Similarly, the ethanol oxidation reaction showed a stability of 96.4% after 20 hours and an oxidative current 235 $mA/cm^2$ at a peak potential 0.76 V.

Other application of 2D NC include sensor for the recognition of various analytes. Because carbon-based nanomaterials can be easily functionalized, they possess good conductivity along with faster electron transport. NC of Zn–Cu MOF/rGO was used as an electrochemical sensor for the electrochemical detection of oxytocin (OXY). Based on the cyclic voltammetry measurements, the electro-catalytic activity of the Zn–Cu MOF/rGO NC increased due to the synergistic interactions. As a result, the Zn–Cu MOF/rGO NC was able to detect lower OXY levels (limit of detection (LOD) of 1.1 nM (S/N = 3)) with a linear range of 40–400 nM. These composite materials can be included into developed devices for on-site application.

## 2.7 CONCLUSION

The swift advancement of nanotechnology in recent years has rendered the exploration of NCs increasingly significant for the creation of innovative materials tailored for advanced applications. To address the rising demand for multifunctional materials, NCs emerge as an optimal solution due to their versatility and high degree of integrated association. This field is inherently multidisciplinary, encompassing both scientific knowledge and technological expertise to develop macroscopic engineered materials derived from nanoscale structures. Such materials are well-suited to satisfy the evolving requirements stemming from scientific and technological progress. The remarkable potential of NCs is underscored by substantial investments from numerous corporations and governments globally. Consequently, NCs are anticipated to exert a considerable influence on the global economy and business landscape. Notably, they offer significant advantages to various industrial sectors, including electronics, chemicals, transportation, healthcare, and, importantly, environmental protection. Therefore, they are projected to play a crucial role in fostering a cleaner, greener, and safer environment in the years ahead.

## REFERENCES

Bin, J., Wu, M. L., Cong, X., Liu, H. N., & Tan, P. H. (2018). Raman spectroscopy of graphene-based materials and its applications in related devices. *Chemical Society Reviews*, 47, 1822–1873.

Chinnaiah, K., Kannan, K., Krishnamoorthi, R., Palko, N., & Gurushankar, K. (2024a). Nanostructured Ag/NiO composites for supercapacitor and antibacterial applications, and in-silico theoretical investigation. *Journal of Physics and Chemistry of Solids*, 184, 111730.

Chinnaiah, K., Kannan, K., Palko, N., Grishina, M., Chen, Y. S., & Gurushankar, K. (2024b). Exploring the potential of Withania somnifera-mediated Ag/$Mn_3O_4$ nanocomposites as electrode material for high-performance supercapattery device. *Journal of the Taiwan Institute of Chemical Engineers*, 157, 105441.

Chinnaiah, K., Kannan, K., Palko, N., Grishina, M., Guganathan, L., & Gurushankar, K. (2024c). Bioengineered Ag/NiO nanocomposites as advanced battery-supercapacitor electrodes for highly efficient symmetric hybrid devices. *Ionics*, 30, 1691.

Ghidiu, M., Naguib, M., Shi, C., Mashtalir, O., Pan, L. M., Zhang, B., Yang, J., Gogotsi, Y., Billinge, S. J. L., & Barsoum, M. W. (2014). Synthesis and characterization of two-dimensional $Nb_4C_3$ (MXene). *Chemical Communications*, 50, 9517–9520.

Gupta, V., Chandra, N., Tari, V., & Kannan, K. (2024). A sustainable approach for the green synthesis of silver nanoparticles using *Brassica oleracea* subsp. *botrytis* (L.) leaves and its antibacterial activity. *Physics and Chemistry of Solids*, 25(2), 303–310. https://doi.org/10.15330/pcss.25.2.303-310

Gurushankar, K., Karuppaiya, C., Tari, V., & Kannan, K. (2024). Exploring MXenes and MBenes in metal sulfur and selenium battery technologies. In K. Kannan & V. Tari (Eds.), *Novel energy storage and conversion technologies for two-dimensional MXenes and MBenes*. IGI Global. https://doi.org/10.4018/979-8-3693-9546-2.ch005

Henrique, P., Camargo, C., Satyanarayana, K. G., & Wypych, F. (2009). Nanocomposites: Synthesis, structure, properties and new application opportunities. *Materials Research*, 12(1), 1–39.

Jadoon, J. K., & Pham, P. V. (2024). Hybrid $TiO_2$–RGO nanocomposite as high specific capacitance electrode for supercapacitor. *Nanotechnology*, 35, 435706.

Jun, B. M., et al. (2019). Review of MXenes as new nanomaterials for energy storage/delivery and selected environmental applications. *Nano Research*, 12(3), 471–487.

Kannan, K., & Tari, V. (Eds.). (2025). *Novel energy storage and conversion technologies for two-dimensional MXenes and MBenes*. https://www.amazon.com/dp/9798369395462

Khandoker, N., Hawkins, S. C., Ibrahim, R., Huynh, C. P., & Deng, F. (2011). Tensile strength of spinnable multiwall carbon nanotubes. *Procedia Engineering*, 10, 2572–2578.

Li, T., Yao, L., Liu, Q., Gu, J., Luo, R., et al. (2018a). Fluorine-free synthesis of high-purity $Ti_3C_2Tx$ (T = OH, O) via alkali treatment. *Angewandte Chemie International Edition*, 57(21), 6115–6119.

Li, X., Xin, Z. Y., & Li, J. S. (2018b). *Growth of graphene from solid carbon sources. Textile Bioengineering and Informatics Symposium Proceedings*, vol. 2018.

Liu, J., Jiang, X., Zhang, R., Zhang, Y., Wu, L., et al. (2019). MXene-enabled electrochemical microfluidic biosensor: Applications toward multicomponent continuous monitoring in whole blood. *Advanced Functional Materials*, 29(6), 1807326.

Liu, W. W., Chai, S. P., Mohamed, A. R., & Hashim, U. (2014). Synthesis and characterization of graphene and carbon nanotubes: A review on the past and recent developments. *Journal of Industrial and Engineering Chemistry*, 20(4), 1171–1185.

Ma, L., Ting, L. R. L., Molinari, V., Giordano, C., & Yeo, B. S. (2015). Efficient hydrogen evolution reaction catalyzed by molybdenum carbide and molybdenum nitride nanocatalysts synthesized via the urea glass route. *Journal of Materials Chemistry A*, 3(16), 8361–8368.

Mbayachi, V. B., Ndayiragije, E., Sammani, T., Taj, S., Mbuta, E. R., & Khan, A. U. (2021). Graphene synthesis, characterization and its applications: A review. *Results in Chemistry*, 3, 100163.

Mishra, N., Boeckl, J., Motta, N., & Iacopi, F. (2016). Graphene growth on silicon carbide: A review. *Physica Status Solidi A*, 213(9), 2277–2289.

Mu, X., Wang, D., Du, F., Chen, G., Wang, C., Wei, Y., Gogotsi, Y., Gao, Y., & Dall'Agnese, Y. (2019). Revealing the pseudo-intercalation charge storage mechanism of MXenes in acidic electrolyte. *Advanced Functional Materials*, 29(29), 1902953.

Naguib, M., Come, J., Dyatkin, B., Presser, V., Taberna, P., Simon, P., Barsoum, M. W., & Gogotsi, Y. (2012). MXene: A promising transition metal carbide anode for lithium-ion batteries. *Electrochemistry Communications*, 16, 61–64.

Parveen, N., Ansari, S. A., Alnahdi, K. M., Hammud, H. H., Aljamhi, W. A., Alam, M. W., Adil, S. F., & Zoubi, W. A. (2025). Eco-friendly synthesis and applications of graphene-titanium dioxide nanocomposites for pollutant degradation and energy storage. *Journal of Photochemistry and Photobiology A: Chemistry*, 459, 116096.

Prabhu, S., Gowdhaman, A., Harish, S., Navaneetham, M., & Ramesh, R. (2021). Synthesis of petal-like $CoMoO_4$/r-GO composites for high performances hybrid supercapacitor. *Materials Letters*, 295, 129821.

Radha, S. K., Crowley, K., Holler, B. A., Gao, X. P. A., Lambrecht, W. R. L., Volkova, H., Berger, M. H., Pentzer, E., Pachuta, K. G., & Sehirlioglu, A. (2021). Ultrathin 2D-oxides: A perspective on fabrication, structure, defect, transport, electron, and phonon properties. *Journal of Applied Physics*, 129, 220903.

Ronchi, R. M., Arantes, J. T., & Santos, S. F. (2019). Synthesis, structure, properties and applications of MXenes: Current status and perspectives. *Ceramics International*, 45(15), 18167–18188.

Rostami, M., Ziarani, G. M., Badiei, A., Ghasemi, J. B., Khazaei, M., Jourshabani, M., & Lee, B. K. (2025). Harnessing Z-scheme charge transfer in ultrathin $C_3N_5$/$TiO_2$@$Ti_3C_2Cl_2$ MXene nanosheets for sustainable water purification, hydrogen evolution, and biocompatibility. *Journal of Science: Advanced Materials and Devices*, 10(1), 100846.

Salim, O., Mahmoud, K. A., Pant, K. K., & Joshi, R. K. (2019). Introduction to MXenes: Synthesis and characteristics. *Materials Today Chemistry*, 14, 100191.

Saraee, H., Tanzifi, M., Mansouri, M., Saraee, M., & Yaraki, M. T. (2025). Fabrication of a novel Z-scheme 2D/1D g-$C_3N_4$/polyaniline decorated on activated carbon with effective visible-light photocatalytic activity. *Environmental Research*, 267, 120658.

Sasikumar, K., Rajamanikandan, R., & Ju, H. (2025). Z-scheme charge transfer heterostructure with $NiV2O_6$@$gC_3N_4$ nanocomposite: A competent photocatalyst for boosting the photodegradation of antibiotics. *Journal of the Taiwan Institute of Chemical Engineers*, 168, 105960.

Shekhirev, M., Shuck, C. E., Sarycheva, A., & Gogotsi, Y. (2021). Characterization of MXenes at every step, from their precursors to single flakes and assembled films. *Progress in Materials Science*, 120, 100757.

Shi, C., Beidaghi, M., Naguib, M., Mashtalir, O., Gogotsi, Y., & Billinge, S. J. L. (2014). Structure of nanocrystalline $Ti_3C_2$ MXene using atomic pair distribution function. *Physical Review Letters*, 112, 125501.

Sucharitakul, S., Ye, G., Lambrecht, W. R. L., Bhandari, C., Gross, A., He, R., Poelman, H., & Gao, X. P. A. (2017). $V_2O_5$, A 2D van der Waals oxide with strong in-plane electrical and optical anisotropy. *ACS Applied Materials & Interfaces*, 9(28), 23949–23956.

Sun, W., Shah, S., Chen, Y., Tan, Z., Gao, H., et al. (2017). Electrochemical etching of Ti2AlC to Ti2CTx (MXene) in low-concentration hydrochloric acid solution. *Journal of Materials Chemistry A*, 5(41), 21663–21668.

Sun, W., Wang, H. W., Vlcek, L., Peng, J., Brady, A. B., Osti, N. C., Mamontov, E., Tyagi, M., Nanda, J., Greenbaum, S. G., Kent, P. R. C., & Naguib, M. (2020). Multiscale and multimodal characterization of 2D titanium carbonitride MXene. *Advanced Materials Interfaces*, 7(11), 1902207.

Sun, W., Xie, Y., & Kent, P. R. C. (2018). Double transition metal MXenes with wide band gaps and novel magnetic properties. *Nanoscale*, 10(25), 11962–11968.

Tari, V., Gupta, V., Joshi, A., Kurve, P., & Chandra, N. (2024). Biodiversity loss due to forest diversion. In A. Rathore (Ed.), *Impact of societal development and infrastructure on biodiversity decline* (pp. 137–165). IGI Global. https://doi.org/10.4018/979-8-3693-6950-0.ch010

Tari, V., Kannan, K., Koliar, J., Gurushankar, K., & Wahyuni, D. K. (2025). Influence of zinc oxide nanomaterials on crop plants: A prospective analysis. In D. Jabbrova, D. Sarkar, Rakesh, S. Singh, & R. Datta (Eds.), *Smart nanomaterials in agroforestry systems*. Springer. https://doi.org/10.1007/978-981-96-1337-3_8

Tari, V., Kannan, K., & Vishwakarma, V. (2022). Emerging antiviral technologies. In P. Nunthavarawong, S. M. Rangappa, S. Siengchin, & M. Thoppil-Mathew (Eds.), *Antimicrobial and antiviral materials: Polymers, metals, ceramics, and applications* (1st ed., pp. 137–150). CRC Press. https://doi.org/10.1201/9781003143093-8

Telychko, M., Berger, J., Majzik, Z., Jelínek, P., & Švec, M. (2015). Graphene on SiC(0001) inspected by dynamic atomic force microscopy at room temperature. *Beilstein Journal of Nanotechnology*, 6, 901–906.

Urbankowski, P., Anasori, B., Makaryan, T., Er, D., Kota, S., et al. (2016). Synthesis of two-dimensional titanium nitride Ti4N3 (MXene). *Nanoscale*, 8(22), 11385.

Vacik, J., et al. (2021). Ion-beam-induced crystallization of radiation-resistant MAX phase nanostructures. *Radiation Effects and Defects in Solids*, 176(1–2), 119–137.

Varadarajan, S., Kavitha, A., Selvaraju, P., Muthu, S. E., Gurushankar, K., Shanmugan, S., & Kannan, K. (2024). Enhanced photocatalytic hydrogen evolution by TiO2: A synergistic approach with defect-rich SnS2 and Ti3C2 MXene cocatalysts. *Hydrogen*, 5(4), 940–957.

Verger, L., Xu, C., Natu, V., Cheng, H. M., Ren, W., & Barsoum, M. W. (2019). Overview of the synthesis of MXenes and other ultrathin 2D transition metal carbides and nitrides. *Current Opinion in Solid State and Materials Science*, 23(3), 149–163.

Vinyas, M., Athul, S. J., Harursampath, D., Loja, M., & Thoi, T. N. (2019). A comprehensive review on analysis of nanocomposites: From manufacturing to properties characterization. *Materials Research Express*, 6, 092002.

Wang, C., Chen, S., & Song, L. (2020). Tuning 2D MXenes by surface controlling and interlayer engineering: Methods, properties, and synchrotron radiation characterizations. *Advanced Functional Materials*, 30(47), 2000869.

Wang, H. W., Naguib, M., Page, K., Wesolowski, D. J., & Gogotsi, Y. (2016). Resolving the structure of Ti3C2Tx MXenes through multilevel structural modeling of the atomic pair distribution function. *Chemistry of Materials*, 28(1), 349–359.

Xie, H., Li, Z., Cheng, L., Haidry, A. A., Tao, J., Xu, Y., Xu, K., & Ou, J. Z. (2022). Recent advances in the fabrication of 2D metal oxides. *iScience*, 25(1), 103598.

Xu, C., Wang, L., Liu, Z., Chen, L., Guo, J., et al. (2015). Large-area high-quality 2D ultrathin Mo2C superconducting crystals. *Nature Materials*, 14, 1135–1141.

Xu, H., Wu, X., Li, X., Luo, C., Liang, F., Orignac, E., Zhang, J., & Chu, J. (2018). Properties of graphene-metal contacts probed by Raman spectroscopy. *Carbon*, 127, 491–497.

Zahra, T., Gassoumi, G., Gouadria, S., Abdullah, M., Alharbi, F. F., Aman, S., & Farid, H. M. T. (2024). Facile fabrication of BiFeO3/g-C3N4 nanohybrid as efficient electrode materials for supercapacitor application. *Diamond and Related Materials*, 144, 110927.

Zhan, X., Si, C., Zhou, J., & Sun, Z. (2020). MXene and MXene-based composites: Synthesis, properties and environment-related applications. *Nanoscale Horizons*, 5(2), 235–258.

Zhong, J., Ji, Z., Gao, X., Lin, Z., Lu, X., Zhu, G., Xu, K., Song, P., & Shen, X. (2025). Engineering phosphorus vacancies in reduced graphene oxide anchored Co2P nanoparticles toward optimal supercapacitive properties. *Fuel*, 386, 134281.

Chapter 3

# Two-dimensional hybrid composite electrode materials for supercapacitors

## Advances, prospects, and sustainability

*Dev Kumar, Yu-Wu Wang, Vanshika Bhardwaj, Durvesh Gautam, Ekta Singh, Karthik Kannan, and Aman Kumar*

### 3.1 INTRODUCTION

There has never been a greater need for reliable and sustainable energy storage due to the global movement toward cleaner energy sources. As the use of renewable energy sources like solar and wind power grows, there is an even more need to develop systems that can efficiently store surplus energy and discharge it when needed. This change in the energy framework has significantly accelerated supercapacitor research and development (Dissanayake & Kularatna-Abeywardana, 2024). Supercapacitors are preferred over traditional batteries due to their quick charging and power-delivery capabilities, which make them ideal for a variety of applications, including renewable energy systems and portable devices. On that note, however, one of their greatest weaknesses—the limited energy density—remains an impediment for extensive applications (Kannan & Tari, 2025; Gurushankar et al., 2024; Dissanayake & Kularatna-Abeywardana, 2024; Muralee Gopi et al., 2025).

In recent development of high-performance energy storage devices such as supercapacitors and batteries has driven intensive research into advanced electrode materials. Among them, two-dimensional (2D) materials notably graphene, transition metal dichalcogenides (TMDs), and MXenes have garnered significant attention due to their unique physicochemical properties, including high surface area, excellent electrical conductivity, and tunable chemical functionalities (Tari et al., 2022; Gupta et al., 2024; Tari et al., 2025). However, there are commonly inherent restrictions with explicit 2D materials. In contrast to TMDs, which may show poor conductivity and mechanical instability, graphene may experience restacking and low pseudocapacitance. To conquer such obstacles, researchers are increasingly using hybrid composite techniques, which integrate 2D materials with other nanostructures (such as metal oxides, polymers, or carbon-based materials) or combine two or more types of 2D materials. These 2D hybrid composite materials are engineered to synergistically enhance electrochemical

DOI: 10.1201/9781003584742-3

performance by leveraging the complementary strengths of their constituents, improving ion transport, electronic conductivity, structural stability, and charge storage capacity. Thus, the rational design of 2D hybrid composites represents a promising strategy to create next-generation electrode materials with superior energy and power densities, cycling stability, and rate capability (Hafeez, 2022).

### 3.1.1 Overview of energy storage challenges

The world's transition to carbon-neutral energy sources by 2060, as pledged under the Paris Agreement to limit global warming to 1.75°C by 2100, invites renewable energy to provide ~74% of the energy mix (UN Environment and International Energy Agency, 2017). UN Environment Program and International Energy Agency (IEA) reports (UN Environment and International Energy Agency, 2017; Energy Technology Policy Division-IEA, 2017) emphasize the central role of wind, solar, and bioenergy with carbon capture and storage (BECCS). The global transition toward carbon-neutral energy systems by 2060, as stipulated by the Paris Agreement to limit global warming to 1.75°C by 2100, necessitates renewable energy sources constituting ~74% of the energy mix (Abergel et al., n.d.). Analyses by the UN Environment Program and International Energy Agency (IEA) (Energy Technology Policy Division-IEA, 2017; IEA, 2024; UN Environment and International Energy Agency, 2017) underscore the centrality of wind, solar, and BECCS. However, the inherent intermittency of these sources introduces destabilizing power fluctuations, compromising grid reliability, a challenge extensively detailed by Rugolo and Aziz (2012) in the context of electricity storage. Energy Storage Systems (ESS) are thus critical, enabling surplus energy storage during low-demand periods and discharge during peak demand to ensure grid stability. Xie et al. (UN Environment and International Energy Agency, 2017) and L. Yang et al. (2016b) emphasize that ESS must achieve scalable deployment, operational autonomy, and seamless multigrid integration to regulate power and frequency effectively. Though central, ESS are plagued by several challenges: high capital costs hinder mass deployment (Giovanelli et al., 2019), while microgrid integration islanded or grid-connected, necessitates advanced architectures to provide power quality and ancillary services like EV charging (Mazgut et al., 2015). Performance bottlenecks like energy conversion efficiency, safe charge-discharge cycling, and system lifetime are unsolved (Parra et al., 2017). Battery ESS, critical for electric transport, is plagued by cost and energy density limitations, necessitating ongoing innovation (Sheikhi et al., 2013). Environment and application-specific optimization renders deployment even more challenging (Azeem et al., 2021). Decentralized generation boosts ESS's peak shaving capability, minimizing capacity expansion needs (Haakana et al., 2017; Nguyen et al., 2015). Mass deployment of ESS and innovative charging technologies (Cheng et al., 2017; Mou et al., 2014) offer load-balancing opportunities, but technical and economic challenges need to be overcome through ongoing R&D for a clean energy future.

### 3.1.2 Importance of 2D hybrid composite electrodes in supercapacitors

The ever-growing global population, coupled with increased industrial activity, has significantly enhanced the requirement for more effective energy storage materials. For this purpose, 2D hybrid composite electrodes have emerged as a revolution in supercapacitor technology. The materials are progressively being hailed as having the capacity to fuel a wide range of technologies—electric vehicles to large-scale stationary storage systems—based on their excellent power delivery, high charge-discharge rate, long lifespan, and consistent capacitance behavior (Adedoja et al., 2023).

Improvements in the materials science sector have facilitated the introduction of novel 2D hybrid electrode materials, with new dimensions being opened for enhancing the performance of supercapacitors (Philip & Ruban Kumar, 2023). Graphene (Wei et al., 2023), transition metal dichalcogenides (TMDs) (Ali et al., 2022), and MXenes (Nasrin et al., 2022) have better electrochemical properties such as high surface area, high conductivity, and high mechanical flexibility (Forouzandeh & Pillai, 2021). Schematic depiction of their synthesis routes, primary characteristics, and probable applications are depicted in Figure 3.1.

Through hybridization of various materials into hybrid composites, researchers have been able to improve key performance indicators like specific capacitance, energy density, and cycling stability to a great extent (Dar et al., 2024). Researchers like Chu and Majumdar (Chu & Majumdar, 2012) have reported the effect of these technologies in addressing global sustainable energy challenges, while Telaretti and Dusonchet (Telaretti & Dusonchet, 2017) mention their practicality, particularly in the U.S. energy sector. One of the greatest challenges, however, still persists: the achievement of high energy and power densities simultaneously, a challenge that Horn et al. (2019) explored, particularly in the context of electric vehicles. Xie et al. (2016) argue that the development of advanced electrode materials is critical to overcoming this challenge, with carbon-based materials like graphene and carbon nanotubes, transition metal oxides like $MnO_2$ and $RuO_2$, and conducting polymers being the foundation of current designs. Of these, 2D hybrid composites, particularly those with MXenes, are particularly promising due to their greater electrical conductivity, large surface area, and structural stability.

MXenes, originally synthesized in 2011 by the group of Yury Gogotsi via exfoliation of $Ti_3AlC_2$ (Naguib et al., 2011), have transformed electrode performance. Owing to metallic conductivity, $Ti_3C_2$ MXenes possess 260 S/cm, compared to reduced graphene oxide's ~1 S/cm (Wu et al., 2017), and owing to hydrophilic character, intercalation of ions is favored (S.

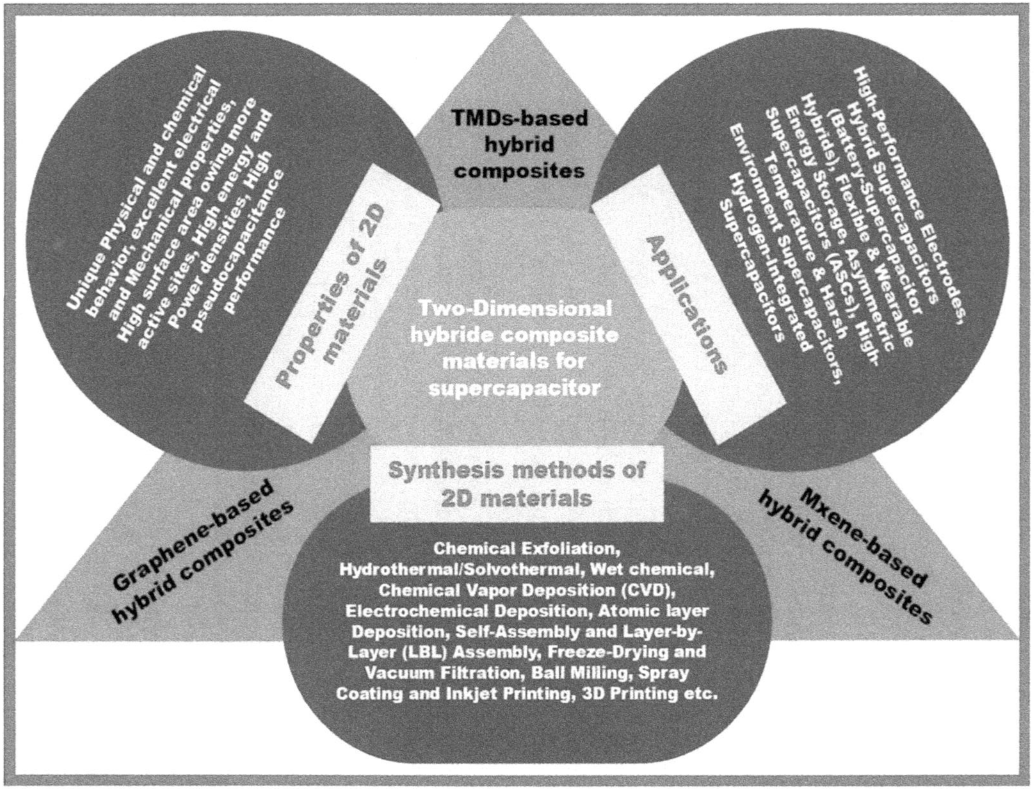


*Figure 3.1* The methods of synthesis, materials properties, and potential applications of 2D materials (Ali et al., 2023; Javed et al., 2023).

Zhang et al., 2015). Wu et al. (2017) also showed them to be stable along with carbon nanoplates, while Zhang (S. Zhang et al., 2015) emphasized the breadth of application in ultrathin 2D nanomaterials. Hybrid electrodes based on the combination of the pairing of $Ti_3C_2T_x$ or $V_4C_3$ with $MnO_2$ or carbon nanotubes (CNT) demonstrate such potential. An example is presented by Zhou et al. (2019) with an MXene/$MnO_2$ composite on carbon cloth, where it recorded a specific capacitance of 511.2 F/g, outperforming the performance of the standalone $MnO_2$ nanorod at 302.3 F/g. This has been explained by the synergy of electric double-layer capacitance and pseudocapacitance by Cao et al. (2022). Gogotsi and Pomerantseva (Pomerantseva & Gogotsi, 2017) also reported that smaller pores (less than 2 nm) can be used to boost capacitance and energy density, a characteristic optimized by 2D hybrids without sacrificing larger pores for power density in high-current applications (Xie et al., 2016). Apart from such performance enhancement, such composites are also remarkable in ion transport and stability, considerations that are important in practical applications. For instance, multivalent ion intercalation, such as $Mg^{2+}$ or $Zn^{2+}$, into MXene layers can lower resistance and maximize power output, as shown by Pomerantseva and Gogotsi (Pomerantseva & Gogotsi, 2017). Hybrid materials like $Ti_3C_2T_x$/MWCNT or $MnO_2$/$Ti_3C_2T_x$ have been reported to maintain about 88% of their capacitance after 10,000 cycles, which is superior to $MnO_2$-only electrodes, as shown by Rakhi et al. (2016). Mechanically, $Ti_3C_2$ MXene's remarkable Young's modulus of 330 GPa (Gao et al., 2018) is an extremely strong performance, while Lu et al., 2020 reported that coupling it with $Ni_2Co$-LDH nanoarrays can deliver battery-type supercapacitors with a rate capability of 126 mAh/g. Economically and environmentally, their adaptability reduces production costs (Xie et al., 2016), making them sustainable solutions. By bridging the gap between energy and high-power densities, MXene-based 2D hybrid composite electrodes are transforming supercapacitors into highly efficient and versatile energy storage devices (Cao et al., 2022). Hybrid composites, consisting of the union of 2D materials with other functional components like conducting polymers, metal oxides, or carbon materials, take advantage of the unique synergies between components. This approach optimizes electrochemical properties to produce substantial improvement in specific capacitance, energy density, and cycling stability. Graphene-based composites, for example, can achieve excellent charge storage capacities when coupled with transition metal oxides, while MXenes significantly enhance conductivity and electrochemical performance when coupled with polymers or other 2D materials.

This chapter explores the recent advancements in 2D hybrid composite electrodes for supercapacitors, with a focus on how the structure of these materials relates to their properties and how electrochemical performance can be enhanced. New fabrication techniques are improving charge storage efficiency, while optimizing interfacial interactions to boost energy density and cycling stability. Scalable production methods and multifunctional design approaches are key to integrating these composites into renewable energy systems. However, sustainability issues such as resource efficiency and managing the lifecycle of these materials need to be addressed to ensure the eco-friendly development of supercapacitors. Ultimately, 2D hybrid composites offer a promising foundation for the next generation of energy storage technologies.

### 3.1.3 Fabrication techniques

The design of high-performance hybrid composite electrode materials is at the core of next-generation supercapacitor technology because these materials synergistically merge the extremely high electrical conductivity (~$10^6$ S/m) of graphene, bandgap tunability, and TMD redox sites with the model pseudocapacitive performance and mechanical robustness of MXenes (Choi et al., 2017; Shao et al., 2015; Yu et al., 2018). Through ordered nanostructure design with optimal porosity (e.g., pore diameters in 2–50 nm porosity mesoporous networks) and interfacial interactions, these composites maximize charge storage capacity (up to more than 500 F/g in some systems) and facilitate stable cycling durability (>10,000 cycles) (Dissanayake & Kularatna-Abeywardana, 2024; Forouzandeh et al., 2020). Next-generation supercapacitors can be fabricated universally using the following fabricating methods by taking advantage of the very high electrical conductivity, high surface area, and improved electrochemical functions of graphene, TMDs, and MXenes based hybrid composites.

## 3.2 SOLUTION-BASED SYNTHESIS METHODS

### 3.2.1 Hydrothermal/solvothermal method

A popular technique to develop hybrid electrode materials is the hydrothermal/solvothermal method, which combines various nanostructures with 2D materials including graphene, TMDs, and MXenes. This method allows for the controlled development of homogeneous nanostructures and close-knit component interfaces by using chemical processes in sealed, in high-temperature conditions (typically 120–250°C), high-pressure autoclave environment using water (hydrothermal) or organic solvents (solvothermal) (Gan et al., 2020; O'Hare, 2001). It provides well-dispersed, hierarchical hybrid composites with numerous active sites, greater structural stability, and higher electrical conductivity by enabling the in-situ production of TMD nanosheets or metal oxides directly on graphene or MXene platforms. As a result, hydrothermal/solvothermal synthesis produces electrodes with high electrochemical performance, such as high capacitance, rate capability, and cycle durability, making this process a flexible and successful approach for next-generation energy storage materials. This process allows the material properties to be controlled with high precision using parameters such as reaction time, chemical composition, and temperature.

**Applications:**

- Hydrothermal/solvothermal methods enable in-situ growth of uniform and hierarchical metal oxide nanostructures on graphene, TMDs, or MXenes, enhancing redox activity and structural stability for improved energy storage.
- These methods facilitate the controlled deposition or polymerization of conductive polymers onto 2D materials, improving mechanical flexibility and electrochemical performance of hybrid electrodes.
- It is generally employed to fabricate TMD/Graphene and MXene/TMD hybrid architectures by enabling in-situ growth of TMDs or MXenes on graphene films.
- This method can produce hybrid materials with high energy storage capacity, high charging rates, and long lifetime, key characteristics of next-generation energy storage devices.

**Advantages:**

- The main benefit of this method is its ability to produce materials with a high surface area and tailored porosity, both of which are essential for optimal supercapacitor performance.
- Produces highly crystalline nanomaterials.
- Allows for controlled morphology and composition.
- Compared to other synthesis methods, this approach is relatively simple, scalable, and energy-efficient, making it attractive for industrial applications, i.e., Scalable for large-scale production.

### 3.2.2 Chemical precipitation

Chemical precipitation is a straightforward and efficient synthesis method widely employed to fabricate hybrid electrode materials by combining graphene, TMDs like $MoS_2$, MXenes, with other nanostructures such as metal oxides, conductive polymers, or carbon-based materials. This technique involves the controlled formation of precursor compounds from solution, which subsequently precipitate onto the surfaces of 2D materials, enabling uniform growth and intimate contact between different components. By precisely tuning reaction conditions such as pH, temperature, and precursor concentration, chemical precipitation allows the formation of well-dispersed, hierarchical hybrid composites with enhanced electrochemical properties. These hybrids leverage the high conductivity of graphene and MXenes, the rich redox activity of TMDs and metal oxides, and the mechanical flexibility of polymers to achieve improved charge storage through synergistic electric double-layer and pseudocapacitive mechanisms. Consequently, electrodes prepared via chemical precipitation exhibit increased specific capacitance, excellent rate capability, and long-term cycling stability, making them highly promising for advanced supercapacitor applications. The beauty of this method lies in its simplicity; the materials essentially "self-assemble" through controlled chemical reactions. It involves the co-precipitation of metal precursors to form metal oxides or sulfides on graphene or MXene nanosheets (Chaudhuri & Yun, 2022).

**Applications:**

- Used to synthesize MXene/TMD and Graphene/TMD composites by controlled nucleation and growth of the secondary material.
- Hybrid materials synthesized by chemical precipitation offer high specific capacitance and energy density by combining electric double-layer and pseudocapacitive behavior.
- Incorporation of polymers and carbon-based materials enhances mechanical flexibility, making the hybrids suitable for bendable and wearable energy storage devices.

**Advantages:**

- Simple and cost-effective.
- Works at mild temperatures (usually below 100°C).
- Allows for precise tuning of composition and morphology.
- Easy to adapt for large-scale production.
- Allows for easy doping with other elements to enhance performance.

**Challenges:**

- Requires careful control of reaction conditions to prevent clumping.
- May produce less crystalline materials compared to high-temperature methods.
- Needs additional processing steps (like annealing) to optimize performance.

### 3.2.3 Chemical vapor deposition (CVD)

Chemical Vapor Deposition (CVD) is a high-temperature synthesis technique that facilitates the direct growth of TMDs and graphene on conductive substrates, yielding high-quality films with well-defined morphologies. This approach is particularly beneficial for supercapacitor electrodes, as it enables the formation of ultrathin, defect-free, and high-surface-area hybrid materials, enhancing electrochemical performance. In this method, precursor gases such as $CH_4$ and $C_2H_2$ for graphene, along with metal oxides like $MoO_3$ and $WO_3$ for TMDs, are introduced into a high-temperature furnace, typically operating between 600 and 1000°C under controlled gas flow conditions. Chemical reactions involved enable controlled growth of TMD monolayers or graphene films on substrates like Ni, $SiO_2$, and Cu foils. By controlling conditions like gas flow rate, temperature, and deposition time, ultrathin layers of $MoS_2$, $WS_2$, or graphene/MXene can be synthesized with controlled structures (Dahmen, 2003; Q. Zhang et al., 2023a).

**Applications:**

- CVD enables the synthesis of high-quality, uniform thin films and layered structures with precise control, making it ideal for micro-scale, high-power supercapacitors used in portable and on-chip electronics.
- Employed for the direct growth of TMDs ($MoS_2$, $WS_2$, etc.) on graphene sheets.
- Can be used to apply ultra-thin TMD layers onto graphene or MXenes.

**Advantages:**

- This method has several advantages, such as the formation of crystalline and high-purity defect-free films.
- It also possesses very precise thickness control, and accordingly monolayers or multilayers of specified properties can be built.
- The highly ordered nature of CVD-grown materials facilitates charge transfer efficiency and ion diffusion, both of which are necessary to obtain optimal supercapacitor performance.

**Challenges:**

- It is expensive and requires specialized equipment.
- This method is not necessarily suitable for large-scale use.
- It is similarly expensive and energy-consuming, which includes hot furnaces and vacuum equipment.
- In addition, scalability is an enormous challenge, as mass production is difficult to practice.
- Optimal electrode properties are obtained by maintaining tight control over growth conditions, which makes the process more difficult.

### 3.2.4 Electrochemical deposition

Electrochemical techniques utilize direct electrodeposition to deposit materials onto a conductive substrate through voltage. The technique allows controlled synthesis of nanostructured materials via electrochemical reactions to create uniform, highly conductive, and electrochemically active films. Controlled parameters like voltage, current density, electrolyte composition, and deposition time can be utilized to synthesize customized hybrid nanostructures with enhanced properties for energy storage (Kumar et al., 2024).

In this process, metal precursor salts (e.g., W, Ti, or Mo for MXenes and TMDs) and graphene oxide (GO) are dissolved in an electrolyte solution. When an external voltage is applied, metal cations are reduced and deposited on the working electrode, and nanostructured layers with good adhesion are formed. At the same time, GO can be reduced to supply reduced graphene oxide (rGO) to improve electrical conductivity (Kumar et al., 2024). This way, direct synthesis of hybrid structures, for example, MXene/graphene, Ni-Co, or $MoS_2$/rGO layered double hydroxides (LDHs), on flexible or rigid substrates can be used as supercapacitor electrodes.

**Applications:**

- It is utilized to deposit TMDs or MXenes onto graphene-based electrodes, modifying the configuration.
- Enables you to create materials with amazing precision, either as thin films or tiny nanoparticles, such as doing something perfectly by hand.

**Advantages:**

- Low-cost and scalable.
- Excels with flexible substrates, so it is extremely versatile.
- Forms tightly clinging nanostructured films that are the perfect thickness and possess a gigantic surface area—key to maximizing charge storage.
- Permits simultaneous mixing of several materials, creating teamwork effects that enhance charge mobility and energy storage.

### 3.2.5 Self-assembly and layer-by-layer (LBL) assembly

This approach builds composite layers by leveraging electrostatic attractions, hydrogen bonds, or Van der Waals forces to allow materials to self-assemble like puzzle pieces. It enables high precision in the control of nanostructure growth at the molecular level, to allow the formation of complex, layered architectures with high electrochemical performance. By leveraging these molecular interactions to its benefit, self-assembled materials show high uniformity, tunable porosity, and controllable surface chemistry, and are a top contender for high-performance energy storage devices (Wang et al., 2025). Materials like graphene, TMDs, and MXenes spontaneously arrange themselves into ordered structures under provided conditions, under the control of internal molecular forces and external conditions like pH, temperature, or solvent properties. For instance, in a solution-based self-assembly process, positively charged TMD nanosheets (like $MoS_2$ or $WS_2$) aggregate with negatively charged GO to form layered heterostructures that show improved conductivity, improved ion accessibility, and high mechanical stability—appropriate for supercapacitor electrodes (Tian et al., 2019). The Layer-by-Layer (LBL) assembly method, a sequential layer-by-layer process, builds multilayered thin films by alternating the deposition of oppositely charged materials, anchored in place by electrostatic attractions, hydrogen bonds, or other molecular forces. When applied in the construction of supercapacitor electrodes, LBL assembly enables the controlled placement of materials like graphene, MXenes, metal oxides, and conductive polymers to form highly ordered structures with optimized electrochemical performance. As an example, an LBL-assembled graphene/MXene hybrid electrode can show improved charge storage, due to improved ion transport pathways and improved electrochemical stability (Sneha et al., 2025).

**Applications:**

- Graphene/TMD, MXene/TMD, and MXene/Graphene, hybrid films can be formed.
- Allows the creation of super-thin, precisely organized multilayer structures.

**Advantages:**

- These assembly methods are most effective at creating ultra-uniform coatings that may be thickness-adjusted, defect-free, and nanoscale precisely conformal.
- Self-assembly and LBL methods are ideally suited for flexible, wearable energy storage devices, which render electrodes light and strong.
- These methods allow for the fine-tuning of material properties by adjusting the number of layers, composition, and interaction strength between components.
- Compatible with various substrates (flexible or rigid).

**Challenges:**

- Challenges remain in optimizing these techniques for large-scale production.
- The LBL method requires multiple deposition steps, which can slow down manufacturing throughput.
- Achieving strong interfacial adhesion between different material layers is critical for long-term electrochemical stability.
- Requires precise control over surface interactions.

## 3.2.6 Freeze-drying and vacuum filtration

### *3.2.6.1 Freeze-drying*

In this method, a liquid suspension of materials such as graphene, MXenes, or TMDs is first frozen and then subjected to sublimation under low pressure. This process removes the solvent without collapsing the material structure, resulting in a highly porous architecture with interconnected nanostructures. The key advantages of freeze-drying include enhanced ion diffusion, large surface area, and improved mechanical stability, making it highly beneficial for energy storage applications. However, the process can be time-consuming and requires careful optimization of freezing conditions to ensure uniform pore formation (Yeon et al., 2022).

**Applications:**

- Produces porous, aerogel-like structures for high-performance supercapacitors.
- Used in MXene/Graphene and Graphene/TMD composites.

**Advantages:**

- Results in highly porous, three-dimensional (3D) architectures with excellent conductivity.
- Retains a high surface area for electrolyte accessibility.

#### 3.2.6.2 *Vacuum filtration*

This technique involves filtering a suspension of nanomaterials through a porous membrane under vacuum, leading to the deposition of a uniform and compact thin film. The method allows precise thickness control, excellent film uniformity, and strong interfacial adhesion between layers, enhancing the electrode's electrical conductivity and mechanical flexibility. Although vacuum filtration is scalable and widely applied in the fabrication of flexible supercapacitor electrodes, it may cause restacking of the nanosheets, and this can restrict ion accessibility and overall electrochemical performance. Combining this process with post-treatment processes, such as thermal annealing or chemical modification, can restrict such limitation and further improve supercapacitor performance (Xing et al., 2023).

**Applications:**

- Used to whip up bendy MXene/graphene electrodes that flex with ease.
- Crafts neatly layered or stacked setups perfect for supercapacitor electrodes.

**Advantages:**

- Low-cost, scalable, and simple.
- Yields highly dense, mechanically durable films with robust structural integrity.

## 3.3 MECHANICAL AND PHYSICAL METHODS

### 3.3.1 Ball milling

A high-energy mechanical milling process breaks bulk materials into nanosheets and enables in situ hybridization. During the process, grinding media (steel, zirconia, or tungsten carbide balls) exert mechanical forces inside a rotating chamber, breaking down bulk materials into ultrathin nanosheets. Wet ball milling prevents agglomeration and improves dispersion and structural integrity. This method enhances the surface area, electron transport, and ion diffusion, all critical for optimizing supercapacitor performance.

Ball milling also enables in situ hybridization, where different nanomaterials are milled together to form composite structures with synergistic properties. For example, $MoS_2$/GO hybrids and MXene/graphene composites produced through ball milling exhibit improved conductivity, electrochemical stability, and charge storage capabilities (Pradeep et al., 2022).

**Applications:**

- Used for the preparation of MXene/TMD composites and graphene/TMD.
- Enables defect engineering and nanostructure formation.

**Advantages:**

- Scalable and cost-effective.
- Enables one-step hybrid composite synthesis.

### 3.3.2 Spray coating and inkjet printing

Nanomaterials such as MXene, graphene, or TMD nanosheets, dissolved in a liquid solvent, are deposited onto substrate materials by spraying or printing. Spray coating and inkjet printing are new approaches to the fabrication of 2D hybrid composite supercapacitor electrodes, with uniform, scalable, and high-resolution deposition of material. In spray coating, a nanomaterial suspension, for example, graphene, MXenes, or TMDs, is atomized and sprayed onto a substrate under compressed gas to deposit thin, uniform films of adjustable thickness. The method is best suited for large-area supercapacitor electrodes due to cost-effectiveness and compatibility with flexible substrates but possibly requiring post-treatments for enhanced adhesion and conductivity. In contrast, inkjet printing is high-resolution, mask-free digital-based inkjet printing process for the printing of functional inks containing nanomaterials, for example, rGO or MXenes, into controlled patterns onto various substrates. The method is best suited for micro-scale patterning with low material loss, suitable for compact, wearable, and flexible supercapacitors. Post-processing treatments, for example, annealing, and challenges in optimizing ink formulation, controlling the viscosity limit, have to be overcome to achieve maximum electrochemical performance. While spray coating is more suitable for large-scale coatings, inkjet printing offers high-resolution patterning for advanced energy storage applications, making both techniques promising for next-generation supercapacitor electrodes in portable electronics and smart devices (Moridi et al., 2014; Sundriyal & Bhattacharya, 2017).

**Applications:**

- Used for fabricating thin-film electrodes.
- Suitable for wearable and flexible energy storage devices.

**Advantages:**

- Simple and customizable for large-area deposition.
- Compatible with printed electronics.

## 3.4 HYBRID AND ADVANCED METHODS

### 3.4.1 Atomic layer deposition (ALD)

Atomic Layer Deposition (ALD) is a precise technique that deposits thin layers of material on a substrate through sequential, self-limiting chemical reactions. Alternating between pulses of precursor gases and exposure of the substrate, a monolayer is deposited per cycle with good control of film growth. Carried out under low temperatures (typically 100–300°C), ALD is compatible to coat temperature-sensitive substrates, such as flexible substrates, without damage (Ansari et al., 2024).

For the material deposition on 2D hybrid composite supercapacitor electrodes' substrates, ALD has optimal uses for the deposition of materials such as transition metal oxides (e.g., $TiO_2$, $Al_2O_3$), conductive polymers, and TMDs. With proper precursor gases and adjustment of parameters like temperature, pressure, and pulse time, ALD can allow the preparation of controlled film thickness, porosity, and composition. Such control is necessary in order to create the electrochemical performance of supercapacitor electrodes for high capacitance, high-rate charge/discharge operations, and durability (Nahirniak et al., 2023

**Applications:**

- Deposits extremely thin coatings of metal oxides or sulfides on graphene or MXene surfaces.

**Advantages:**

- Provides uniformity and high precision.
- Improves the electrochemical processes stability.

### 3.4.2 3D printing

3D printing, or additive manufacturing, is a novel fabrication technique that allows for scalable and precise construction of 3D objects from digital blueprints. In the case of supercapacitors, 3D printing possesses a distinct advantage by allowing the design and construction of extremely intricate, optimized, and tailored electrode architectures, especially when coupled with 2D hybrid composite materials like graphene, MXenes, and TMDs (Saeed et al., 2024).

The process of 3D printing involves deposition of materials layer by layer, for instance, conductive inks or pastes with 2D nanomaterials, to fabricate complex geometries. FDM, SLA, and direct ink writing (DIW) are common methods used to fabricate supercapacitor electrodes. Such methods allow for precise control of material location, which can be employed to fabricate structures with engineered microstructures, porosity, and surface areas that significantly enhance the electrochemical performance of supercapacitors (Gómez et al., 2021).

In DIW, for instance, a conductive paste or ink that includes 2D materials such as graphene, MXenes, or TMDs is pushed through a nozzle in a layer-by-layer design to form complex 3D geometries (Hou et al., 2024). The method permits the formation of highly porous structures that facilitate ion diffusion and increase charge storage efficiency, both important to high-performance supercapacitor electrodes. By facilitating precise control of micro- and macro-scale geometry, such as interconnected pores and surface texture, 3D printing enhances electrode performance by boosting capacitance, rate capability, and cycling stability.

**Applications:**

- Produces tailored, high-efficiency supercapacitor electrodes.
- Supports inks formulated with graphene and MXenes.

**Advantages:**

- Allows customized electrode designs for enhanced performance.
- Supports consistent, large-scale manufacturing.

The fabrication technique for hybrid composites incorporating MXenes, graphene, and TMDs, is vital for boosting the electrochemical capabilities of supercapacitors. Hydrothermal, CVD, electrodeposition, vacuum filtration, and self-assembly are widely employed based on material compatibility, cost, and scalability. Advancements in hybrid fabrication techniques further allow the development of next-generation supercapacitors with high energy density, power density, and long cycle life.

We also give a table and a paragraph summarizing different types of 2D composite materials, synthesis methods, and advantages, which are relevant for supercapacitors, batteries, and sensor applications (Table 3.1).

*Table 3.1* Overview of 2D materials their composites, synthesis methods & advantages

| *2D material type* | *Composite type* | *Synthesis method* | *Advantages* | *[Ref.]* |
|---|---|---|---|---|
| Graphene-based | Graphene/Metal Oxide (e.g., $MnO_2$, $Fe_3O_4$) | Hydrothermal + Sol-Gel | High surface area, improved redox activity | Li et al. (2017) |
| | rGO/Polyaniline (PANI), Polypyrrole (PPy) | In-situ Chemical Polymerization | Enhanced conductivity and pseudo-capacitance | Chauhan et al. (2016) |
| | rGO/CNT (Carbon Nanotube) | Vacuum Filtration or Layer-by-Layer | Flexible film, enhanced mechanical and electrical properties | Peng et al. (2020) |
| | Graphene/Transition Metal Carbide (e.g., $Mo_2C$) | Chemical Vapor Deposition (CVD) | Strong coupling, thermal stability | Gogotsi (2015) |
| | Graphene/Metal-Organic Frameworks (MOFs) | Self-assembly + Thermal Annealing | High porosity, tunable structure | Jayaramulu et al. (2022) |
| TMDs-based | $MoS_2$/Graphene | Hydrothermal + Annealing | Improved conductivity and capacity retention | Salarizadeh et al. (2022) |
| | $MoS_2$/CNT | CVD or Solution Processing | Controlled growth, synergistic effect | Cheng et al. (2012) |
| | $WS_2$/Carbon Composite | Solvothermal + Freeze Drying | High electrochemical stability | Xu et al. (2013) |
| | $MoSe_2$/PANI | In-situ Polymerization | Improved charge storage capability | Wang et al. (2023) |
| | $MoS_2/Ni(OH)_2$ | Electrochemical Deposition + Annealing | High pseudocapacitance | Choudhary et al. (2015) |

| | | | | |
|---|---|---|---|---|
| MXene-based | MXene/$TiO_2$, $MnO_2$ | HF Etching + Sol-Gel + Filtration | Pseudocapacitive contribution + fast ion transport | Paul et al. (2025) |
| | MXene/PANI, PPy | Electrochemical or In-situ Polymerization | High flexibility, good mechanical integrity | Awan et al. (2024) |
| | MXene/Graphene | Layer-by-Layer or Vacuum Filtration | Synergistic enhancement of capacitance and stability | T. Zhou et al. (2020) |
| | MXene/Carbon Nanofiber | Electrospinning + Carbonization | High surface area, continuous conductive pathways | Y. Zhang et al. (2023b) |
| | MXene/MOFs (e.g., ZIF-67) | Self-assembly + Pyrolysis | High porosity, enhanced ion transport | Y. Chen et al. (2025) |
| | MXene/Bimetallic Oxide (e.g., $NiCo_2O_4$) | Hydrothermal + Electrochemical Deposition | Excellent electrochemical performance | Lakhan et al. (2024) |
| | MXene/Layered Double Hydroxides (LDHs) | Co-precipitation + Filtration | Multi-redox-active layers, high specific capacitance | Lin et al. (2023) |
| | MXene/Boron Nitride (BN) | Vacuum-Assisted Assembly | Thermally stable, wide potential window | Chen et al. (2019) |

## 3.5 PROGRESS IN TWO-DIMENSIONAL (2D) COMPOSITE MATERIALS FOR SUPERCAPACITOR ELECTRODES

Supercapacitors, ultracapacitors, or electrochemical capacitors, have drawn immense interest as highly effective electrochemical energy storage devices. Their unique characteristics like fast charge-discharge, long cycle life, high power density, high capacitance, safety, and environmental compatibility make them valuable assets in the development of energy storage technology (Liu et al., 2010; Rakhi et al., 2012). They find broad applications, particularly in electric and hybrid electric vehicles, where they improve acceleration, regenerative braking, and efficiency in energy (Dhakal et al., 2021; Hussain et al., 2019). However, supercapacitors are plagued by issues like low energy density, high self-discharge, narrow voltage windows, and dielectric absorption, which hinder their practical application (Yadlapalli et al., 2022). Due to the challenges identified above, numerous studies have focused on the search for optimal electrode materials, which are pivotal in maximizing electrochemical performance (Lai et al., 2012). The optimal electrode materials for supercapacitors should be highly conductive, thermally stable, and chemically stable, have large surface areas, be corrosion-resistant, low cost, and environmentally friendly. They should also facilitate efficient faradic charge transfer to maximize capacitance (Forouzandeh et al., 2020; Frackowiak & Béguin, 2001; Xie et al., 2016).

In recent developments, 2D materials have emerged as extremely promising candidates for supercapacitor electrodes. The materials are known for their gigantic surface areas that ensure maximum exposure of active sites, leading to high specific capacitance, as well as high mechanical strength, flexibility, and electronic property tunability. Their layered structures allow easy ion intercalation, supporting ion access to active sites and efficient charge storage (Hegazy et al., 2024; Joseph et al., 2020). A highly diverse range of 2D materials, such as graphene, TMDs, metal-organic frameworks (MOFs), hexagonal boron nitride (hBN), transition metal oxides/hydroxides, and black phosphorus, have been found to perform extremely well as supercapacitor materials (S. Ali et al., 2023; Javed et al., 2023). In addition, MXenes, a recently discovered family of 2D materials, have also been reported to be of interest due to their exceptional electrochemical properties, such as high specific energy and power, high conductivity, and high rates of charge storage (Wang et al., 2020). In addition to the performance capability of their pure forms, composites of 2D materials have become highly popular as supercapacitor electrode materials due to their significantly enhanced electrochemical properties and overall performance. Through strategic combination and hybridization, these composites leverage the individual electrochemical strengths of every component, experiencing synergistic benefits that optimize charge storage performance (Chaudhary et al., 2019, 2021; Cong et al., 2017; Zhu et al., 2017). Ongoing studies continue to highlight the gargantuan potential of 2D material-based composites as supercapacitor electrode materials.

### 3.5.1 Graphene-composite electrodes for supercapacitors

Graphene, a 2D atom-thick material, is made up of $sp^2$-hybridized carbon atoms organized in a hexagonal honeycomb lattice (Allen et al., 2010; Gupta et al., 2015). Since its discovery to date, graphene has been the most researched 2D material due to its remarkable properties, including a predicted surface area of 2600 $m^2/g$ (Stoller et al., 2008), outstanding mechanical strength (tensile strength of 130 GPa and elastic modulus up to 1000 GPa) (Lee et al., 2012), excellent electron mobility ($2 \times 10^5$ $cm^2/Vs$), and excellent thermal conductivity (4840–5300 W/mK) (Balandin et al., 2008). GO, a functional derivative of graphene, can be prepared by chemically oxidizing graphite, then dispersing and exfoliation it into aqueous solvents (Chen et al., 2012). Due to the extensive surface area of graphene, its high adsorbability of the electrolyte ions at the electrode-electrolyte interface makes it an ideal capacitive material that can increase the capacitance of supercapacitors (Iqbal et al., 2022). However, single graphene nanosheets are prone to aggregating and restacking during production due to strong $\pi$–$\pi$ interactions and weak van der Waals interactions, which nullifies the use of the surface area, hinders ion mobility, and compromises supercapacitive performance. To address these, scientists fabricated graphene-based composites and hybrids by mixing graphene with highly conducting materials like conductive polymers or metal oxides. These composites leverage the synergy of their components, which addresses the ion transfer limitation of the carbon-based compounds and prevents nanosheet aggregation, making available a greater surface area and facilitating better electrolyte diffusion for enhanced electrochemical performance (Han et al., 2018; Palaniselvam & Baek, 2015; Shao et al., 2015). For instance, Wang and co-authors prepared a $MnCo_2O_4$/graphene composite using a two-step solid-phase synthesis and hydrothermal route. The composite exhibited a specific capacitance of 503 F/g at a current density of 1 A/g, with a satisfactory 97.4% capacitance retention after 5000 cycles, and is thus facile to employ in energy storage devices (H. Wang et al., 2019). Similarly, Jana and co-authors prepared a $Co_9S_8$/rGO hybrid via a hydrothermal route. As an electrode material, the hybrid exhibited a satisfactory specific capacitance of 1349 F/g at a current density of 2.2 A/g, with an energy density of 68.6 Wh/kg at a power density of 1319 W/kg, and maintained 96% of its capacitance after 1000 cycles, and is thus apt for long-term energy storage devices (Jana et al., 2015).

### 3.5.2 Transition metal dichalcogenide (TMD) composites for supercapacitor electrodes

TMDs are a 2D family of materials with structural homologues in graphene, represented by the general formula $MX_2$. Here, M is a transition metal such as titanium (Ti), molybdenum (Mo), vanadium (V), tungsten (W), rhenium (Re), or tantalum (Ta), and X is a chalcogen element such as sulfur (S), selenium (Se), or tellurium (Te). TMD monolayers are bound by van der Waals forces, each being compositionally three atomic sheets: a transition metal sheet sandwiched between two sheets of chalcogens (Li & Zhu, 2015). $TiS_2$, $MoS_2$, $MoSe_2$, $WS_2$, $TaS_2$, and $NbSe_2$ are the most significant TMDs that form an essential class of 2D materials having immense potential in a wide range of applications. TMDs have emerged as electrode materials due to their excellent physicochemical characteristics such as high surface areas, high catalytic activity, good semiconducting properties, low cost, high chemical stability, good mechanical strength, and ease of synthesis (Forouzandeh & Pillai, 2021). The composition of interlayer distance in TMDs makes them highly appealing for electrode applications, as it offers facile access for electrolyte ions between layers. The transition metal atoms can also participate in redox processes, resulting in high capacitance (Hegazy et al., 2024; Liu & Peng, 2017). However, the high surface energy of 2D TMDs and interlayer interactions result in restacking of the layers and decreased accessibility of active sites (Leng et al., 2016). To address this issue, researchers have synthesized hybrid materials by incorporating TMDs into conductive substrates, enhancing performance (Choi et al., 2017; Jiang & Park, 2014; Yousaf et al., 2016). For example, a $MoS_2$/PANI@C composite electrode exhibited a superior specific capacitance of 678 F/g at a scan rate of 1 mV/s, and an impressive capacitance retention of 80% after 10,000 cycles, which reflects its superior energy storage potential (C. Yang et al., 2016a). In another investigation, Chaudhary and Khanuja successfully synthesized a $MoS_2$/PbS hybrid via a chemical route, which demonstrated a notable double-layer capacitance of 205.50 F/g at a scan rate of 10 mV/s (Chaudhary & Khanuja, 2022).

### 3.5.3 MXenes-based composites as electrode materials

MXene, a burgeoning class of materials, encompasses transition metal carbides, carbonitrides, and nitrides, characterized by the general formula $M_{n+1}X_nT_x$ (n = 1–3). In this context, M represents a range of transition metals (including Sc, Ti, Cr, V, Hf, Zr, Mo, Nb, W, and Ta), X denotes carbon and/or nitrogen, and T signifies the terminal surface functional group (–O, –OH, or –F) (Z. Wang et al., 2020). MXenes are rapidly emerging as promising electrode materials for supercapacitors, boasting exceptional electrochemical performance attributed to their remarkable properties, including large interlayer spacing, exceptional hardness, good electrical conductivity, tunable bandgap, high melting point, inherent surface area, excellent thermal stability, and hydrophilic nature (Garg et al., 2020; Hu et al., 2020; Venkateshalu & Grace, 2020). Although having superior properties, pure MXenes suffer from a chain of issues, including tendency to restack, susceptibility to oxidation, low electrical conductivity, and poor mechanical flexibility. In addition, pure MXene-based electrodes suffer from poor potential window and low reversible capacity, which limit their broader application in supercapacitors (Nahirniak et al., 2023; Tomy et al., 2021). These issues can be addressed by synthesizing composite materials in which MXenes are mixed with complementary materials to take advantage of their synergistic effects. Through the incorporation of additives between the MXene layers, composites prevent restacking, increase accessible surface area, and enhance electrochemical performance. As a result, MXene-based composite electrodes have attracted high interest because of their superior performance characteristics (Nahirniak et al., 2023; Yu et al., 2018). In one recent work, Ayman and coauthors synthesized a CoF/MXene composite that had superior electrochemical characteristics, with a specific capacitance of 1268.75 F/g at a current density of 1 A/g and retaining around 97% of its capacitance after 5000 cycles (Ayman et al., 2020). In another work, researchers synthesized an MXene/PANI composite through a hydrothermal method, which had a specific capacitance of 563 F/g at a current density of 0.5 A/g. A device employing this composite as an electrode had superior energy and power densities values of 22.67 Wh/kg and 0.217 kW/kg, respectively (Y. Li et al., 2021). The energy density and power density of the electrode materials are the primary factors that determine the electrochemical performance of supercapacitors. 2D composites have demonstrated great potential in improving these performance measures by combining synergistic components with the high surface area, customizable electrical characteristics, and structural plasticity of 2D materials as mentioned above. Comparing the energy and power densities for various 2D composites can provide a thorough understanding of their capabilities. Key findings value of energy and power densities from current research are compiled in the Table 3.2, which sheds light on the structure-performance links of these novel electrode materials.

## 3.6 PROSPECTS FOR FUTURE DEVELOPMENT

The future development of hybrid composite materials for supercapacitor applications lies at the intersection of material innovation, synthesis advancement, and the growing demand for efficient, sustainable, and multifunctional energy storage systems. One of the most promising candidates is hybrids that combine graphene, TMDs such as $MoS_2$, MXenes (e.g., $Ti_3C_2T_2$), with other such as nanostructures such as metal oxides, conductive polymers, and

*Table 3.2* Comparison of energy and power densities of 2D composites as electrode materials for supercapacitors

| *S.No.* | *2D composite* | *Energy density Wh/kg* | *Power density W/kg* | *Reference* |
|---|---|---|---|---|
| 1. | $Co_9S_8$/rGO | 68.6 | 1319 | Jana et al. (2015) |
| 2. | PDA-rGO@ $NiMoO_4$ | 48 | 800 | Chen et al. (2020) |
| 3. | $Fe_3O_4$/RGO | 39.1 | 1800 | Jana et al. (2015) |
| 4. | $MoS_2$/PPy | 83.3 | 3332 | Tang et al. (2015) |
| 5. | C-$MoS_2$ | 32 | 224.58 | Luo et al. (2018) |
| 6. | $WS_2$/PANI | 50 | 800 | Mohan & Rakhi (2024) |
| 7. | ZnSe@$MoSe_2$ | 43 | 740 | Sangeetha Vidhya et al. (2022) |
| 8. | MXene/$CoS_2$ //rGO | 28.8 | 800 | H. Liu et al. (2020) |
| 9. | MXene/PANI | 22.67 | 217 | Li et al. (2021) |
| 10. | MCS-MXene //AC | 25.6 | 6400 | Nasrin et al. (2021) |
| 11. | $MnO_2$//$Fe_2O_3$// MXene | 32.2 | 900.6 | Shi et al. (2021) |

carbon-based materials (e.g., CNTs). Each of these materials offers unique advantages: graphene provides exceptional electrical conductivity and flexibility; TMDs provide layered structures favorable for redox activity and ion intercalation; MXenes exhibit high capacitance, surface functionalization, and excellent electrical properties. Looking ahead, future prospects focus heavily on advanced interface engineering and structural optimization of these hybrid materials. The goal is to develop 3D hierarchical porous architectures that facilitate rapid ion and electron transport, increase the accessible surface area, and provide mechanical stability. Precise control over the interfacial interactions between graphene, MXene, TMDs, and other additives will be crucial to maintain electrochemical integrity and enhance energy storage behavior under operational stress. Equally important is the development of scalable, environmentally friendly, and cost-effective synthesis techniques. Traditional methods such as chemical precipitation, hydrothermal/solvothermal synthesis, and chemical vapor deposition (CVD) have proven effective for laboratory-scale fabrication, but often lack scalability or environmental friendliness. For example, hydrothermal and solvothermal methods are widely used to grow TMD nanosheets or deposit metal oxides on graphene or MXene substrates due to their ability to control crystal growth and morphology. Chemical precipitation is attractive for large-scale, low-temperature processing and is particularly effective for integrating metal oxides and polymers with 2D materials. On the other hand, CVD produces high-quality, uniform thin films and is suitable for vertical stacking or heterostructure fabrication, although it can be energy-intensive and less compatible with MXenes due to thermal sensitivity. Moving forward, innovation in synthesis methods should aim to combine scalability with precision examples include microwave-assisted synthesis, in-situ polymerization, ALD, and electrochemical deposition. Emerging printing technologies such as inkjet printing, spray coating, and roll-to-roll processing will also play an important role in the large-scale production of flexible, wearable supercapacitor electrodes. Another promising direction is the use of green chemistry and water-based processes to reduce the environmental impact and improve the sustainability of electrode manufacturing.

In response to growing technological demands, future hybrid materials must address several key performance criteria: high energy and power density, mechanical flexibility, environmental stability, and long operational life. As consumer electronics, electric vehicles, and wearable technologies develop, supercapacitors must not only provide fast energy storage and delivery, but also maintain performance when bent, stretched, and exposed to changing environmental conditions. This is particularly important for flexible and wearable supercapacitors, where the integration of soft materials such as polymers or textiles with 2D hybrids can enable stretchable and foldable energy storage platforms. Solid-state and gel-based electrolytes combined with these flexible electrodes will enable safer and more compact devices suitable for next-generation portable applications. In addition, the development of micro-supercapacitors—miniaturized devices with high areal capacitance—will benefit from CVD-based thin-film techniques and precise patterning techniques that allow integration directly onto chips or circuits. Beyond materials and methods, the role of computational tools and artificial intelligence (AI) in accelerating discovery and optimization cannot be underestimated. Machine learning algorithms, data-driven design, and density functional theory (DFT) simulations will increasingly guide material selection, predict electrochemical behavior and optimize structural configurations. These tools can substantially reduce the experimental workload, allowing researchers to focus on the most promising material combinations and fabrication routes. This approach is essential given the vast parameter space involved in hybrid.

### 3.6.1 Sustainability considerations

Sustainable development of 2D hybrid composite electrodes for supercapacitors necessitates environmentally benign and scalable synthesis strategies (Gupta et al., 2024). Conventional production of graphene, TMDs, and MXenes often relies on energy-intensive processes or toxic reagents, underscoring the need for greener alternatives such as electrochemical exfoliation, bio-derived precursors, and low-temperature solvothermal methods. Incorporating earth-abundant, non-toxic materials (e.g., Fe-/Mn-based compounds) can mitigate dependence on scarce or hazardous elements (e.g., Co, V). Closed-loop strategies, including electrode recycling and biodegradable binders/separators, further reduce ecological impact (Yalçın et al., 2022). Life cycle assessment (LCA) must guide material design to quantify energy use, emissions, and disposal implications. By harmonizing sustainable synthesis, eco-compatible components, and circular economy frameworks, 2D hybrid composites can achieve high performance while aligning with planetary sustainability objectives.

## 3.7 CONCLUSION

Hybrid composite materials combining graphene, TMDs (such as $MoS_2$), MXenes, and other nanostructures like metal oxides, conductive polymers, and carbon-based materials represent a transformative approach to designing next-generation supercapacitor electrode materials. By leveraging the synergistic properties of these components'—high conductivity, mechanical flexibility, redox activity, and structural tunability—these hybrids overcome the individual limitations of each material, enabling enhanced energy and power densities, improved cycling life, and multifunctionality. These systems achieve ultrahigh specific capacitance, superior rate capability, and prolonged cycling stability by integrating electric double-layer capacitance with Faradaic pseudocapacitance. However, their commercialization necessitates overcoming critical barriers in scalability, sustainability, and cost-efficiency. Future advancements must prioritize eco-friendly synthesis techniques, including solvent-free exfoliation, bio-derived precursors, and plasma-assisted processing, to mitigate environmental impacts. The replacement of scarce or toxic elements (e.g., Co, V) with earth-abundant alternatives (e.g., Fe-, Mn-, or Ni-based compounds) is imperative for sustainable scale-up. The continued advancement of synthesis techniques such as chemical precipitation, hydrothermal/solvothermal processes, and chemical vapor deposition, along with the emerging use of green, scalable, and low-temperature fabrication methods, is key to translating these materials from lab to market. Looking forward, innovations in interface engineering, 3D hierarchical structuring, and device integration will play critical roles in maximizing electrochemical performance. Additionally, the integration of machine learning and computational modeling will accelerate the discovery and optimization of material combinations tailored to specific performance requirements. Meeting the growing demands for flexible, wearable, and high-performance energy storage systems will depend not only on materials innovation but also on the ability to manufacture robust, scalable, and environmentally sustainable devices. As such, hybrid composite materials offer a promising pathway toward realizing compact, efficient, and reliable supercapacitors for future energy storage technologies. Ultimately, the viability of these systems hinges not only on their electrochemical efficacy but also on their adherence to sustainability objectives, ensuring a dual contribution to technological progress and ecological preservation.

## REFERENCES


Abergel, T., Dean, B., & Dulac, J. (n.d.). *Towards a zero-emission, efficient, and resilient buildings and construction sector*. www.globalabc.org

Adedoja, O. S., Sadiku, E. R., & Hamam, Y. (2023). An overview of the emerging technologies and composite materials for supercapacitors in energy storage applications. *Polymers*, *15*(10), 2272. https://doi.org/10.3390/polym15102272

Ali, M., Afzal, A. M., Iqbal, M. W., Mumtaz, S., Imran, M., Ashraf, F., Ur Rehman, A., & Muhammad, F. (2022). 2D-TMDs based electrode material for supercapacitor applications. *International Journal of Energy Research*, *46*(15), 22336–22364. https://doi.org/10.1002/er.8698

Ali, S., Ahmad, T., Tahir, M. Y., Usman, M., Chhattal, M., Hussain, I., Khan, S., Hassan, A. M., Assiri, M. A., Rosaiah, P., Javed, M. S., Akkinepally, B., & Qi, J. (2023). The emergence of density functional theory for supercapacitors: Recent progress and advances. *Journal of Energy Storage*, *73*, 109100. https://doi.org/10.1016/j.est.2023.109100

Allen, M. J., Tung, V. C., & Kaner, R. B. (2010). Honeycomb carbon: A review of graphene. *Chemical Reviews*, *110*(1), 132–145. https://doi.org/10.1021/cr900070d

Ansari, M. Z., Hussain, I., Mohapatra, D., Ansari, S. A., Rahighi, R., Nandi, D. K., Song, W., & Kim, S. (2024). Atomic layer deposition—A versatile toolbox for designing/engineering electrodes for advanced supercapacitors. *Advanced Science*, *11*(1). https://doi.org/10.1002/advs.202303055

Awan, H. T. A., Abdah, M. A. A. M., Mehar, M., Walvekar, R., Chaudhary, V., Khalid, M., & Khosla, A. (2024). MXene-polymer hybrid composites for advanced energy storage: Insights into supercapacitors and batteries. *Journal of Energy Storage*, *95*, 112449. https://doi.org/10.1016/j.est.2024.112449

Ayman, I., Rasheed, A., Ajmal, S., Rehman, A., Ali, A., Shakir, I., & Warsi, M. F. (2020). $CoFe_2O_4$ nanoparticle-decorated 2D MXene: A novel hybrid material for supercapacitor applications. *Energy & Fuels*, *34*(6), 7622–7630. https://doi.org/10.1021/acs.energyfuels.0c00959

Azeem, O., Ali, M., Abbas, G., Uzair, M., Qahmash, A., Algarni, A., & Hussain, M. R. (2021). A comprehensive review on integration challenges, optimization techniques and control strategies of hybrid AC/DC microgrid. *Applied Sciences*, *11*(14), 6242. https://doi.org/10.3390/app11146242

Balandin, A. A., Ghosh, S., Bao, W., Calizo, I., Teweldebrhan, D., Miao, F., & Lau, C. N. (2008). Superior thermal conductivity of single-layer graphene. *Nano Letters*, *8*(3), 902–907. https://doi.org/10.1021/nl0731872

Cao, F., Zhang, Y., Wang, H., Khan, K., Tareen, A. K., Qian, W., Zhang, H., & Ågren, H. (2022). Recent advances in oxidation stable chemistry of 2D MXenes. *Advanced Materials*, *34*(13). https://doi.org/10.1002/adma.202107554

Chaudhary, N., & Khanuja, M. (2022). High-performance supercapacitor electrode material based on the two-dimensional/three-dimensional architecture of $MoS_2$ – PbS hybrid material. *Energy & Fuels*, *36*(2), 1034–1042. https://doi.org/10.1021/acs.energyfuels.1c02685

Chaudhary, N., Khanuja, M., & Islam, S. S. (2019). Broadband photodetector based on 3D architect of $MoS_2$-PANI hybrid structure for high photoresponsive properties. *Polymer*, *165*, 168–173. https://doi.org/10.1016/j.polymer.2019.01.028

Chaudhary, N., Kumar, A., Imtiyaz, S., & Khanuja, M. (2021). Architectural design of electrode material for supercapacitor application based on a $MoS_2/CeO_2$ Heterostructure synthesized by facile hydrothermal technique. *ECS Journal of Solid State Science and Technology*, *10*(5), 053005. https://doi.org/10.1149/2162-8777/ac02a3

Chaudhuri, H., & Yun, Y.-S. (2022). Synthesis and environmental applications of graphene oxide/layered double hydroxides and graphene oxide/MXenes: A critical review. *Separation and Purification Technology*, *297*, 121518. https://doi.org/10.1016/j.seppur.2022.121518

Chauhan, N. P. S., Mozafari, M., Chundawat, N. S., Meghwal, K., Ameta, R., & Ameta, S. C. (2016). High-performance supercapacitors based on polyaniline–graphene nanocomposites: Some approaches, challenges and opportunities. *Journal of Industrial and Engineering Chemistry*, *36*, 13–29. https://doi.org/10.1016/j.jiec.2016.03.003

Chen, C., Qin, S., Liu, D., Wang, J., Yang, G., Su, Y., Zhang, L., Cao, W., Ma, M., Qian, Y., Liu, Y., Liu, J. Z., & Lei, W. (2019). Ultrafast, stable ionic and molecular sieving through functionalized boron nitride membranes. *ACS Applied Materials & Interfaces*, *11*(33), 30430–30436. https://doi.org/10.1021/acsami.9b08296

Chen, D., Feng, H., & Li, J. (2012). Graphene oxide: Preparation, functionalization, and electrochemical applications. *Chemical Reviews*, *112*(11), 6027–6053. https://doi.org/10.1021/cr300115g

Chen, Y., Bai, J., Yang, D., Sun, P., & Li, X. (2020). Excellent performance of flexible supercapacitor based on the ternary composites of reduced graphene oxide/molybdenum disulfide/poly (3,4-ethylenedioxythiophene). *Electrochimica Acta*, *330*, 135205. https://doi.org/10.1016/j.electacta.2019.135205

Chen, Y., Shi, Y., Song, G., Yang, B., & Pang, H. (2025). Compositing MXene with organic coordination frameworks (MOFs, COFs, and HOFs) for enhanced electrochemical performance. *Next Materials*, *6*, 100323. https://doi.org/10.1016/j.nxmate.2024.100323

Cheng, P.-H., Huang, T.-H., Chien, Y.-W., Wu, C.-L., & Fu, L.-C. (2017). *Demand-side management in residential community realizing sharing economy with bidirectional PEV. 2017 IEEE International Conference on Systems, Man, and Cybernetics (SMC)*, 1615–1620. https://doi.org/10.1109/SMC.2017.8122846

Cheng, Y., Lu, S., Zhang, H., Varanasi, C. V., & Liu, J. (2012). Synergistic effects from graphene and carbon nanotubes enable flexible and robust electrodes for high-performance supercapacitors. *Nano Letters*, *12*(8), 4206–4211. https://doi.org/10.1021/nl301804c

Choi, W., Choudhary, N., Han, G. H., Park, J., Akinwande, D., & Lee, Y. H. (2017). Recent development of two-dimensional transition metal dichalcogenides and their applications. *Materials Today*, *20*(3), 116–130. https://doi.org/10.1016/j.mattod.2016.10.002

Choudhary, N., Patel, M., Ho, Y.-H., Dahotre, N. B., Lee, W., Hwang, J. Y., & Choi, W. (2015). Directly deposited $MoS_2$ thin film electrodes for high performance supercapacitors. *Journal of Materials Chemistry A*, *3*(47), 24049–24054. https://doi.org/10.1039/C5TA08095A

Chu, S., & Majumdar, A. (2012). Opportunities and challenges for a sustainable energy future. *Nature*, *488*(7411), 294–303. https://doi.org/10.1038/nature11475

Cong, L., Xie, H., & Li, J. (2017). Hierarchical structures based on two-dimensional nanomaterials for rechargeable lithium batteries. *Advanced Energy Materials*, *7*(12). https://doi.org/10.1002/aenm.201601906

Dahmen, K.-H. (2003). Chemical Vapor Deposition. In *Encyclopedia of Physical Science and Technology* (pp. 787–808). Elsevier. https://doi.org/10.1016/B0-12-227410-5/00102-2

Dar, M. A., Majid, S. R., Satgunam, M., Siva, C., Ansari, S., Arularasan, P., & Rafi Ahamed, S. (2024). Advancements in Supercapacitor electrodes and perspectives for future energy storage technologies. *International Journal of Hydrogen Energy*, *70*, 10–28. https://doi.org/10.1016/j.ijhydene.2024.05.191

Dhakal, G., Mohapatra, D., Tamang, T. L., Lee, M., Lee, Y. R., & Shim, J.-J. (2021). Redox-additive electrolyte–driven enhancement of the electrochemical energy storage performance of asymmetric $Co_3O_4$//carbon nano-onions supercapacitors. *Energy*, *218*, 119436. https://doi.org/10.1016/j.energy.2020.119436

Dissanayake, K., & Kularatna-Abeywardana, D. (2024). A review of supercapacitors: Materials, technology, challenges, and renewable energy applications. *Journal of Energy Storage*, *96*, 112563. https://doi.org/10.1016/j.est.2024.112563

Energy Technology Policy Division-IEA. (2017). *Energy technology perspectives 2017*. Energy Technology Policy Division-IEA.

Forouzandeh, P., Kumaravel, V., & Pillai, S. C. (2020). Electrode materials for supercapacitors: A review of recent advances. *Catalysts*, *10*(9), 969. https://doi.org/10.3390/catal10090969

Forouzandeh, P., & Pillai, S. C. (2021). Two-dimensional (2D) electrode materials for supercapacitors. *Materials Today: Proceedings*, *41*, 498–505. https://doi.org/10.1016/j.matpr.2020.05.233

Frackowiak, E., & Béguin, F. (2001). Carbon materials for the electrochemical storage of energy in capacitors. *Carbon*, *39*(6), 937–950. https://doi.org/10.1016/S0008-6223(00)00183-4

Gan, Y. X., Jayatissa, A. H., Yu, Z., Chen, X., & Li, M. (2020). Hydrothermal synthesis of nanomaterials. *Journal of Nanomaterials*, *2020*, 1–3. https://doi.org/10.1155/2020/8917013

Gao, L., Zhang, H., Surjadi, J. U., Li, P., Han, Y., Sun, D., & Lu, Y. (2018). Mechanically stable ternary heterogeneous electrodes for energy storage and conversion. *Nanoscale*, *10*(5), 2613–2622. https://doi.org/10.1039/C7NR07789K

Garg, R., Agarwal, A., & Agarwal, M. (2020). A review on MXene for energy storage application: effect of interlayer distance. *Materials Research Express*, 7(2), 022001. https://doi.org/10.1088/2053-1591/ab750d

Giovanelli, C., Kilkki, O., Sierla, S., Seilonen, I., & Vyatkin, V. (2019). Task allocation algorithm for energy resources providing frequency containment reserves. *IEEE Transactions on Industrial Informatics*, *15*(2), 677–688. https://doi.org/10.1109/TII.2018.2821676

Gogotsi, Y. (2015). Transition metal carbides go 2D. *Nature Materials*, *14*(11), 1079–1080. https://doi.org/10.1038/nmat4386

Gómez, I. J., Alegret, N., Dominguez-Alfaro, A., & Vázquez Sulleiro, M. (2021). Recent Advances on 2D Materials towards 3D Printing. *Chemistry*, *3*(4), 1314–1343. https://doi.org/10.3390/chemistry3040095

Gupta, A., Sakthivel, T., & Seal, S. (2015). Recent development in 2D materials beyond graphene. *Progress in Materials Science*, *73*, 44–126. https://doi.org/10.1016/j.pmatsci.2015.02.002

Gupta, V., Chandra, N., Tari, V., & Kannan, K. (2024). A sustainable approach for the green synthesis of silver nanoparticles using *Brassica oleracea* subsp. *botrytis* (L.) leaves and its antibacterial activity. *Physics and Chemistry of Solids*, *25*(2), 303–310. https://doi.org/10.15330/pcss.25.2.303-310

Gurushankar, K., Karuppaiya, C., Tari, V., & Kannan, K. (2024). Exploring MXenes and MBenes in metal sulfur and selenium battery technologies. In K. Kannan & V. Tari (Eds.), *Novel energy storage and conversion technologies for two-dimensional MXenes and MBenes*. IGI Global. https://doi.org/10.4018/979-8-3693-9546-2.ch005

Haakana, J., Haapaniemi, J., Tikka, V., Lassila, J., & Partanen, J. (2017). Risk or benefit on the electricity grid: distributed energy storages in system services. *CIRED - Open Access Proceedings Journal*, *2017*(1), 1971–1974. https://doi.org/10.1049/oap-cired.2017.0833

Hafeez, M. (2022). *Recent progress and overview of nanocomposites*. In *Nanocomposite Materials for Biomedical and Energy Storage Applications*. IntechOpen. https://doi.org/10.5772/intechopen.102469

Han, Y., Ge, Y., Chao, Y., Wang, C., & Wallace, G. G. (2018). Recent progress in 2D materials for flexible supercapacitors. *Journal of Energy Chemistry*, *27*(1), 57–72. https://doi.org/10.1016/j.jechem.2017.10.033

Hegazy, H. H., Khan, J., Shakeel, N., Alabdullkarem, E. A., Saleem, M. I., Alrobei, H., & Yahia, I. S. (2024). 2D-based electrode materials for supercapacitors – Status, challenges, and prospects. *RSC Advances*, *14*(45), 32958–32977. https://doi.org/10.1039/D4RA05473C

Horn, M., MacLeod, J., Liu, M., Webb, J., & Motta, N. (2019). Supercapacitors: A new source of power for electric cars? *Economic Analysis and Policy*, *61*, 93–103. https://doi.org/10.1016/j.eap.2018.08.003

Hou, Y., Baig, M. M., Lu, J., Zhang, H., Liu, P., Zhu, G., Ge, X., Pang, H., & Zhang, Y. (2024). Direct ink writing 3D printing of low-dimensional nanomaterials for micro-supercapacitors. *Nanoscale*, *16*(26), 12380–12396. https://doi.org/10.1039/D4NR01590H

Hu, M., Zhang, H., Hu, T., Fan, B., Wang, X., & Li, Z. (2020). Emerging 2D MXenes for supercapacitors: Status, challenges and prospects. *Chemical Society Reviews*, *49*(18), 6666–6693. https://doi.org/10.1039/D0CS00175A

Hussain, I., Lamiel, C., Mohamed, S. G., Vijayakumar, S., Ali, A., & Shim, J.-J. (2019). Controlled synthesis and growth mechanism of zinc cobalt sulfide rods on Ni-foam for high-performance supercapacitors. *Journal of Industrial and Engineering Chemistry*, *71*, 250–259. https://doi.org/10.1016/j.jiec.2018.11.033

IEA. (2024). *Energy technology perspectives 2024*. www.iea.org

Iqbal, M. Z., Alam, F., Khan, J., Alam, S., Alhassan, N., & Alzaid, M. (2022). Reduced graphene oxide/cobalt phosphate composites with improved electrochemical performance for supercapattery devices. *International Journal of Energy Research*, *46*(15), 23377–23387. https://doi.org/10.1002/er.8636

Jana, M., Saha, S., Samanta, P., Murmu, N. C., Kim, N. H., Kuila, T., & Lee, J. H. (2015). Development of high energy density supercapacitor through hydrothermal synthesis of RGO/nano-structured cobalt sulphide composites. *Nanotechnology*, *26*(7), 075402. https://doi.org/10.1088/0957-4484/26/7/075402

Javed, M. S., Mateen, A., Hussain, I., Ali, S., Asim, S., Ahmad, A., tag Eldin, E., Bajaber, M. A., Najam, T., & Han, W. (2023). The quest for negative electrode materials for Supercapacitors: 2D materials as a promising family. *Chemical Engineering Journal*, *452*, 139455. https://doi.org/10.1016/j.cej.2022.139455

Jayaramulu, K., Mukherjee, S., Morales, D. M., Dubal, D. P., Nanjundan, A. K., Schneemann, A., Masa, J., Kment, S., Schuhmann, W., Otyepka, M., Zbořil, R., & Fischer, R. A. (2022). Graphene-based metal–organic framework hybrids for applications in catalysis, environmental, and energy technologies. *Chemical Reviews*, *122*(24), 17241–17338. https://doi.org/10.1021/acs.chemrev.2c00270

Jiang, J.-W., & Park, H. S. (2014). Mechanical properties of MoS2/graphene heterostructures. *Applied Physics Letters*, *105*(3). https://doi.org/10.1063/1.4891342

Joseph, N., Shafi, P. M., & Bose, A. C. (2020). Recent advances in 2D-$MoS_2$ and its composite nanostructures for supercapacitor electrode application. *Energy & Fuels*, *34*(6), 6558–6597. https://doi.org/10.1021/acs.energyfuels.0c00430

Kannan, K., & Tari, V. (Eds.). (2025). *Novel energy storage and conversion technologies for two-dimensional MXenes and MBenes*. https://www.amazon.com/dp/9798369395462

Kumar, S. A., Sahoo, S., Laxminarayana, G. K., & Rout, C. S. (2024). Electrochemical deposition for cultivating nano- and microstructured electroactive materials for supercapacitors: Recent developments and future perspectives. *Small*. https://doi.org/10.1002/smll.202402087

Lai, L., Yang, H., Wang, L., Teh, B. K., Zhong, J., Chou, H., Chen, L., Chen, W., Shen, Z., Ruoff, R. S., & Lin, J. (2012). Preparation of supercapacitor electrodes through selection of graphene surface functionalities. *ACS Nano*, *6*(7), 5941–5951. https://doi.org/10.1021/nn3008096

Lakhan, M. N., Hanan, A., Wang, Y., Lee, H. K., & Arandiyan, H. (2024). Integrated MXene and metal oxide electrocatalysts for the oxygen evolution reaction: synthesis, mechanisms, and advances. *Chemical Science*, *15*(38), 15540–15564. https://doi.org/10.1039/D4SC04141K

Lee, J.-U., Yoon, D., & Cheong, H. (2012). Estimation of young's modulus of graphene by Raman spectroscopy. *Nano Letters*, *12*(9), 4444–4448. https://doi.org/10.1021/nl301073q

Leng, K., Chen, Z., Zhao, X., Tang, W., Tian, B., Nai, C. T., Zhou, W., & Loh, K. P. (2016). Phase restructuring in transition metal dichalcogenides for highly stable energy storage. *ACS Nano*, *10*(10), 9208–9215. https://doi.org/10.1021/acsnano.6b05746

Li, X., & Zhu, H. (2015). Two-dimensional $MoS_2$: Properties, preparation, and applications. *Journal of Materiomics*, *1*(1), 33–44. https://doi.org/10.1016/j.jmat.2015.03.003

Li, Y., Kamdem, P., & Jin, X.-J. (2021). Hierarchical architecture of MXene/PANI hybrid electrode for advanced asymmetric supercapacitors. *Journal of Alloys and Compounds*, *850*, 156608. https://doi.org/10.1016/j.jallcom.2020.156608

Li, Y., Xu, L., Gao, J., & Jin, X. (2017). Hydrothermal fabrication of reduced graphene oxide/activated carbon/$MnO_2$ hybrids with excellent electrochemical performance for supercapacitors. *RSC Advances*, 7(62), 39024–39033. https://doi.org/10.1039/C7RA07056J

Lin, Z., Li, X., Zhang, H., Xu, B. Bin, Wasnik, P., Li, H., Singh, M. V., Ma, Y., Li, T., & Guo, Z. (2023). Research progress of MXenes and layered double hydroxides for supercapacitors. *Inorganic Chemistry Frontiers*, *10*(15), 4358–4392. https://doi.org/10.1039/D3QI00819C

Liu, C., Li, F., Ma, L., & Cheng, H. (2010). Advanced materials for energy storage. *Advanced Materials*, *22*(8). https://doi.org/10.1002/adma.200903328

Liu, H., Hu, R., Qi, J., Sui, Y., He, Y., Meng, Q., Wei, F., Ren, Y., Zhao, Y., & Wei, W. (2020). One-step synthesis of nanostructured $CoS_2$ grown on titanium carbide MXene for high-performance asymmetrical supercapacitors. *Advanced Materials Interfaces*, *7*(6). https://doi.org/10.1002/admi.201901659

Liu, Y., & Peng, X. (2017). Recent advances of supercapacitors based on two-dimensional materials. *Applied Materials Today*, *8*, 104–115. https://doi.org/10.1016/j.apmt.2017.05.002

Lu, C., Li, A., Zhai, T., Niu, C., Duan, H., Guo, L., & Zhou, W. (2020). Interface design based on $Ti_3C_2$ MXene atomic layers of advanced battery-type material for supercapacitors. *Energy Storage Materials*, *26*, 472–482. https://doi.org/10.1016/j.ensm.2019.11.021

Luo, W., Zhang, G., Cui, Y., Liu, Y., Jin, C., Hao, J., Zhang, J., & Zheng, W. (2018). One-pot synthesis of highly stable carbon–$MoS_2$ nanosphere electrodes using a co-growth mechanism for supercapacitors. *New Journal of Chemistry*, *42*(12), 10111–10117. https://doi.org/10.1039/C8NJ01387J

Mazgut, R., Kascak, S., Drgona, P., Radvan, R., & Cubon, P. (2015). Supply structure with an additional energy storage for electric vehicle. *2015 International Conference on Electrical Drives and Power Electronics (EDPE)*, 409–412. https://doi.org/10.1109/EDPE.2015.7325329

Mohan, V. V., & Rakhi, R. B. (2024). $WS_2$/Conducting polymer nanocomposite-based flexible and binder-free electrodes for high-performance supercapacitors. *Electrochimica Acta*, *498*, 144657. https://doi.org/10.1016/j.electacta.2024.144657

Moridi, A., Hassani-Gangaraj, S. M., Guagliano, M., & Dao, M. (2014). Cold spray coating: Review of material systems and future perspectives. *Surface Engineering*, *30*(6), 369–395. https://doi.org/10.1179/1743294414Y.0000000270

Mou, Y., Xing, H., Lin, Z., & Fu, M. (2014). *A new approach to distributed charging control for plug-in hybrid electric vehicles. Proceedings of the 33rd Chinese Control Conference*, 8118–8123. https://doi.org/10.1109/ChiCC.2014.6896359

Muralee Gopi, C. V. V., Alzahmi, S., Narayanaswamy, V., Vinodh, R., Issa, B., & Obaidat, I. M. (2025). Supercapacitors: A promising solution for sustainable energy storage and diverse applications. *Journal of Energy Storage*, *114*, 115729. https://doi.org/10.1016/j.est.2025.115729

Naguib, M., Kurtoglu, M., Presser, V., Lu, J., Niu, J., Heon, M., Hultman, L., Gogotsi, Y., & Barsoum, M. W. (2011). Two-Dimensional Nanocrystals: Two-Dimensional Nanocrystals Produced by Exfoliation of $Ti_3$ AlC $_2$ (Adv. Mater. 37/2011). *Advanced Materials*, *23*(37), 4207. https://doi.org/10.1002/adma.201190147

Nahirniak, S., Ray, A., & Saruhan, B. (2023). Challenges and future prospects of the MXene-based materials for energy storage applications. *Batteries*, *9*(2), 126. https://doi.org/10.3390/batteries9020126

Nasrin, K., Subramani, K., Karnan, M., & Sathish, M. (2021). $MnCo_2S_4$ – MXene: A novel hybrid electrode material for high performance long-life asymmetric supercapattery. *Journal of Colloid and Interface Science*, *600*, 264–277. https://doi.org/10.1016/j.jcis.2021.05.037

Nasrin, K., Sudharshan, V., Subramani, K., & Sathish, M. (2022). Insights into 2D/2D MXene heterostructures for improved synergy in structure toward next-generation supercapacitors: A review. *Advanced Functional Materials*, *32*(18). https://doi.org/10.1002/adfm.202110267

Nguyen, H. K., Song, J. Bin, & Han, Z. (2015). Distributed demand side management with energy storage in smart grid. *IEEE Transactions on Parallel and Distributed Systems*, *26*(12), 3346–3357. https://doi.org/10.1109/TPDS.2014.2372781

O'Hare, D. (2001). Hydrothermal Synthesis. In *Encyclopedia of Materials: Science and Technology* (pp. 3989–3992). Elsevier. https://doi.org/10.1016/B0-08-043152-6/00701-4

Palaniselvam, T., & Baek, J.-B. (2015). Graphene based 2D-materials for supercapacitors. *2D Materials*, *2*(3), 032002. https://doi.org/10.1088/2053-1583/2/3/032002

Parra, D., Swierczynski, M., Stroe, D. I., Norman, S. A., Abdon, A., Worlitschek, J., O'Doherty, T., Rodrigues, L., Gillott, M., Zhang, X., Bauer, C., & Patel, M. K. (2017). An interdisciplinary review of energy storage for communities: Challenges and perspectives. *Renewable and Sustainable Energy Reviews*, *79*, 730–749. https://doi.org/10.1016/j.rser.2017.05.003

Paul, T. K., Khaleque, Md. A., Ali, Md. R., Aly Saad Aly, M., Bacchu, Md. S., Rahman, S., & Khan, Md. Z. H. (2025). MXenes from MAX phases: synthesis, hybridization, and advances in supercapacitor applications. *RSC Advances*, *15*(12), 8948–8976. https://doi.org/10.1039/D5RA00271K

Peng, S., Wang, L., Zhu, Z., & Han, K. (2020). Electrochemical performance of reduced graphene oxide/carbon nanotube hybrid papers as binder-free anodes for potassium-ion batteries. *Journal of Physics and Chemistry of Solids*, *138*, 109296. https://doi.org/10.1016/j.jpcs.2019.109296

Philip, A., & Ruban Kumar, A. (2023). Recent advancements and developments employing 2D-materials in enhancing the performance of electrochemical supercapacitors: A review. *Renewable and Sustainable Energy Reviews*, *182*, 113423. https://doi.org/10.1016/j.rser.2023.113423

Pomerantseva, E., & Gogotsi, Y. (2017). Two-dimensional heterostructures for energy storage. *Nature Energy*, *2*(7), 17089. https://doi.org/10.1038/nenergy.2017.89

Pradeep, N. B., Hegde, M. M. R., Rajendrachari, S., & Surendranathan, A. O. (2022). Investigation of microstructure and mechanical properties of microwave consolidated TiMgSr alloy prepared by high energy ball milling. *Powder Technology*, *408*, 117715. https://doi.org/10.1016/j.powtec.2022.117715

Rakhi, R. B., Ahmed, B., Anjum, D., & Alshareef, H. N. (2016). Direct chemical synthesis of $MnO_2$ nanowhiskers on transition-metal carbide surfaces for supercapacitor applications. *ACS Applied Materials & Interfaces*, *8*(29), 18806–18814. https://doi.org/10.1021/acsami.6b04481

Rakhi, R. B., Chen, W., Cha, D., & Alshareef, H. N. (2012). Nanostructured ternary electrodes for energy-storage applications. *Advanced Energy Materials*, *2*(3), 381–389. https://doi.org/10.1002/aenm.201100609

Rugolo, J., & Aziz, M. J. (2012). Electricity storage for intermittent renewable sources. *Energy & Environmental Science*, *5*(5), 7151. https://doi.org/10.1039/c2ee02542f

Saeed, G., Kang, T., Byun, J. S., Min, D., Kim, J. S., Sadavar, S. V., & Park, H. S. (2024). Two-dimensional (2D) materials for 3D printed micro-supercapacitors and micro-batteries. *Energy Materials*, *4*(2). https://doi.org/10.20517/energymater.2023.81

Salarizadeh, P., Askari, M. B., & Di Bartolomeo, A. (2022). $MoS_2$/$Ni_3S_2$/reduced graphene oxide nanostructure as an electrocatalyst for alcohol fuel cells. *ACS Applied Nano Materials*, *5*(3), 3361–3373. https://doi.org/10.1021/acsanm.1c03946

Sangeetha Vidhya, M., Yuvakkumar, R., Senthil Kumar, P., Ravi, G., Velauthapillai, D., & Bijad, M. (2022). Recent progression of flower like ZnSe@$MoSe_2$ designed as an electrocatalyst for enhanced supercapacitor performance. *Topics in Catalysis*, *65*(5–6), 684–693. https://doi.org/10.1007/s11244-021-01538-6

Shao, Y., El-Kady, M. F., Wang, L. J., Zhang, Q., Li, Y., Wang, H., Mousavi, M. F., & Kaner, R. B. (2015). Graphene-based materials for flexible supercapacitors. *Chemical Society Reviews*, *44*(11), 3639–3665. https://doi.org/10.1039/C4CS00316K

Sheikhi, A., Maani, A., & Ranjbar, A. M. (2013). Evaluation of intelligent distribution network response to plug-in hybrid electric vehicles. *2013 IEEE Grenoble Conference*, 1–6. https://doi.org/10.1109/PTC.2013.6652184

Shi, T.-Z., Feng, Y.-L., Peng, T., & Yuan, B.-G. (2021). Sea urchin-shaped $Fe_2O_3$ coupled with 2D MXene nanosheets as negative electrode for high-performance asymmetric supercapacitors. *Electrochimica Acta*, *381*, 138245. https://doi.org/10.1016/j.electacta.2021.138245

Sneha, N., Shakthivel, K. R., & Kiruthika, S. (2025). Flexible and durable conducting fabric electrodes for next-generation wearable supercapacitors. *ACS Applied Materials & Interfaces*, *17*(5), 7568–7580. https://doi.org/10.1021/acsami.4c15201

Stoller, M. D., Park, S., Zhu, Y., An, J., & Ruoff, R. S. (2008). Graphene-based ultra capacitors. *Nano Letters*, *8*(10), 3498–3502. https://doi.org/10.1021/nl802558y

Sundriyal, P., & Bhattacharya, S. (2017). Inkjet-printed electrodes on A4 paper substrates for low-cost, disposable, and flexible asymmetric supercapacitors. *ACS Applied Materials and Interfaces*, *9*(44), 38507–38521. https://doi.org/10.1021/acsami.7b11262

Tang, H., Wang, J., Yin, H., Zhao, H., Wang, D., & Tang, Z. (2015). Growth of polypyrrole ultrathin films on $MoS_2$ monolayers as high-performance supercapacitor electrodes. *Advanced Materials*, *27*(6), 1117–1123. https://doi.org/10.1002/adma.201404622

Tari, V., Kannan, K., Koliar, J., Gurushankar, K., & Wahyuni, D. K. (2025). Influence of zinc oxide nanomaterials on crop plants: A prospective analysis. In D. Jabbrova, D. Sarkar, Rakesh, S. Singh, & R. Datta (Eds.), *Smart nanomaterials in agroforestry systems*. Springer. https://doi.org/10.1007/978-981-96-1337-3_8

Tari, V., Kannan, K., & Vishwakarma, V. (2022). Chapter 8: Emerging antiviral technologies. In P. Nunthavarawong, S. M. Rangappa, S. Siengchin, & M. Thoppil-Mathew (Eds.), *Antimicrobial and antiviral materials: Polymers, metals, ceramics, and applications* (1st ed., pp. 137–150). CRC Press. https://doi.org/10.1201/9781003143093-8

Telaretti, E., & Dusonchet, L. (2017). Stationary battery technologies in the U.S.: Development Trends and prospects. *Renewable and Sustainable Energy Reviews*, *75*, 380–392. https://doi.org/10.1016/j.rser.2016.11.003

Tian, W., VahidMohammadi, A., Wang, Z., Ouyang, L., Beidaghi, M., & Hamedi, M. M. (2019). Layer-by-layer self-assembly of pillared two-dimensional multilayers. *Nature Communications*, *10*(1), 2558. https://doi.org/10.1038/s41467-019-10631-0

Tomy, M., Ambika Rajappan, A., Vimuna, V. M., & Thankappan Suryabai, X. (2021). Emergence of novel 2D materials for high-performance supercapacitor electrode applications: A brief review. *Energy & Fuels*, *35*(24), 19881–19900. https://doi.org/10.1021/acs.energyfuels.1c02743

UN Environment and International Energy Agency. (2017). *Towards a zero-emission, efficient, and resilient buildings and construction sector: G. Status Report 2017*.

Venkateshalu, S., & Grace, A. N. (2020). MXenes—A new class of 2D layered materials: Synthesis, properties, applications as supercapacitor electrode and beyond. *Applied Materials Today*, *18*, 100509. https://doi.org/10.1016/j.apmt.2019.100509

Wang, D., Yang, F., Wang, C., Chu, F., Nan, J., & Chen, R. (2023). In-situ polymerization of PANI on hydrogel electrolyte enabling all-in-one supercapacitors mechanically stable at low temperatures. *Chemical Engineering Journal*, *455*, 140949. https://doi.org/10.1016/j.cej.2022.140949

Wang, H., Shen, C., Liu, J., Zhang, W., & Yao, S. (2019). Three-dimensional $MnCo_2O_4$/graphene composites for super capacitor with promising electrochemical properties. *Journal of Alloys and Compounds*, *792*, 122–129. https://doi.org/10.1016/j.jallcom.2019.03.405

Wang, J., Qu, X., Li, Y., Yang, G., Zhou, S., Wang, Y., Yu, X., Qiu, Y., & Yang, Y. (2025). In situ oxidized $Mo_2CT$ MXene film via electrochemical activation for smart electrochromic supercapacitors. *Journal of Colloid and Interface Science*, *684*, 170–179. https://doi.org/10.1016/j.jcis.2025.01.008

Wang, Z., Xu, Z., Huang, H., Chu, X., Xie, Y., Xiong, D., Yan, C., Zhao, H., Zhang, H., & Yang, W. (2020). Unraveling and regulating self-discharge behavior of $Ti_3C_2T_x$ MXene-based supercapacitors. *ACS Nano*, *14*(4), 4916–4924. https://doi.org/10.1021/acsnano.0c01056

Wei, J., Sajjad, M., Zhang, J., Li, D., & Mao, Z. (2023). The rise of novel 2D materials beyond graphene: A comprehensive review of their potential as supercapacitor electrodes. *Surfaces and Interfaces*, *42*, 103334. https://doi.org/10.1016/j.surfin.2023.103334

Wu, X., Wang, Z., Yu, M., Xiu, L., & Qiu, J. (2017). Stabilizing the MXenes by carbon nanoplating for developing hierarchical nanohybrids with efficient lithium storage and hydrogen evolution capability. *Advanced Materials*, *29*(24). https://doi.org/10.1002/adma.201607017

Xie, L., Sun, G., Su, F., Guo, X., Kong, Q., Li, X., Huang, X., Wan, L., Song, W., Li, K., Lv, C., & Chen, C.-M. (2016). Hierarchical porous carbon microtubes derived from willow catkins for supercapacitor applications. *Journal of Materials Chemistry A*, *4*(5), 1637–1646. https://doi.org/10.1039/C5TA09043A

Xing, Y., Jia, H., Wang, Z., Xie, L., Liu, D., Wang, Z., Li, M., & Kong, Q. (2023). Vacuum-filtration-assisted ice-templated freeze drying for preparing capacitive graphene aerogel for thermal management. *Crystals*, *13*(3), 458. https://doi.org/10.3390/cryst13030458

Xu, X., Rout, C. S., Yang, J., Cao, R., Oh, P., Shin, H. S., & Cho, J. (2013). Freeze-dried WS2 composites with low content of graphene as high-rate lithium storage materials. *Journal of Materials Chemistry A*, *1*(46), 14548. https://doi.org/10.1039/c3ta13329j

Yadlapalli, R. T., Alla, R. R., Kandipati, R., & Kotapati, A. (2022). Super capacitors for energy storage: Progress, applications and challenges. *Journal of Energy Storage*, *49*, 104194. https://doi.org/10.1016/j.est.2022.104194

Yalçın, M., Turgut, N., Gökbulut, C., Mermer, S., Sofuoğlu, S. C., Tari, V., & Turgut, C. (2022). Removal of pesticide residues from apple and tomato cuticle. *Environmental Science and Pollution Research*. https://doi.org/10.1007/s11356-022-23269-1

Yang, C., Chen, Z., Shakir, I., Xu, Y., & Lu, H. (2016a). Rational synthesis of carbon shell coated polyaniline/$MoS_2$ monolayer composites for high-performance supercapacitors. *Nano Research*, *9*(4), 951–962. https://doi.org/10.1007/s12274-016-0983-3

Yang, L., Tai, N., Fan, C., & Meng, Y. (2016b). Energy regulating and fluctuation stabilizing by air source heat pump and battery energy storage system in microgrid. *Renewable Energy*, *95*, 202–212. https://doi.org/10.1016/j.renene.2016.04.019

Yeon, J. S., Gupta, N., Bhattacharya, P., & Park, H. S. (2022). A new era of integrative ice frozen assembly into multiscale architecturing of energy materials. *Advanced Functional Materials*, *32*(19). https://doi.org/10.1002/adfm.202112509

Yousaf, M., Mahmood, A., Wang, Y., Chen, Y., Ma, Z., & Han, R. P. S.. (2016). Advancement in layered transition metal dichalcogenide composites for lithium and sodium ion batteries. *Journal of Electrical Engineering*, *4*(2). https://doi.org/10.17265/2328-2223/2016.02.003

Yu, L., Hu, L., Anasori, B., Liu, Y.-T., Zhu, Q., Zhang, P., Gogotsi, Y., & Xu, B. (2018). MXene-bonded activated carbon as a flexible electrode for high-performance supercapacitors. *ACS Energy Letters*, *3*(7), 1597–1603. https://doi.org/10.1021/acsenergylett.8b00718

Zhang, Q., Geng, D., & Hu, W. (2023a). Chemical vapor deposition for few-layer two-dimensional materials. *SmartMat*, *4*(3). https://doi.org/10.1002/smm2.1177

Zhang, S., Yin, B., Jiang, H., Qu, F., Umar, A., & Wu, X. (2015). Hybrid ZnO/ZnS nanoforests as the electrode materials for high performance supercapacitor application. *Dalton Transactions*, *44*(5), 2409–2415. https://doi.org/10.1039/C4DT03270E

Zhang, Y., Feng, Z., Wang, X., Hu, H., & Wu, M. (2023b). MXene/carbon composites for electrochemical energy storage and conversion. *Materials Today Sustainability*, *22*, 100350. https://doi.org/10.1016/j.mtsust.2023.100350

Zhou, H., Lu, Y., Wu, F., Fang, L., Luo, H., Zhang, Y., & Zhou, M. (2019). $MnO_2$ nanorods/MXene/CC composite electrode for flexible supercapacitors with enhanced electrochemical performance. *Journal of Alloys and Compounds*, *802*, 259–268. https://doi.org/10.1016/j.jallcom.2019.06.173

Zhou, T., Wu, C., Wang, Y., Tomsia, A. P., Li, M., Saiz, E., Fang, S., Baughman, R. H., Jiang, L., & Cheng, Q. (2020). Super-tough MXene-functionalized graphene sheets. *Nature Communications*, *11*(1), 2077. https://doi.org/10.1038/s41467-020-15991-6

Zhu, L., Nuo Peh, C. K., Zhu, T., Lim, Y.-F., & Ho, G. W. (2017). Bifunctional 2D-on-2D $MoO_3$ nanobelt/$Ni(OH)_2$ nanosheets for supercapacitor-driven electrochromic energy storage. *Journal of Materials Chemistry A*, *5*(18), 8343–8351. https://doi.org/10.1039/C7TA01858D

Chapter 4

# Research direction and challenges of MXene and graphene-based components in Metal-Sulfur and Selenium batteries

*Vinotha Krishnasamy*

## 4.1 INTRODUCTION

Innovative battery technology is progressing at an accelerated pace, focusing on meeting the energy storage demands through incorporating smart materials and different chemical compositions. Although lithium-ion batteries (LIBs) have been extensively utilized, their efficacy has stagnated because of constraints related to energy density, intercalation-driven electrochemical processes, and cycle longevity. Alternative battery chemistries like metal-sulfur and metal-selenium have surfaced as viable alternatives, owing to their elevated energy densities and their potential for economically advantageous implementations. Modifications to electrodes are made to improve the performance of batteries (Tari et al., 2022, 2025; Chinnaiah et al., 2024a, 2024b, 2024c).

Lithium–sulfur (Li–S) batteries could be a promising solution to next-generation energy needs. They come with a theoretical specific energy of 2600 Wh $kg^{-1}$, and a specific capacity of 1675 mAh $g^{-1}$. Such high values are enabled by multi-electron transfer electrochemistry (Zhou et al., 2021), which relies on the reversible redox reactions between lithium and sulfur. The theoretical capacity of the sulfur (S) cathode is ten times that of Li-ion cathode materials (He et al., 2018). Also, sulfur as a cathode material is abundant, non-toxic, and cheap; all these features make it very appealing for new battery technology. However, disadvantages such as volume expansion during discharge, the polysulfide shuttling effect, and the insulating nature of sulfur give rise to capacity decay, low utilization, and poor cyclability. Studies focus on the development and advancement of Li–S batteries in terms of sulfur-based composite cathode design, separator modification, binder improvement, electrolyte optimization, and lithium metal protection (Chen et al., 2021). The addition of kinetic promoters to attain practical high energy density is being explored (Zhao et al., 2023). Likewise, lithium–selenium (Li–Se) batteries are generally observed to have a high volumetric energy density and an intermediate output voltage but suffer from such obvious shortcomings as polyselenide dissolution, poor cycle life, and the high price of selenium (Se).

DOI: 10.1201/9781003584742-4

Both Li–S and Li–Se batteries encounter common obstacles, including the shuttling of polysulfides or polyselenides, leading to quick capacity degradation and low coulombic efficiency. For Li–S batteries, sulfur's low volumetric density and limited cyclability further hinder widespread adoption, despite its availability as a byproduct of oil refining. On the other hand, the eco-friendly benefits of selenium and its high energy density are offset by the challenges involved in achieving reproducible performance, which is mostly accounted for by the poor utilization of active materials and high production costs.

Customizing energy storage devices according to specific application requirements, such as power density, volumetric and gravimetric energy densities, and cycle life, is essential for optimization (Fang et al., 2017). The functionality of the sulfur and selenium batteries can be enhanced by incorporating new materials in the design that could help overcome their problems. A great deal of research nowadays points toward materials with exceptional energy storage potential, most notably, MXene and graphene. These two-dimensional (2D) materials hold the promise of improving conductivity, energy density, and structural stability, which are key elements in the future of battery systems.

The addition of advanced materials, like MXene and graphene, can help deal with the challenges. They exhibit significant properties that are beneficial in many applications. The arrangement of carbon atoms in graphene enables the electrons to move freely at high speeds, thus saving energy. It has been discovered that graphene has a self-repair property. Their 2D structure allows the atoms to react from two sides. Special chemical reactivity is seen in the carbon atoms present at the edge of the graphene sheet (Denissova et al., 2022). The unique properties of MXene include high electrical conductivity, mechanical properties, and functionalized surfaces. They are used in many applications (Varadarajan et al., 2024). In water, they can form a stable colloidal solution with a high zeta potential (Gogotsi & Anasori, 2019). MXene can be coupled with different metal ions, which alter the interlayer spacing and influence conductivity. Electrochemical intercalation of MXenes makes them useful in the energy storage applications of metal-ion batteries (Long et al., 2022).

MXene and graphene enhance the performance of the cathode by increasing its conductivity, relieving the shuttling phenomena, and promoting stable cycling. Their integration into Li–S, and Li–Se batteries can introduce energy storage devices with increased energy densities, long cycle life, and improved safety. Preliminary research shows that metal-sulfur and metal-selenium batteries, based on advanced materials, will likely play a significant part in meeting the global demand for efficient and sustainable energy storage systems. Generally, metallic anodes of Li, Mg, and Na are coupled with the high-energy cathode materials from group 4. There are multiple studies on other metal-based sulfur and selenium batteries. Among group 4's primary elements, oxygen and sulfur are considered effective cathode materials for battery applications (Gu et al., 2019).

This chapter discusses the prospects and challenges of metal-sulfur and selenium batteries. Lithium-based sulfur and selenium batteries are widely discussed. MXene and graphene synthesis methods and their contribution to overcoming the challenges in these batteries are elaborated. The focus of the chapter is to understand the role of MXene and graphene in reducing or removing the problems that hinder performance. They can be utilized to further enhance the battery design and performance in sulfur and selenium batteries.

## 4.2 SULFUR AND SELENIUM BATTERIES

The basic chemistry in Li–S and Li–Se batteries is quite similar. The two main requirements for a battery to be compatible with grid energy storage, portable electronics, electric transportation, and other applications are lower price and enhanced discharge capacity. The discharge capacity of a battery is a significant factor in determining the operation duration of electronic devices. Energy-intensive applications like transportation call for high-energy-capacity batteries. Sulfur remains a suitable option in this regard and is expected to dominate the market in the post-Li-ion battery era (Ould Ely et al., 2018).

Sulfur has many allotropes, with the stable and substantial form being orthorhombic a-S8 (8 sulfur atoms in a puckered ring). Heating causes the S-S bond to break, resulting in linear S8 chains and longer ones like S16, S24, and so on, adding a viscosity change to sulfur. Also, oxidation causes the intertwining of the long chains, which increases viscosity and affects the operation of the battery.

### 4.2.1 Lithium–sulfur battery electrochemistry

The rechargeable Li–S battery has a theoretical energy density and capacity higher than commercial LIBs. The sulfur cathode exhibits high theoretical capacity because of the two-electron redox process (Zhou et al., 2021). They are also comparatively low in cost, widely available, and less toxic, so they are considered better alternatives to traditional batteries.

The battery configuration of Li–S consists of a Li metal anode, an organic electrolyte, and an elemental sulfur cathode. $Li_2S$ is formed when elemental sulfur reacts with lithium during discharge. Li and S are produced when lithium sulfide decomposes during the charging process. The cathode (sulfur) charge transfer reaction is,

$$S_8 + 16Li^+ + 16e^- \rightarrow 8Li_2S \tag{4.1}$$

In Li–S, the battery cathode is made of octa-sulfur S8. During discharge, it is reduced by producing several soluble polysulfides with high solubility. Li-ions move from the anode (negative electrode) to the cathode (positive electrode) and react with the polysulfides, forming $Li_2S$ and $Li_2S_2$ (Figure 4.1). The gravimetric energy expected from a Li–S battery is 500 Wh $kg^{-1}$, against that of a Li-ion cell, which is approximately 243 Wh $kg^{-1}$.

Li–S has promising prospects as a next-generation battery; however, the major challenges have to be addressed. Finding an appropriate sulfur host to retain the polysulfides is required. A suitable sulfur host to effectively anchor the polysulfide has not been identified. By combining different host materials, the desired effect could be achieved. Despite the drawbacks, it is highly anticipated that Li-S batteries will be widely commercialized by overcoming the hurdles through rigorous theoretical and experimental studies.

### 4.2.2 Lithium–selenium battery electrochemistry

The metal selenium batteries consist of a selenium cathode, an organic electrolyte, a separator, and a metal anode (Li, Na, K, Mg). Selenium has higher

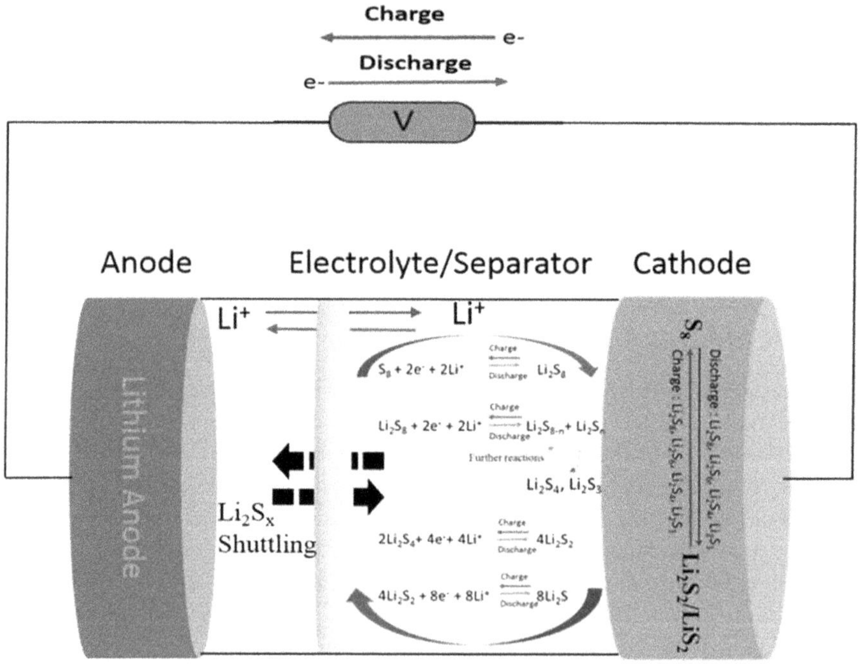


*Figure 4.1* Lithium-sulfur battery electrochemistry.

**(Reprinted with permission from Ould Ely et al. (2018). Copyright © 2018, American Chemical Society.)**

conductivity than sulfur because of the smaller bandgap. The Se cathode can operate over a wide temperature range. The reaction in the Li–Se battery is as follows,

$$Se + 2Li + 2e^{-} \leftrightarrow Li_2Se \tag{4.2}$$

The formation of polyselenides results in approximately 98% volume expansion in the cathode. The polyselenides are thermodynamically unstable and soluble in various electrolytes, and their solubility in different electrolytes is quite variable. Typically, ether-based electrolytes or carbonate-based electrolytes can be used in the Li–Se battery.

Since they undergo two-electron reactions, both sulfur and selenium batteries have more electrons participating in the reaction and migrating, so they can attain good specific capacity and performance.

## 4.3 CHALLENGES

### 4.3.1 Lithium–sulfur batteries

Li–S batteries face challenges in achieving optimal performance. Another important concern in realizing the Li-S battery technology is polysulfide shuttling. For Li–S batteries to be a practical substitution for lithium-ion, it is foremost to solve the technical issues. Key issues include the inhibition of polysulfides, the increase in sulfur loading to 4–6 mg $cm^{-2}$, the increase in sulfur fraction to at least 70%, the improvement of sulfur utilization to at least 80%, the reduction of the electrolyte/sulfur weight ratio to 3:1 or less, and the stability of the lithium anode material (Ould Ely et al., 2018). Limitations of the sulfur cathode include insulation, PS shuttle, and volume change. Measures to suppress dendrite formation in the Li anode must be taken to make them practical (Wei et al., 2020). Addressing these issues effectively will enable the realization of the full potential of Li–S batteries.

#### *4.3.1.1 Volume expansion*

Sulfur cathode is prone to volume expansion. In a Li–S battery, the transition from S8 to $Li_2S$ causes volume expansion in the cathode, damaging the structure (Zhao et al., 2021). The density of sulfur is more than that of $Li_2S$ (1.67 g $cm^{-3}$). There is a volume expansion of about 80% (Wang et al., 2013).

#### *4.3.1.2 The shuttle effect*

The shuttle effect refers to the dissolution-migration of polyselenides/polysulfides in the electrolyte, which hinders the electrochemical performance (Figure 4.2). The solubility of polysulfide ions in the electrolyte in

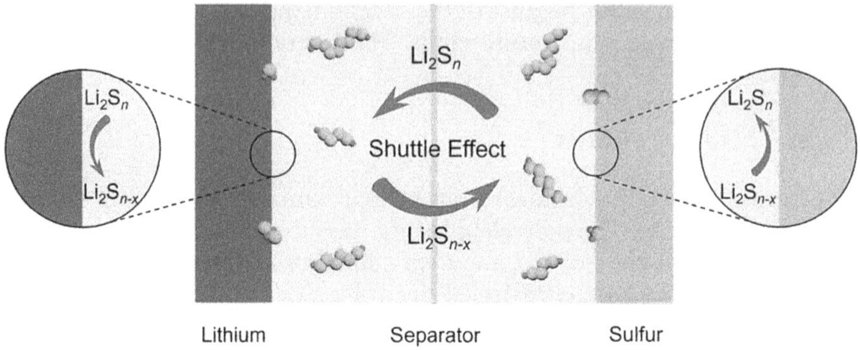


*Figure 4.2* Lithium-sulfur battery showing the shuttle effect.

**(Reproduced from (Zhou et al., 2022.)**

Li–S batteries is a major issue. Due to a concentration gradient, the dissolved ions move from the cathode to the anode. Now, the insoluble $Li_2S_2$/ $Li_2S$ in the Li anode reacts with the ions, forming medium-chain ions that return to the sulfur cathode. This is termed the shuttle effect. The major drawbacks of the shuttle effect are,

- Decrease in coulomb efficiency
- Reduction of capacity during cycling
- Active mass loss at the cathode
- The cathode becomes insulated by the deposition (i.e. hindrance to ion and electron transfer)

## 4.3.2 Lithium–selenium batteries

Li–Se batteries are an emerging battery chemistry with high volume capacity and specific capacity. Challenges in developing the battery include volume change, shuttle effect, sluggish redox kinetics of polyselenide, and low reactivity of bulk Se (W. Qiu et al., 2023). The sodium-selenium (Na-Se) battery has recently been researched because selenium has a very good theoretical capacity and conductivity higher than sulfur (>20 times), which makes it an attractive option for energy storage. Challenges in the widespread application of Li–Se batteries are listed below.

### *4.3.2.1 Cost and safety*

Selenium metal is more expensive than sulfur and is highly toxic. Due to these factors, the widespread commercialization is hindered. In order to make it suitable for everyday use, it needs to be sealed properly and taken care of. All these would further add to the cost of battery manufacturing. Cost is a major issue when it comes to large-scale applications.

#### 4.3.2.2 Selenium loading rate

The volume expansion in the selenium battery can be mitigated by loading it onto materials that provide support. The Se loading capacity has an impact on the capacity of the positive electrode. Hence, the battery design should incorporate support materials that increase the Se loading capacity and also provide support (Feng et al., 2022).

#### 4.3.2.3 Volume expansion

During cycling, the electrode's crystal structure changes, causing volume expansion that is detrimental to battery operation. The Se crystals accommodate the lithium, resulting in volume expansion, and there is mechanical stress between the particles, and the cathode undergoes structural damage after discharge. In the long run, the loss of Se active materials will lead to a short lifetime, capacity fading, and ultimately to the pulverization of electrodes.

#### 4.3.2.4 Slow reaction kinetics

The average conductivity of Se (trigonal, with the highest conductivity among other Se forms) is only close to that of a semiconductor. The Li-ion diffusion and electron transfer are slow, and they affect the kinetics of the reaction, the utilization of Se, and the rate performance. It is also observed that the transformation of Se to $Li_2Se$ is slow because of sluggish Li-ion diffusion kinetics and charge transformation issues (Li et al., 2016).

#### 4.3.2.5 Shuttle effect of polyselenides

In ether-based electrolytes, the soluble polyselenides cause a loss in active material. The continuous diffusion of polyselenide with the electrodes leads to the corrosion of the solid electrolyte interphase (SEI) film, deposition of insulators on the anode, the consumption of electrolytes, and self-discharge of the battery. The polyselenide shuttle effect affects the cycle stability and the coulombic efficiency of the Li–Se battery (Energy Materials, 2023).

Different novel strategies are adopted to address each of the above issues. The strategies discussed here could be categorized into two basic strategies: (1) containment strategies, to confine polysulfides with a suitable coating or interlayer separator, and (2) destruction strategies, which use an electrocatalyst to facilitate the autocatalytic decomposition or precipitation of polysulfides so that they can no longer perform their shuttling functions. Electrolytes with suitable additives can resolve the polysulfide problem.

Studies on Li–Se batteries should concentrate on alleviating the above issues effectively for the commercial application of the battery. Different techniques are explored for achieving better performance. In the following sections, the contribution of 2D materials MXene and graphene, in enhancing sulfur- and selenium-based metal battery performance is discussed.

## 4.4 MXENE AND GRAPHENE-BASED COMPOSITES

Different methods to attenuate the shuttle effect include

- Physical confinement using porous carbon (M. Wang et al., 2019b)
- Chemical bonding of the polysulfides with polar metal compounds (Hong et al., 2020)
- Using a conductive layer between the cathode and the separator (Huang et al., 2015).

However, the high electrical conductivity and mechanical strength of MXenes can prevent the shuttle effect by providing sufficient anchoring (Jayan & Islam, 2020; Kannan and Tari, 2025; Gurushankar et al., 2024).

Suitable host materials are required for polysulfide anchoring. Here, we discuss the MXene- and graphene-based composites as host materials. In physical confinement, physical barriers can be created using porous and layered (shelled) structures of sulfur hosts in the electrode. Chemical confinement involves four types of interaction with the sulfur host.

1) redox interactions
2) Lewis acid–base interactions
3) polar–polar interactions
4) covalent binding interactions.

Carbon-selenium (C/Se) composite cathodes were studied to alleviate volume expansion, inhibit polyselenide dissolution, and accelerate electron transport. Carbon materials can be useful host materials for Se because of their porous structure and adsorption effect (Wang et al., 2023).

### 4.4.1 MXene in metal-sulfur battery and metal-selenium battery

MXenes are 2D materials that are transition metal carbides, nitrides, and carbonitrides. Different compositions of MXenes are currently available. The applications are attributed to their unique 2D structure with enormous surface chemistry and exceptional physical and chemical properties. Though they are applicable in different fields, this chapter deals with the potential in battery chemistry, specifically in metal-sulfur and selenium batteries.

The distinctive properties of MXenes are enhanced by advanced synthesis methods. MXenes are utilized in different components of the battery with some modifications. The modifications can be incorporated during the synthesis procedure. MXenes can also be used along with other materials like carbonaceous materials and metal oxides to suit the requirements and enhance their performance. In metal-sulfur batteries (Li, Na, K), they can be used in electrodes, metal anode protective layers, separators, S hosts, and other components (Liu et al., 2022).

2D materials are seen to provide vast space for metal ion shuttling, together with impressive sustainability of volume change, which leads to enhanced storage capacity and extraordinary cycling stability (Naguib & Gogotsi, 2015). MXenes are described by the general formula $M_{n+1} X_nT_x$.

M—early transition metal,
X—carbon or nitrogen or both, and
n = 1, 2, or 3.

$T_x$ can be any surface functional group such as -OH, =O, or -F. The electrochemical performance of MXene can be improved either by the decoration of the MXene surface or by chemical treatment.

In the synthesis of MXenes, different terminations can be added to the MXene sheets that influence the properties as well as the application in different fields. MXenes can have enhanced optoelectronic (Zhou et al., 2018), electronic (Liu et al., 2016), and magnetic properties (Khazaei et al., 2013) based on the terminations usually decided during the synthetic process or controlled through post-treatment (Kamysbayev et al., 2020; Li et al., 2019).

There are various advantages to incorporating MXenes in battery design as shown in Figure 4.3. It is observed that MXene with different terminations can be used as a sulfur host or as an interlayer material between the cathode and the separator to inhibit the shuttle effect and improve cyclic stability (Rao et al., 2017; D. Wang et al., 2019a). So, in LiS MXene, different terminations can be added that can contribute to strong chemical interaction with lithium polysulfides. The capacity of Li–S batteries can be improved by using MXene because it has good metallic conductivity, thereby enabling quick electron transfer. This, in turn, ensures better utilization of sulfur (Liang et al., 2015).

MXenes effectively control Li dendrites by enabling uniform nucleation and growth. MXene@Cellulose nanofiber (CNF) film was investigated and found to have good flexibility and improved mechanical strength. It has more Li nucleation sites, which inhibit dendrites while providing better cycling stability and high capacity (Wang et al., 2020).

MXene has a structural advantage in that it can host sulfur or Li and mitigate the effect of volume expansion. Another notable issue with Li-ion batteries is dendrite formation. By using horizontally aligned MXenes, it is possible to guide the nucleation and growth to obtain horizontal-growth Li

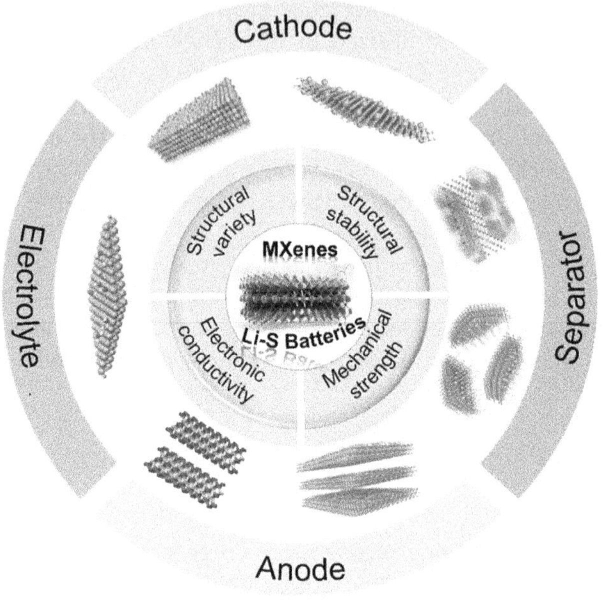


*Figure 4.3* MXene in Lithium Sulfur battery.

**(Reproduced from Liu et al., 2024a, https://creativecommons.org/licenses/by/4.0/.)**

anodes. Moreover, the fluoride terminations in MXene form an electrolyte interface. So, MXene has a major contribution in increasing the cycle stability of Li-ion battery (Zhang et al., 2019). This can be advantageous in lithium-based battery systems.

MXenes and polysulfides exhibit a Lewis acid–base interaction that suppresses the polysulfide diffusion and improves the lifecycle of the sulfur cathode (Liang et al., 2015). Though lithium metal anodes are an extraordinary choice for high-energy-density batteries, they pose safety risks during cycling. The further development of the anode is restrained. MXenes are introduced as anode complements to improve the safety features and overall performance of the battery. MXene characteristics like large specific surface area, ultrafast ion diffusion, excellent conductivity, and flexibility make them a good choice for the intended purposes.

Rechargeable Li–Se batteries have chemical properties identical to those of lithium-sulfur batteries. They also react to form polyselenides ($Li_2Se$) (Jayan & Islam, 2020). Various studies on the utilization of MXene in Se batteries are discussed here.

The unique nanoarchitecture of metal-organic framework (MOF)-carbon nanotube (CNT)/MXene is effective in acting as a suitable sulfur host in Li–Se batteries and also enhances the performance of the battery. The incorporation of catalytic advantages associated with this nanoarchitecture has significantly improved the conductivity of the cathode, the movement of ions, and the overall electrochemical performance of lithium-selenium batteries. The selenium cathode featuring a Fe-CNT/MXene nanoarchitecture demonstrated exceptional stability at a charge rate of 1C, achieving a capacity retention of 95.8% after 2000 cycles. Furthermore, it maintained its integrity at a charge rate of 10 C/1 C, with a capacity retention of 97.9% after 450 cycles (Vallem et al., 2024).

In another study, the Se@MXene composite was synthesized using a straightforward melt diffusion method. Trigonal selenium nanocrystals, with a mass loading of 63.3 wt%, are distributed both on the surface and within the interlayer spaces of the conductive MXene. The Se@MXene cathode exhibited a substantial specific capacity of 632 $mAhg^{-1}$, which corresponds to 93.6% of its theoretical capacity, at a current density of 50 mA $g^{-1}$. Furthermore, it maintains a reversible capacity of 274 mAh $g^{-1}$ even at an elevated current density of 4000 mA $g^{-1}$. Notably, the Se@MXene cathode demonstrates stable cycling performance over 300 cycles (Liu et al., 2024b).

The hybrid separator (CCNT/MXene/PP) cetrimonium bromide (CTAB)/ CNT/$Ti_3C_2T_x$ MXene hybrid modified polypropylene (PP) effectively immobilizes polyselenides through enhanced Lewis acid-base interactions (CTAB/ MXene and polyselenides). The addition of carbon nanotubes (CNT) enhances electrolyte infiltration and facilitates ionic transport. For the first time, in situ permeation experiments have been conducted to visually examine the behavior of polyselenides, demonstrating the suppression of the shuttle effect and the protection of the lithium anode from corrosion when utilizing CCNT/MXene/PP separators (Zhang et al., 2020).

### 4.4.2 Graphene in metal-sulfur battery and selenium battery

Graphene is an important 2D material that is utilized in several applications. The extraordinary physical and chemical characteristics, along with the excellent electronic conductivity of graphene, have the potential to address some of the issues that inhibit the large-scale commercialization of Li–S batteries. The porous nature of graphene, the high conductivity, and flexibility can be useful in suppressing the polysulfide shuttling (J. Zhang et al., 2022).

The polysulfide dissolution is controlled by immobilizing sulfur and polysulfides. The functional groups on the surface of graphene oxide (GO) enabled the trapping of sulfur and polysulfides (Ji et al., 2011). The electrochemical performance of the as-prepared cathode composite was tested with different electrolytes under various conditions, which exhibited a stable high capacity. The GO–S nanocomposite was efficient in suppressing polysulfide dissolution and relieving the stress caused by sulfur expansion (Ould Ely et al., 2018).

The performance of graphene-sulfur composites is enhanced through the process of nitrogen doping (Qiu et al., 2014). The resulting nitrogen-doped graphene demonstrates significant electrical conductivity and an ability to effectively capture sulfur and polysulfides. The developed sulfur/nitrogen-doped graphene (S–NG) composite achieved a cycling performance of 120 cycles (Ould Ely et al., 2018).

Graphene sheets as an additive in the cathode reported a reversible capacity (570 mAh $g^{-1}$) after 100 cycles (Lin et al., 2013). A sulfur/3D-graphene composite exhibited an initial discharge capacity (1260 mAh $g^{-1}$) and a reversible capacity (700 mAh $g^{-1}$) after 100 cycles. The highly porous three-dimensional structure of graphene facilitated efficient pathways for the transport of electrons and ions (Xu et al., 2015).

Unstacked nitrogen-doped graphene (NG) with a significantly higher sulfur loading (5 mg $cm^{-2}$) reached an energy density (comparable to lithium-ion batteries) (Song et al., 2016). The sulfur/NG composite, which contained 80% sulfur loading, demonstrated a high discharge capacity. The remarkable reversible capacities observed at various current densities can be attributed to the capacity of the unstacked porous nitrogen-doped graphene to adsorb polysulfides, thereby facilitating a high sulfur loading of up to 90 wt% (Ould Ely et al., 2018).

To enhance the ion conduction of graphene across the layers, it is activated, providing more sites for sulfur accommodation. Potassium hydroxide (KOH)-activated graphene has mesopores large enough to transport Li ions and small enough to hold the polysulfides (Ding et al., 2013). Also, the presence of residual oxygen groups in the graphene sheets influences the adsorption of sulfur and polysulfide ions (Ji et al., 2011).

Graphene can be a suitable cathode modification material for Li–S batteries because of its high electrical conductivity and hardness. It complements the sulfur cathode, which is insulating in nature with its high conductivity. The hardness of graphene resists sulfur expansion during the discharge process. Different strategies can be adopted to enhance the performance of metal-sulfur batteries using graphene.

A novel sulfur cathode design with sulfur sandwiched between two graphene layers localizes the active sulfur species at the cathode (Zhou & Zhou, 2017). A combination of reduced graphene oxide and graphene foam is reported as an effective sulfur host (Hu et al., 2015).

A high-conductivity and porous nanoribbon of vanadium nitride and graphene sulfur host was tested for the cathode of a Li–S battery and was used to adsorb the polysulfides from the catholyte. The charge transfer and electrolyte absorption were made easy by the graphene network and the nanoribbons, with the mesopores facilitating fast ion transportation and adsorption of the polysulfides. This composite was able to achieve an initial capacity of 1471 mAh $g^{-1}$ at 0.2 C and mitigated polarization (Sun et al., 2017).

The diffusion of polysulfides can be reduced by coating a shell or layer on the sulfur particles and suppressing the shuttle effect (Wu et al., 2016). Using a flexible layer-by-layer self-assembly strategy, polyelectrolyte multilayers

(PEMs) and graphene sheets are coated on the surface of a hollow carbon/sulfur composite (Figure 4.4a). The graphene sheets stabilize the PEMs and reduce the inner resistance of the electrode. This composite exhibited cycle stability of over 200 cycles at 1 $Ag^{-1}$ and coulombic efficiency of 99% (Figure 4.4b).

In a study, Li–S cathode, consisting of a freestanding composite thin film assembled from S nanoparticles, reduced graphene oxide (rGO), and a multifunctional additive poly(anthraquinonyl sulfide) (PAQS) (Figure 4.4c). The S nanoparticles provide a high initial specific capacity, and the layered and porous rGO structure provides electron and ion transport paths and restricts polysulfide shuttling. PAQS is not only a highly efficient sulfide trapping agent but also an excellent Li+ conductor, which benefits the battery reaction kinetics at a high rate. The resulting cathode exhibits an initial specific capacity of 1255 mAh $g^{-1}$ with a decay rate as low as 0.046% per cycle over 1200 cycles. Importantly, it displays a reversible capacity of 615 mAh $g^{-1}$ when discharged at a high rate of 8 C (13.744 $Ag^{-1}$) (Chen et al., 2015) (Figure 4.4d).

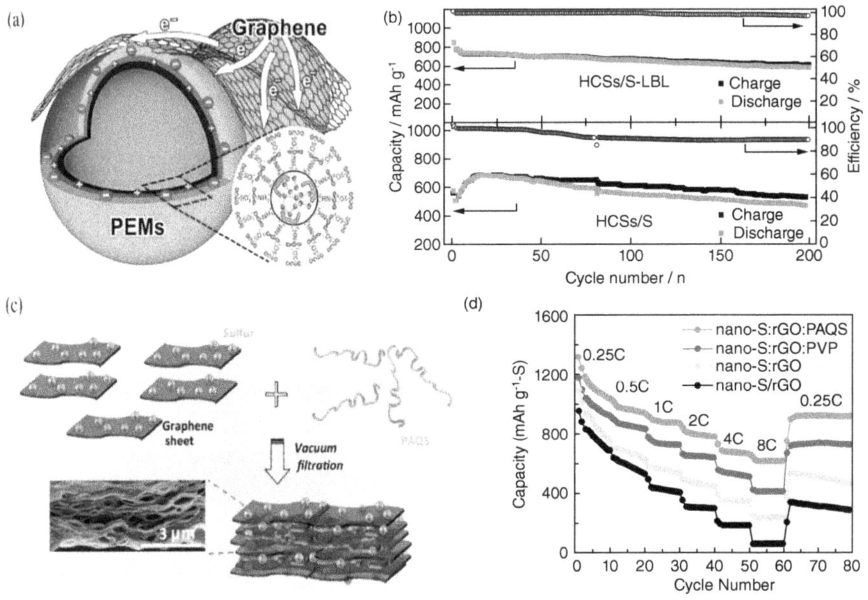


*Figure 4.4* (a) Layer-by-layer film-coated hollow carbon/sulfur composites. (b) Cycle life performance and coulombic efficiency of the composite at 1 A $g^{-1}$. (Reprinted with permission from Wu et al. (2016). Copyright © 2016, American Chemical Society.) (c) Schematic illustration and (d) rate performance of a nano-S:rGO: PAQS thin film cathode.

**(Reprinted with permission from H. Chen et al. (2015). Copyright © 2015, American Chemical Society.)**

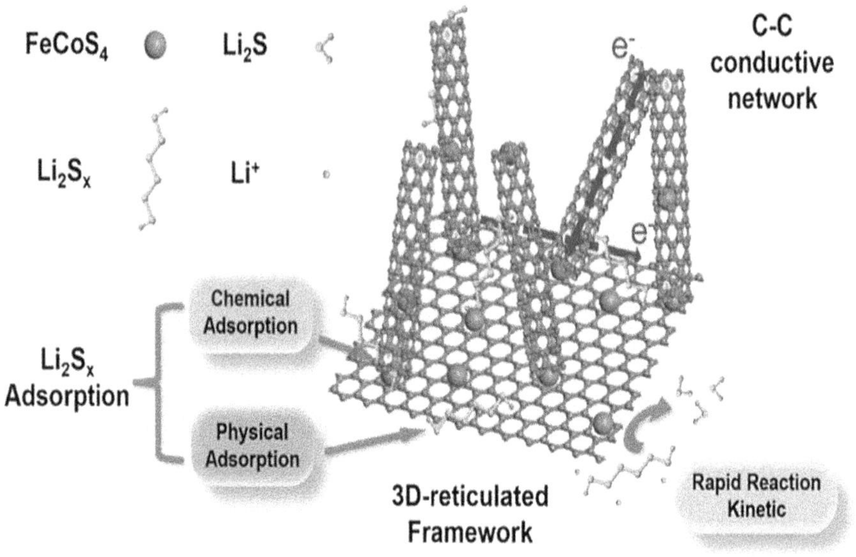


*Figure 4.5* 3D Framework of Fe–Co sulfide/graphene/CNTs (FCS@GC) to mitigate the shuttle effect in LiS battery.

**(Reprinted with permission from Ying et al. (2024). Copyright © 2024, American Chemical Society.)**

In Li–S batteries, the Fe–Co sulfide/graphene/CNTs (FCS@GC) functional interlayer composite slows down the shuttle effect at many levels and enhances the kinetics (Figure 4.5). A stable specific capacity of 787.5 mA h $g^{-1}$ after 200 charge/discharge cycles at 0.5 C has been reported; the capacity decay rate per cycle is only 0.036% after 800 cycles. After 110 cycles at a high sulfur loading of 5 mg $cm^{-2}$, a capacity of 750.5 mA h $g^{-1}$ is maintained. Hence, the composite exhibits good potential for the commercialization of Li–S batteries (Ying et al., 2024).

High-energy ball milling process is used to construct Se-decorated graphene cathodes, and this helps in binding the Se atoms to the bulk graphene support (Sha et al., 2018). There is research work on the electrochemical performance of the battery by using graphene with CNTs, porous CNFs, and porous carbon.

### 4.4.3 DFT study of MXene and graphene

Through density functional theory (DFT) studies on the interaction of lithium polysulfides with MXene and surface functions, it was found that the shuttle effect is mitigated by the terminations in MXene, which form a strong metal-S bond (Rao et al., 2017). Surface-functionalized MXene retains the metallic conductivity of MXene. However, $Ti_3C_2S_2$ and $Ti_3C_2O_2$, and $Ti_3C_2N_2$ display fair adsorption, which can restrain polysulfide dissolution and shuttling. $Ti_3C_2S_2$ and $Ti_3C_2O_2$ can be good host materials for sulfur at the cathode, as these surface-functionalized MXenes reduced the $Li_2S$ decomposition barrier considerably (D. Wang et al., 2019a).

The structural stability, electronic properties, and adsorption behavior of MXene and graphene were determined using density functional theory. The polyselenide bonding strength with MXene was evaluated by analyzing the van der Waals interactions using the DFT-D3 method. Using DFT calculations, the interaction of MXene and graphene as host materials in the Se cathode was investigated. Graphene is used as a reference material to determine the anchoring property of MXene. It is observed that graphene is not capable of effectively anchoring the polyselenides to prevent dissolution. It is also inferred that bare $Ti_3C_2$ MXenes have strong binding with polyselenides, but dissociations are occurring at the surface that affect the reversibility of the battery. Adding functional groups to the bare $Ti_3C_2$ can overcome these limitations. The $Ti_3C_2S_2$ and $Ti_3C_2O_2$ were found to be capable of providing sufficient binding for the polyselenides, and no chemical decomposition was observed. This confirms the superior performance of the S- and O-functionalized MXene for the Se cathode. The use of functionalized MXene is considered to be beneficial for the anchoring of the polyselenides, providing an electronic conduction path to overcome the insulating nature of Se (Jayan & Islam, 2020).

## 4.5 CONCLUSION

MXene and graphene-based composites have emerged as highly promising materials for the enhancement of metal-sulfur and metal-selenium batteries, attributable to their remarkable electrical conductivity, extensive surface area, and chemical stability. The synergistic integration of these materials effectively addresses critical challenges such as polysulfide shuttling, volumetric expansion, and slow reaction kinetics. MXenes provide superior conductivity, redox activity, and structural integrity, while graphene contributes to enhanced electron transport and mechanical flexibility.

Recent studies underscore the potential of these composites in the development of high-capacity cathodes, protective interlayers, and conductive hosts for sulfur- and selenium-based batteries. Despite significant advancements, several challenges persist, including the optimization of composite design, scalability improvements, and the assurance of long-term cycling stability. Future research endeavors should prioritize the customization of interfacial chemistry, the exploration of innovative synthesis methodologies, and the establishment of environmentally sustainable production processes. In summary, MXene and graphene-based composites signify a transformative strategy for the next generation of metal-sulfur and metal-selenium batteries. Their ongoing development holds the promise of facilitating high-performance energy storage systems characterized by enhanced capacity, efficiency, and sustainability, thereby propelling progress in renewable energy applications.

## REFERENCES


Chen, H., Wang, C., Dai, Y., Qiu, S., Yang, J., Lu, W., & Chen, L. (2015). Rational design of cathode structure for high-rate performance lithium-sulfur batteries. *Nano Letters*, *15*(8), 5443–5448. https://doi.org/10.1021/acs.nanolett.5b01837

Chen, Y., Wang, T., Tian, H., Su, D., Zhang, Q., & Wang, G. (2021). Advances in lithium–sulfur batteries: From academic research to commercial viability. *Advanced Materials*, *33*(29), 2003666. https://doi.org/10.1002/adma.202003666

Chinnaiah, K., Kannan, K., Krishnamoorthi, R., Palko, N., & Gurushankar, K. (2024a). Nanostructured Ag/NiO composites for supercapacitor and antibacterial applications, and in-silico theoretical investigation. *Journal of Physics and Chemistry of Solids*, *184*, 111730.

Chinnaiah, K., Kannan, K., Palko, N., Grishina, M., Chen, Y.-S., & Gurushankar, K. (2024b). Exploring the potential of Withania somnifera-mediated Ag/$Mn_3O_4$ nanocomposites as electrode material for high-performance supercapattery device. *Journal of the Taiwan Institute of Chemical Engineers*, *157*, 105441.

Chinnaiah, K., Kannan, K., Palko, N., Grishina, M., Guganathan, L., & Gurushankar, K. (2024c). Bioengineered Ag/NiO nanocomposites as advanced battery-supercapacitor electrodes for highly efficient symmetric hybrid devices. *Ionics*, *30*(3), 1691–1707. https://doi.org/10.1007/s11581-023-05361-9

Denissova, A. I., Volokitin, A. V., & Volokitina, I. E. (2022). Prospects of application and global significance of graphene. *Progress in Physics of Metals*, *23*(2), 268–295. https://doi.org/10.15407/ufm.23.02.268

Ding, B., Yuan, C., Shen, L., Xu, G., Nie, P., Lai, Q., & Zhang, X. (2013). Chemically tailoring the nanostructure of graphene nanosheets to confine sulfur for high-performance lithium-sulfur batteries. *Journal of Materials Chemistry A*, *1*(4), 1096–1101.

Energy Materials. (2023). *Recent Progress and Prospect of Li-Se Batteries: A Comprehensive Review*. https://doi.org/10.20517/energymater.2022.91

Fang, R., Zhao, S., Sun, Z., Wang, D. W., Cheng, H. M., & Li, F. (2017). More reliable lithium-sulfur batteries: Status, solutions and prospects. In *Advanced Materials* (Vol. 29, Issue 48). Wiley-VCH Verlag. https://doi.org/10.1002/adma.201606823

Feng, S., Liu, J., Zhang, X., Shi, L., Anderson, C., Lin, Y., Song, M.-K., Liu, J., Xiao, J., & Lu, D. (2022). Rationalizing nitrogen-doped secondary carbon particles for practical lithium-sulfur batteries. *Nano Energy*, *103*, 107794.

Gogotsi, Y., & Anasori, B. (2019). The Rise of MXenes. In *ACS Nano* (Vol. 13, Issue 8, pp. 8491–8494). American Chemical Society. https://doi.org/10.1021/acsnano.9b06394

Gu, X., Tang, T., Liu, X., & Hou, Y. (2019). Rechargeable metal batteries based on selenium cathodes: Progress, challenges and perspectives. In *Journal of Materials Chemistry A* (Vol. 7, Issue 19, pp. 11566–11583). Royal Society of Chemistry. https://doi.org/10.1039/c8ta12537f

Gurushankar, K., Karuppaiya, C., Tari, V., & Kannan, K. (2024). Exploring MXenes and MBenes in metal sulfur and selenium battery technologies. In K. Kannan & V. Tari (Eds.), *Novel Energy Storage and Conversion Technologies for Two-Dimensional MXenes and MBenes*. IGI Global. https://doi.org/10.4018/979-8-3693-9546-2.ch005

He, Y., Chang, Z., Wu, S., & Zhou, H. (2018). Effective strategies for long-cycle life lithium-sulfur batteries. In *Journal of Materials Chemistry A* (Vol. 6, Issue 15, pp. 6155–6182). Royal Society of Chemistry. https://doi.org/10.1039/c8ta01115j

Hong, X., Wang, R., Liu, Y., Fu, J., Liang, J., & Dou, S. (2020). Recent advances in chemical adsorption and catalytic conversion materials for Li–S batteries. *Journal of Energy Chemistry*, *42*, 144–168.

Hu, G., Xu, C., Sun, Z., Wang, S., Cheng, H.-M., Li, F., & Ren, W. (2015). 3D graphene-foam-reduced-graphene-oxide hybrid nested hierarchical networks for high-performance Li-S batteries. *Advanced Materials (Deerfield Beach, Fla.)*, *28*(8), 1603–1609.

Huang, J.-Q., Zhang, Q., & Wei, F. (2015). Multi-functional separator/interlayer system for high-stable lithium-sulfur batteries: Progress and prospects. *Energy Storage Materials*, *1*, 127–145. https://doi.org/10.1016/j.ensm.2015.09.008

Jayan, R., & Islam, M. M. (2020). Functionalized MXenes as effective polyselenide immobilizers for lithium-selenium batteries: A density functional theory (DFT) study. *Nanoscale*, *12*(26), 14087–14095. https://doi.org/10.1039/d0nr02296a

Ji, L., Rao, M., Zheng, H., Zhang, L., Li, Y., Duan, W., Guo, J., Cairns, E. J., & Zhang, Y. (2011). Graphene oxide as a sulfur immobilizer in high performance lithium/sulfur cells. *Journal of the American Chemical Society*, *133*(46), 18522–18525.

Kamysbayev, V., Filatov, A. S., Hu, H., Rui, X., Lagunas, F., Wang, D., Klie, R. F., & Talapin, D. V. (2020). Covalent surface modifications and superconductivity of two-dimensional metal carbide MXenes. *Science*, *369*(6506), 979–983.

Kannan, K., & Tari, V. (Eds.). (2025). *Novel energy storage and conversion technologies for two-dimensional MXenes and MBenes*. https://www.amazon.com/dp/9798369395462

Khazaei, M., Arai, M., Sasaki, T., Chung, C., Venkataramanan, N. S., Estili, M., Sakka, Y., & Kawazoe, Y. (2013). Novel electronic and magnetic properties of two-dimensional transition metal carbides and nitrides. *Advanced Functional Materials*, *23*(17), 2185–2192.

Li, M., Lu, J., Luo, K., Li, Y., Chang, K., Chen, K., Zhou, J., Rosen, J., Hultman, L., & Eklund, P. (2019). Element replacement approach by reaction with Lewis acidic molten salts to synthesize nanolaminated MAX phases and MXenes. *Journal of the American Chemical Society*, *141*(11), 4730–4737.

Li, Q., Liu, H., Yao, Z., Cheng, J., Li, T., Li, Y., Wolverton, C., Wu, J., & Dravid, V. P. (2016). Electrochemistry of selenium with sodium and lithium: Kinetics and reaction mechanism. *ACS Nano*, *10*(9), 8788–8795.

Liang, X., Garsuch, A., & Nazar, L. F. (2015). Sulfur cathodes based on conductive mxene nanosheets for high-performance lithium–sulfur batteries. *Angewandte Chemie*, *127*(13), 3979–3983. https://doi.org/10.1002/ange.201410174

Lin, T., Tang, Y., Wang, Y., Bi, H., Liu, Z., Huang, F., Xie, X., & Jiang, M. (2013). Scotch-tape-like exfoliation of graphite assisted with elemental sulfur and graphene–sulfur composites for high-performance lithium-sulfur batteries. *Energy & Environmental Science*, *6*(4), 1283–1290.

Liu, H., Sun, S., & Jin, B. (2024a). Recent progress in MXenes-based lithium-sulfur batteries. *Energy Materials*, *4*(4). https://doi.org/10.20517/energymater.2023.99

Liu, R., Li, C., Li, Q., Zhang, S., Wang, C., Zhang, Z., Shi, Y., Yang, L., Yin, L., & Wang, R. (2024b). High performance all-solid-state Li–Se battery based on selenium loaded on $Ti_3C_2$ MXene cathode. *Green Energy and Resources*, *2*(1). https://doi.org/10.1016/j.gerr.2024.100058

Liu, S., Song, Z., Jin, X., Mao, R., Zhang, T., & Hu, F. (2022). MXenes for metal-ion and metal-sulfur batteries: Synthesis, properties, and electrochemistry. In *Materials Reports: Energy* (Vol. 2, Issue 1). KeAi Communications Co. https://doi.org/10.1016/j.matre.2021.100077

Liu, Y., Xiao, H., & Goddard III, W. A. (2016). Schottky-barrier-free contacts with two-dimensional semiconductors by surface-engineered MXenes. *Journal of the American Chemical Society*, *138*(49), 15853–15856.

Long, Y., Tao, Y., Shang, T., Yang, H., Sun, Z., Chen, W., & Yang, Q.-H. (2022). Roles of metal ions in MXene synthesis, processing and applications: A perspective. *Advanced Science*, *9*(12), 2200296. https://doi.org/10.1002/advs.202200296

Naguib, M., & Gogotsi, Y. (2015). Synthesis of two-dimensional materials by selective extraction. *Accounts of Chemical Research*, *48*(1), 128–135.

Ould Ely, T., Kamzabek, D., Chakraborty, D., & Doherty, M. F. (2018). Lithium-sulfur batteries: State of the art and future directions. In *ACS Applied Energy Materials* (Vol. 1, Issue 5, pp. 1783–1814). American Chemical Society. https://doi.org/10.1021/acsaem.7b00153

Qiu, W., Huang, X. L., Wang, Y., Feng, C., Ji, H., Liu, H. K., Dou, S. X., & Wang, Z. (2023). Design strategies of performance-enhanced Se cathodes for Li-Se batteries and beyond. *Journal of Energy Chemistry*, *76*, 528–546. https://doi.org/10.1016/j.jechem.2022.09.026

Qiu, Y., Li, W., Zhao, W., Li, G., Hou, Y., Liu, M., Zhou, L., Ye, F., Li, H., & Wei, Z. (2014). High-rate, ultralong cycle-life lithium/sulfur batteries enabled by nitrogen-doped graphene. *Nano Letters*, *14*(8), 4821–4827.

Rao, D., Zhang, L., Wang, Y., Meng, Z., Qian, X., Liu, J., Shen, X., Qiao, G., & Lu, R. (2017). Mechanism on the improved performance of lithium sulfur batteries with MXene-based additives. *The Journal of Physical Chemistry C*, *121*(21), 11047–11054.

Sha, L., Gao, P., Ren, X., Chi, Q., Chen, Y., & Yang, P. (2018). A self-repairing cathode material for lithium–selenium batteries: Se–C chemically bonded selenium–graphene composite. *Chemistry - A European Journal*, *24*(9), 2151–2156. https://doi.org/10.1002/chem.201704079

Song, J., Yu, Z., Gordin, M. L., & Wang, D. (2016). Advanced sulfur cathode enabled by highly crumpled nitrogen-doped graphene sheets for high-energy-density lithium–sulfur batteries. *Nano Letters*, *16*(2), 864–870.

Sun, Z., Zhang, J., Yin, L., Hu, G., Fang, R., Cheng, H.-M., & Li, F. (2017). Conductive porous vanadium nitride/graphene composite as chemical anchor of polysulfides for lithium-sulfur batteries. *Nature Communications*, *8*(1), 14627.

Tari, V., Kannan, K., Koliar, J., Gurushankar, K., & Wahyuni, D. K. (2025). Influence of zinc oxide nanomaterials on crop plants: A prospective analysis. In D. Jabbrova, D. Sarkar, Rakesh, S. Singh, & R. Datta (Eds.), *Smart Nanomaterials in Agroforestry Systems*. Springer. https://doi.org/10.1007/978-981-96-1337-3_8

Tari, V., Kannan, K., & Vishwakarma, V. (2022). Emerging antiviral technologies. In P. Nunthavarawong, S. M. Rangappa, S. Siengchin, & M. Thoppil-Mathew (Eds.),

*Antimicrobial and Antiviral Materials: Polymers, Metals, Ceramics, and Applications* (1st ed., pp. 137–150). CRC Press. https://doi.org/10.1201/9781003143093-8

Vallem, S., Song, S., Oh, Y., Kim, J., Li, M., Li, Y., Cheng, X., & Bae, J. (2024). Designing a Se-intercalated MOF/MXene-derived nanoarchitecture for advancing the performance and durability of lithium-selenium batteries. *Journal of Colloid and Interface Science*, *665*, 1017–1028. https://doi.org/10.1016/j.jcis.2024.03.159

Varadarajan, S., Kavitha, A., Selvaraju, P., Muthu, S. E., Gurushankar, K., Shanmugan, S., & Kannan, K. (2024). Enhanced photocatalytic hydrogen evolution by $TiO_2$: A synergistic approach with defect-rich $SnS_2$ and $Ti_3C_2$ MXene cocatalysts. *Hydrogen*, *5*(4), 940–957. https://doi.org/10.3390/hydrogen5040050

Wang, C. Y., Zheng, Z. J., Feng, Y. Q., Ye, H., Cao, F. F., & Guo, Z. P. (2020). Topological design of ultrastrong MXene paper hosted Li enables ultrathin and fully flexible lithium metal batteries. *Nano Energy*, *74*. https://doi.org/10.1016/j.nanoen.2020.104817

Wang, D., Li, F., Lian, R., Xu, J., Kan, D., Liu, Y., Chen, G., Gogotsi, Y., & Wei, Y. (2019a). A general atomic surface modification strategy for improving anchoring and electrocatalysis behavior of $Ti_3C_2T_2$ MXene in lithium–sulfur batteries. *ACS Nano*, *13*(10), 11078–11086.

Wang, D. W., Zeng, Q., Zhou, G., Yin, L., Li, F., Cheng, H. M., Gentle, I. R., & Lu, G. Q. M. (2013). Carbon-sulfur composites for Li-S batteries: Status and prospects. *Journal of Materials Chemistry A*, *1*(33), 9382–9394. https://doi.org/10.1039/c3ta11045a

Wang, M., Xia, X., Zhong, Y., Wu, J., Xu, R., Yao, Z., Wang, D., Tang, W., Wang, X., & Tu, J. (2019b). Porous carbon hosts for lithium–sulfur batteries. *Chemistry–A European Journal*, *25*(15), 3710–3725.

Wang, R., Wang, D., Dong, Y., Xie, B., Wu, X., Wu, Q., Zhu, S., Diao, G., & Chen, M. (2023). Recent progress of advanced carbon-based cathode in sodium-selenium batteries. *Journal of Alloys and Compounds*, *952*, 169980. https://doi.org/10.1016/j.jallcom.2023.169980

Wei, C., Tao, Y., An, Y., Tian, Y., Zhang, Y., Feng, J., & Qian, Y. (2020). Recent advances of emerging 2D MXene for stable and dendrite-free metal anodes. *Advanced Functional Materials*, *30*(45), 2004613.

Wu, F., Li, J., Su, Y., Wang, J., Yang, W., Li, N., Chen, L., Chen, S., Chen, R., & Bao, L. (2016). Layer-by-layer assembled architecture of polyelectrolyte multilayers and graphene sheets on hollow carbon spheres/sulfur composite for high-performance lithium–sulfur batteries. *Nano Letters*, *16*(9), 5488–5494.

Xu, C., Wu, Y., Zhao, X., Wang, X., Du, G., Zhang, J., & Tu, J. (2015). Sulfur/three-dimensional graphene composite for high performance lithium–sulfur batteries. *Journal of Power Sources*, *275*, 22–25.

Ying, D., Yu, X., Liu, X., Ma, C., Qiao, W., Wang, J., & Ling, L. (2024). Three-dimensional reticulated Fe–Co Sulfide/Graphene/CNTs composites as functional interlayer capturing and converting polysulfide for high-performance lithium–sulfur batteries. *Energy & Fuels*, *38*(10), 9080–9090. https://doi.org/10.1021/acs.energyfuels.4c00599

Zhang, D., Wang, S., Li, B., Gong, Y., & Yang, S. (2019). Horizontal growth of lithium on parallelly aligned MXene layers towards dendrite-free metallic lithium anodes. *Advanced Materials*, *31*(33). https://doi.org/10.1002/adma.201901820

Zhang, F., Guo, X., Xiong, P., Zhang, J., Song, J., Yan, K., Gao, X., Liu, H., & Wang, G. (2020). Interface engineering of MXene composite separator for high-performance Li–Se and Na–Se batteries. *Advanced Energy Materials*, *10*(20). https://doi.org/10.1002/aenm.202000446

Zhang, J., Jia, L., Lin, H., & Wang, J. (2022). Advances and prospects of 2D graphene-based materials/hybrids for lithium metal–sulfur full battery: From intrinsic property to catalysis modification. In *Advanced Energy and Sustainability Research* (Vol. 3, Issue 4). John Wiley and Sons Inc. https://doi.org/10.1002/aesr.202100187

Zhao, M., Peng, H. J., Li, B. Q., & Huang, J. Q. (2023). Kinetic promoters for sulfur cathodes in lithium-sulfur batteries. *Accounts of Chemical Research*. https://doi.org/10.1021/acs.accounts.3c00698

Zhao, Q., Zhu, Q., Liu, Y., & Xu, B. (2021). Status and prospects of MXene-based lithium–sulfur batteries. In *Advanced Functional Materials* (Vol. 31, Issue 21). John Wiley and Sons Inc. https://doi.org/10.1002/adfm.202100457

Zhou, G., & Zhou, G. (2017). Graphene–pure sulfur sandwich structure for ultrafast, long-life lithium-sulfur batteries. In *Design, Fabrication and Electrochemical Performance of Nanostructured Carbon-Based Materials for High-Energy Lithium–Sulfur Batteries: Next-Generation High Performance Lithium–Sulfur Batteries*, 75–94.

Zhou, L., Danilov, D. L., Eichel, R. A., & Notten, P. H. L. (2021). Host materials anchoring polysulfides in Li–S batteries reviewed. In *Advanced Energy Materials* (Vol. 11, Issue 15). John Wiley and Sons Inc. https://doi.org/10.1002/aenm.202001304

Zhou, L., Danilov, D. L., Qiao, F., Wang, J., Li, H., Eichel, R. A., & Notten, P. H. L. (2022). Sulfur reduction reaction in lithium–sulfur batteries: Mechanisms, catalysts, and characterization. In *Advanced Energy Materials* (Vol. 12, Issue 44). John Wiley and Sons Inc. https://doi.org/10.1002/aenm.202202094

Zhou, L., Zhang, Y., Zhuo, Z., Neukirch, A. J., & Tretiak, S. (2018). Interlayer-decoupled Sc-Based MXene with high carrier mobility and strong light-harvesting ability. *The Journal of Physical Chemistry Letters*, *9*(23), 6915–6920. https://doi.org/10.1021/acs.jpclett.8b03077

Chapter 5

# Metal oxide-carbon composite electrode materials for metal-air batteries

## Advances, challenges, and environmental impacts

*Nitika Devi and Yong-Song Chen*

## 5.1 INTRODUCTION

Increasing energy demand is causing a threat to living beings and existing energy resources. Today, the world is largely dependent on fossil fuels, but depletion of these resources suggests a shift toward more sustainable and renewable energy resources (Tari et al., 2024; Tari & Patil, 2017a, 2017b). Recently, scientists have explored various renewable energy resources like solar, wind, bioenergy, etc. However, sporadic nature of renewable energy resources is the biggest constraint in realizing their full potential. Energy storage devices like batteries, fuel cells, capacitors, supercapacitors, etc. could be a solution for long-term use of renewable energy (Capellán-Pérez, Mediavilla, de Castro, Carpintero, & Miguel, 2014; Kalair, Abas, Saleem, Kalair, & Khan, 2021; Gurushankar et al., 2024, Kannan & Tari, 2025). Energy storage devices can work independently and be integrated with the renewable energy resources. Batteries have advantages over supercapacitors and other energy devices such as high energy density, long cycle life, high efficiency, and availability in different sizes (Elalfy, Gouda, Kotb, Bureš, & Sedhom, 2024; He et al., 2024; Tari et al., 2025). Batteries can be of two types, primary and secondary batteries. Primary batteries are non-rechargeable, and secondary batteries are rechargeable batteries like zinc-air, lithium-ion, lead-acid batteries, etc. Metal-air batteries are electrochemical batteries which have 3–30 times more energy density as compared to conventional Li-ion batteries. So, over the years, scientists have tried to develop different metal-air batteries and modify them in order to get efficient batteries. Figure 5.1 gives the historical development of metal-air batteries.

Energy is stored in metal-air batteries through electrochemical reactions in which metal anode undergoes oxidation, and reduction happens to the oxygen at cathode. A separator is used between the anode and cathode which facilitates the ion transportation. Metal-air batteries comprise of anode of metal like Li, Na, Fe, Al, Zn, etc. and a gas diffusion layer used for the oxygen cathode. Although, the theoretical energy density of metal-air batteries is very high, for instance, theoretical energy density of a Li-air

DOI: 10.1201/9781003584742-5

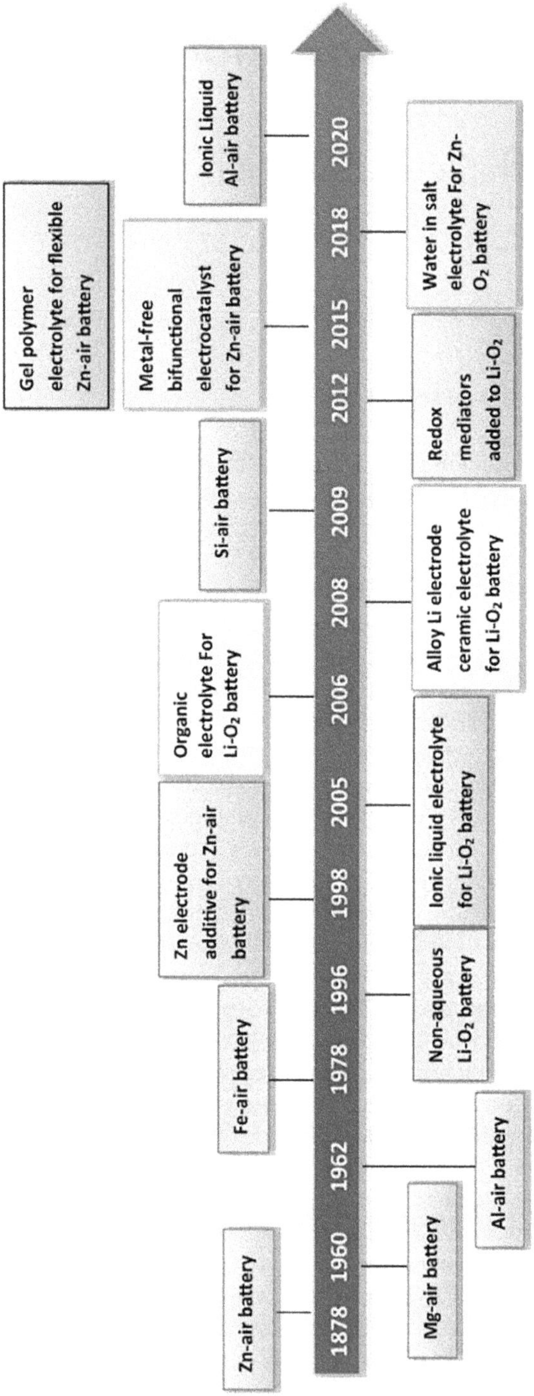


*Figure 5.1* Development and progress of metal-air batteries (Yaqoob, Noor, & Iqbal, 2022).

battery is 11,000 Wh $kg^{-1}$. But experimentally, it is not possible to achieve this value due to many performance-limiting factors. Battery performance can be affected due to slow reaction kinetics at cathode, anode corrosion, and high current density resulting in a dendrite, an ineffective material catalyst to cater excess oxygen and poor cycle life due to unstable discharge product (Li & Lu, 2017; Yaqoob et al., 2022). All these factors contribute to degrading the metal-air battery performance. Scientists have been exploring efficient electrolytes, cathode catalysts, and effective anode materials so that, like theoretical energy density, an efficient battery can also be developed experimentally. Jung et al. (2015) developed one-dimensional manganese-cobalt oxide nanofibers as cathode catalysts for Zn-air batteries. It was concluded that a fibrous spinal catalyst improves the round-trip efficiency of the battery, and catalyst was stable for repeated charge-discharge cycles. An acid–alkaline dual gel was used as an electrolyte for various metal-air batteries such as Zn, Al, and Mg metal-air batteries. Retarded ion diffusion inside the gel contributes to high voltage and longer lifetime (Wang et al., 2022b). Similar to cathode catalyst and electrolyte improvement, an improved anode electrode can also significantly enhance the battery performance. Zhang et al. (2020b) discussed the role of microstructural evolution in an Al-based anode and its effect on the efficiency of the battery. It was reported that structural morphology changes can result in significant improvement in the energy efficiency of the battery.

Metal-air batteries are often consisting of metal or metal-oxide materials as electrodes and catalysts at cathode. Metal-based electrodes suffered from drastic volume expansion and poor electrical conductivity. These structural changes during operation and poor conductivity lead to instability and poor cycling life. This book chapter briefly discusses the fundamentals of metal-air batteries, different kinds of metal-air batteries, and progress and challenges related to the electrodes of metal-air batteries.

## 5.2 FUNDAMENTALS OF METAL-AIR BATTERIES

Metal-air batteries are electrochemical cells comprising three components, an anode, an air cathode, and an electrolyte, which is used as a separator between two electrodes. Different metals such as Li, Na, Al, Fe, Mg, Zn, etc. are utilized as anodes in metal-air batteries. Electrolytes used for metal-air batteries are aqueous electrolytes (protic) and non-aqueous electrolytes (aprotic), hybrid, and solid-state electrolytes. Metal-air batteries can be further classified as aqueous and non-aqueous metal-air batteries depending on the type of electrolyte used. In aqueous media, metals anodes like Zn, Fe, Al, and Mg are stable, while Na, K, and Li anodes are highly stable in non-aqueous media. Thermodynamic instability of metal anodes in different electrolytes can be addressed by various coatings like use of a layer of oxide and hydroxide coating for aqueous media and an ionic conducting film

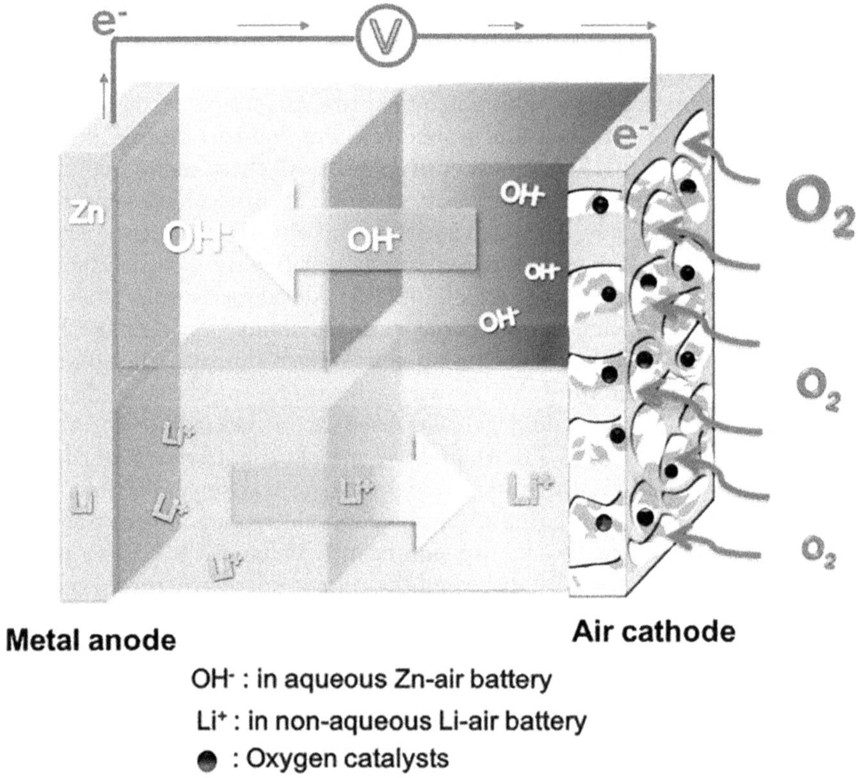


*Figure 5.2* Schematic representation of the working of aqueous and non-aqueous metal-air batteries (Wang, Xu, Xu, & Zhang, 2014).

for non-aqueous electrolytes. Figure 5.2 shows the schematic of metal-air batteries, which illustrates the working of metal-air batteries in both types of electrolytes. Electrochemical reactions for both types of electrolytes are given below:

### Aqueous electrolytes

At anode $M \leftrightarrow M^{n+} + ne^{-}$

At cathode $O_2 + ne^{-} + 2H_2O \leftrightarrow 4OH^{-}$

### Non-aqueous electrolytes

At anode $M \leftrightarrow M^{+} + e^{-}$

At cathode $xM^{+} + O_2 + xe^{-} \leftrightarrow M_xO_2\left(x = 1, 2\right)$

Here, *M* represents the metal ion, and *x* represents the oxidation state of the metal. Electrochemical reactions and working of metal-air batteries are different in aqueous and non-aqueous media. Oxygen reduction reaction (ORR) of oxygen is very similar to the ORR reaction of hydrogen in a fuel cell. Kinetics of ORR are significantly affected by the type of catalyst used on the cathode gas diffusion layer (GDL) (Li & Lu, 2017; Yaqoob et al., 2022).

### 5.2.1 Types of metal-air batteries

Metal-air batteries can be classified into different types depending on the kind of metal anode and types of electrolytes. Li-air batteries, Zn-air batteries, Mg-air batteries, Na-air batteries, Al-air batteries, and Fe-air batteries are some of the most explored metal-air batteries. All these batteries comprise air as a cathode and are designed with different aqueous and non-aqueous electrolytes. Theoretically, the energy density of these batteries is very high, for example, Zn-air batteries have a theoretical energy density of 1353 Wh $kg^{-1}$, Li-air batteries have 11000 Wh $kg^{-1}$ energy density, and likewise, energy density for Na-air batteries is 1601 Wh $kg^{-1}$. But experimentally, it is very difficult to achieve this level of high energy density due to many limitations associated with the battery components. For example, the experimentally reported energy densities of Zn-air batteries are 350–500 Wh $kg^{-1}$. Energy efficiencies of Zn-air batteries are less than 60% due to overpotential at air cathode. Sluggish ORR is the main reason for the low efficiency of metal-air batteries (Li & Lu, 2017; Yaqoob et al., 2022). Figure 5.3 gives an overview of various types of metal-air batteries along with their compatibility with various types of electrolytes.

Following, we have discussed most developed metal-air batteries and along with their merit and demerits.

**Li-air batteries:** Li-air batteries are usually non-aqueous as Li is a highly explosive metal and reacts very rapidly with water. In typical Li-air batteries, $LiO_2$ is oxidized to form Li and $O_2$, and cathode oxidation results in the formation of $Li_2O_2$ and $LiO_2$. The chemical reaction of Li-air batteries is $Li + O_2 \leftrightarrow LiO_2$, E= 2.96 V. Li-air batteries have certain advantages over other metal-air batteries due to small radii of Li; Li-air batteries have high volumetric, gravimetric capacity and power density. Li-air batteries possess uppermost cell potential due to lowest reduction potential of Li ions. Li-air batteries also have some demerits like formation of dendrite at anode, different product formation at cathode, and insoluble reaction by-products at cathode may result into hindering the mobility of Li-ions. Also, sometimes cathode carbon and binder result in lithium carbonate formation. All these factors result in sluggish reactions and low battery performance (Chen, Yang, Huang, & Zhang, 2021; Yaqoob et al., 2022).

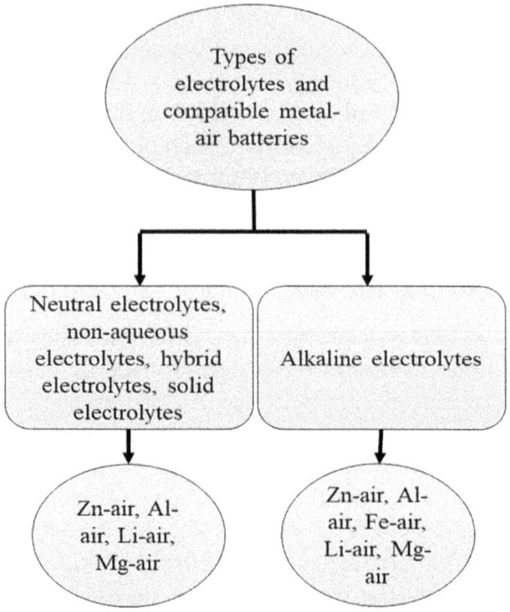


*Figure 5.3* Various types of metal-air batteries along with their compatibility with various type of electrolytes (Rahman, Wang, & Wen, 2013).

**Zn-air batteries:** Like Li-air batteries, the Zn air battery configuration consists of Zn metal at anode and an air cathode. Zn batteries are stable in aqueous electrolytes and can have a specific energy and volumetric energy density of the order of 1218 Wh $kg^{-1}$ and 6136 Wh $L^{-1}$, respectively. The overall electrochemical reaction of a Zn-air battery is $2Zn + O_2 \leftrightarrow 2ZnO$, E=1.66 V. Zn-air batteries are best suited for the small-scale applications like hearing aids because their shelf life and capacities are limited. Advantages of Zn-air batteries are their cost efficiency, lightweight, and environmental friendliness. Demerits of Zn-air batteries are low voltage and corrosive electrolytes (Dobley, 2013; Pan et al., 2018).

**Al-air batteries:** Al-air batteries are attractive for vehicle propulsion due to their high energy density of 4140 Wh $kg^{-1}$. Configuration features of Al-air batteries are the same as previously discussed metal-air batteries except the anode is Al metal. Electrochemical reaction of the electrode in Al-air batteries is $4Al + 3O_2 + 6H_2O \leftrightarrow 4Al(OH)_3$, E = −2.75 V. Advantages associated with the Al-air batteries are light weight of Al anode, high energy density, non-toxic, and safe. But Al-air batteries still suffer from some limitations like Al corrosion in electrolyte, $Al_2O_3$ and $Al(OH)_3$ deposited on the anode and cathode electrodes decrease the electrochemical reaction kinetics. Faster loss of charge on storage leads to limited shelf life of the battery (Goel, Dobhal, & Sharma, 2020; Liu et al., 2017).

**Fe-air batteries:** Fe-air batteries are made of Fe anode and oxygen cathode arranged with suitable electrolyte. The theoretical energy density of Fe-air batteries is around 1200 Wh $kg^{-1}$ with a low-cost advantage. Electrochemically, energy is stored through $3Fe + 2O_2 \leftrightarrow Fe_3O_4$, E = 1.16–1.28 V. Benefits associated with Fe-air batteries are their low cost, non-toxicity, safe operations, and energy density is better as compared to the Li-ion batteries. Limitations of Fe-air batteries are iron corrosion and formation of dendrites on the electrode surface (Figueredo-Rodríguez et al., 2017; Tan, Kawamura, Muto, & Matsuda, 2021).

**Na-air batteries:** Na-air batteries have been introduced in recent times as compared to the much-discussed Li-air batteries, Al-air batteries, and Zn-air batteries. The low cost of Na and better energy density (1600 Wh $kg^{-1}$) and gravimetric capacity (1165.78 mAh $g^{-1}$) as compared to Li-ion batteries make them good alternatives in energy storage devices. Energy is stored through following electrochemical reaction: $Na + O_2 \leftrightarrow NaO_2$, E = 2.27 V, $2Na + O_2 \leftrightarrow Na_2O_2$, E = 2.33 V. As this battery is still in the development stage, several improvements are needed to achieve its potential benefits. Degradation of battery components due to reactive species is a concerning factor in Na-air batteries, which also need to formulate more stable electrolytes toward $O_2^-$. Formation of additional by-products like carbonates, oxalates, etc. on the electrode surface results in sluggish electrochemical reactions (Pozo-Gonzalo & Ortiz-Vitoriano, 2022; Sun, Yang, & Fu, 2012).

**Mg-air batteries:** Mg-air batteries are a combination of Mg anode with air cathode and electrolyte separator. Mg-air batteries are also in the beginning stage, and the biggest challenge in these batteries is to find a compatible electrolyte. The combined reaction occurring in Mg-air batteries is $2Mg + O_2 + 2H_2O \leftrightarrow 2Mg(OH)_2$, E = 3.1 V. Mg-air batteries are getting researchers' attention due to low toxicity, Mg abundance, high reaction rates, and lightweight. But Mg-air batteries have high polarization, and usually their working voltage is much less than the theoretical value, mostly below 1.2 V (Tong, Wei, Chen, & Gao, 2021; Yaqoob et al., 2022).

Other than these mentioned metal-air batteries, there are other newly introduced metal-air batteries. These are vanadium-air batteries, potassium-air batteries, calcium-air batteries, tin-air batteries, silicon-air batteries, and germanium-air batteries. All these are still in the R&D stage and many challenges and optimizations need to be done for the commercialization of these metal-air batteries (Ahuja, Kalpna, & Varshney, 2021).

## 5.3 CRITICAL EVALUATION OF ANODE AND CATHODE IN METAL-AIR BATTERIES

The electrode plays a crucial role in metal-air batteries as electrochemical reactions take place on the surface of the electrode. Electrodes with high surface area, good porosity, and high electrical conductivity are preferably used in energy storage devices. In metal-air batteries, both anode and cathode play an important part as combined good performance can result in highly efficient batteries. Anode in metal-air batteries are usually highly reactive metals like Al, Zn, Fe, Li, etc. Key features like large surface area and porosity of electrodes can provide a large area for electrochemical reactions. In metal-air battery electrochemical reactions, the anode provides the electrons by undergoing oxidation, and electrons further move in the circuit to provide the energy output. The anode metal is responsible for the oxidation part of the electrochemical reaction, so the metal must have good conductivity and surface area. Also, reversibility of the reaction depends upon the nature of the anode metal because usually less reversible reactions result in the formation of dendrites and additional by-products which can decrease the electrode performance and processes like corrosion can damage the anode. Scientists have employed different strategies for improving the performance and mitigating the effects like side reactions and corrosion of metal-air anode. Applying a layer and mixing of nanostructures like graphene, carbon nanotube (CNT), etc. with electrode material can significantly decrease dendrite formation and increase the conductivity of the electrode, which increases the reversibility of the electrochemical reactions. Alloying of different metals has been applied for improving the corrosion resistance of the anode, like exploring Al-alloy for metal-air anode. New electrolytes have also been explored for limiting the side reactions of the metal-air batteries (Bi, Amine, & Lu, 2020; Blurton & Sammells, 1979; Rahman et al., 2013; Wang et al., 2021). Similar to the anode, the cathode is equally important for efficient metal-air batteries. Large surface area catalytic materials are employed for cathode as the cathode permits the atmospheric oxygen for reduction part of electrochemical reactions. Again, a large surface area and porosity will provide a large site for electrochemical reactions as they can enhance the diffusion of oxygen and facilitate the ORR. Cathode catalyst material is also important for handling the oxygen evolution reaction (OER) in recharging metal-air batteries and mitigating the additional product formation such as carbonates, oxalates, etc. Cyclic life and stability of the battery are also dependent on the type of catalysts used for ORR of metal-air batteries. Many strategies have been employed for enhancing the performance of cathode catalysts, such as nanostructure graphene, CNTs, and their composites with various metal-oxides, use of cost-efficient metal catalysts and doping of heteroatoms (Cheng & Chen, 2012; Martin, Neburchilov, Wang, & Qu, 2009; Zhang, Huang, Yan, Shao, & Zhang, 2019). Volume expansion of metal and metal-oxide materials is a common factor for mostly rechargeable metal-air batteries, which decreases the performance and cyclic life of the batteries. It was reported that composite formation of carbon nanostructures like graphene, CNTs, and carbon quantum dots with different metals and metal-oxides can improve the reaction rates and cyclic life of the batteries. So, we will focus on the recent development of these materials' incorporation for the anode and cathode electrodes of metal-air batteries.

## 5.3.1 Metal-oxide/graphene composite electrodes

Metal-oxide/graphene composite electrodes have significantly affected the energy storage area. These electrodes have a synergetic effect of pseudocapacitance behavior and large surface area due to metal-oxide and graphene, respectively. Large surface area and storage capacity of the electrode is significant for its application in energy storage devices. Here, we are discussing how the synergetic properties of metal-oxide and graphene are affecting the performance of metal-air batteries.

### *5.3.1.1 Cathode*

Graphene's high electric conductivity, high surface area, and 2-D structure features make it suitable for cathode catalyst applications. Metal-oxide and graphene composites can give efficient cathode catalysts as their synergetic effect results in improved reaction rates of ORR and OER, and long-cycle life, which ultimately leads to better power density and stability of metal-air batteries (Patra & Tiwari, 2023). Also, it is important to find a replacement for high-cost Pt/C catalyst for developing cost efficient metal-air batteries. Different types of metal-oxide composites have been explored as cathode catalysts for metal-air batteries. Peng et al. (2020) synthesized cobalt oxide/graphene ($Co_3O_4$/GR) nanocomposites by hydrothermal synthesis process. Nanocomposites were further used as cathode catalysts for Li-air batteries and offered very good stability as battery can work stable for 120 h. A ternary metal-oxide and graphene composite were formulated in a thin film using a laser-induced in-situ synthesis process. In this composite, Co/Mn and Ni and Fe species are incorporated with graphene as Co/Mn and Ni and Fe can facilitate ORR and OER reactions. Composites were used for Zn-air batteries, and the battery delivered an energy density and power density of 842 Wh $kg^{-1}$ and 98.9 mW $cm^{-2}$, respectively (M. Ren, Zhang, & Tour, 2019). N. Wang et al. (2022a) explored graphene nanocomposites with Ru-$RuO_2$ Mott-Schottky heterojunctions for cathode electrocatalysts in Zn-air batteries. A quaternary mixed metal sulfide synergized with nitrogen-doped reduced graphene oxide (rGO) bifunctional cathode catalyst was prepared through an anion exchange route. Synergetic effect of nitrogen-doped reduced graphene and uniformly distributed metal sulfide leads to excellent capacity and energy density of Zn-air batteries. Zn-air batteries comprising these composite cathode catalysts displayed 834 mAhg$^{-1}$ capacity and 959 Wh kg$^{-1}$ energy density (Pendashteh, Sanchez, Palma, Anderson, & Marcilla, 2019). Al-Ogaili et al. (2021) developed a graphene-based nanocomposite cathode architecture with palladium and $\alpha$-$MnO_2$ for Li-air batteries. The schematic of the composite development and capacity performance is shown in Figure 5.4. In this work, graphene oxide (GO) is prepared through the modified Hummer's method and $\alpha$-$MnO_2$ nanowire using hydrothermal method. Pd-doped rGO was obtained by sonicating $PdCl_2$ in GO suspension, and then in the final step Pd-doped rGO/$\alpha$-$MnO_2$ composites were prepared by ball milling. The resulting material was utilized as a cathode catalyst for Li-air batteries and showed a discharge capacity of 7500 mA h g$^{-1}$, and the battery could work stably for 50 cycles at a capacity of 800 mA h g$^{-1}$ (Al-Ogaili et al., 2021).

As discussed, many reports have been published on the cathode catalyst for Li-air batteries (Gnana Kumar, Christy, Jang, & Nahm, 2015; Mei et al., 2016), Zn-air batteries (Ruan et al., 2023; Xiang, Li, Wu, Ma, & Sheng, 2022), and recently for Mg-air batteries (Jiang et al., 2017). Not many reports are available on other battery improvements as those are still under developing stage.

#### *5.3.1.2 Anode*

Very few reports are available on improved graphene-based composite anodes for metal-air batteries. Because ORR and OER are main performance-defining reactions in metal-air batteries, researchers have been mainly devoting efforts toward developing cathode catalysts for enhanced ORR and OER. Tan et al. (2020) developed $Fe_3O_4$-embedded rGO composite anodes for FeOx-air battery. In this work, $Fe_3O_4$-embedded rGO was synthesized by chemical precipitation method and then the material was dried in an electric furnace at different temperatures for different duration. A schematic of the synthesis procedure for $Fe_3O_4$-embedded rGO composite is shown in Figure 5.5. Fe-air battery consisting of this electrode was capable of delivering a discharge capacity of 420 mA h g$^{-1}$.

As discussed, dendrite formation during battery operation is a performance-degrading factor in metal-air batteries. Xia et al. (2019) introduced a layer of rGO on a Zn metal anode to reduce the dendrite formation. It was calculated that the rGO-coated Zn electrode showed very low polarization losses and 88.5% capacity retention. Graphene coating can increase the stability of electrodes and significantly enhance the performance of metal-air batteries.

### 5.3.2 Metal-oxide/CNTs composite electrode

Like graphene, CNTs can also be utilized for enhancing the electrode performance in metal-air batteries. CNTs have very good conductivity, high

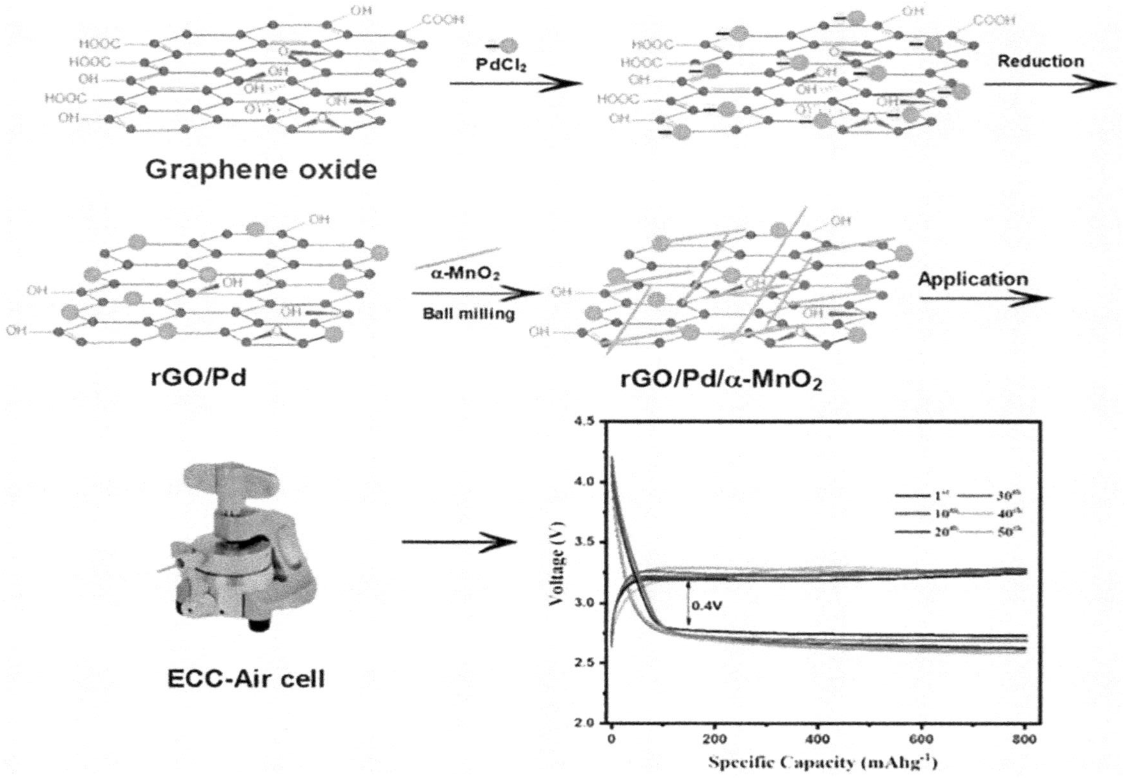


*Figure 5.4* Schematic representation of graphene-based nanocomposite with α-$MnO_2$ and palladium, cathodes for Li-air batteries (Al-Ogaili et al., 2021).

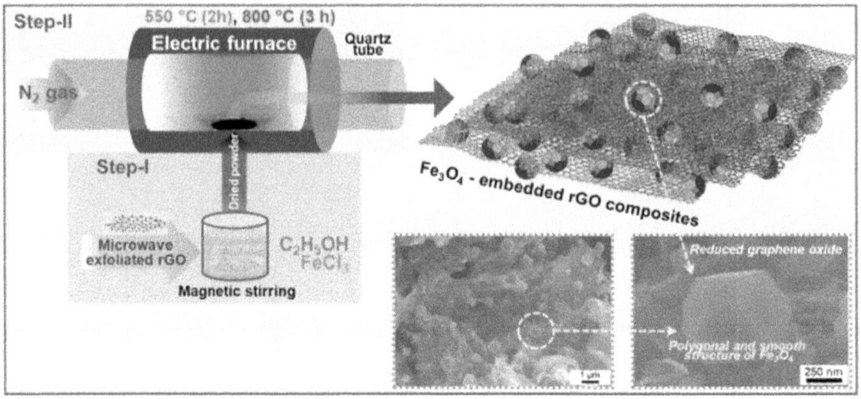


*Figure 5.5* Schematic of synthesis procedure for $Fe_3O_4$-embedded rGO composite formation (Tan et al., 2020).

surface area, and porosity, which contribute to improving the electrode performance. Metal-oxide/CNTs composite synergetic effects have been explored for both cathode and anode electrode applications in metal-air batteries.

#### 5.3.2.1 Cathode

Metal-oxide/CNT composites cathode catalysts have been employed in different types of metal-air batteries like Li-air batteries, Zn-air batteries, etc. Hong et al. (2021) reported the use of a three-dimensional (3D) structure comprising metal-oxides/CNT composite electrocatalysts for Zn-air batteries. In this work, CNT was prepared by using spray pyrolysis method and then using an ultrasonic nebulizer with polystyrene nanobeads, CNT was converted into a 3D nanosphere. $Fe^{3+}$ ions were chemically trapped in a CNT microsphere by treatment with polyvinylpyrrolidone (PVP). FeOOH nanorods were uniformly attached to CNT microsphere using a solution method. In the final step, previously obtained material was further hydrothermally treated with $KMnO_4$ to get the final $MnO_2$-$Fe_2O_3$/CNT microsphere composites. A schematic of above discussed synthesis procedure is given in Figure 5.6. The obtained material showed that it can be used as bifunctional catalyst material as it has capability of improving both ORR and OER reactions. Further, bifunctional catalysts were used for Zn-air cathode catalyst.

Zn-air battery comprising $MnO_2$-$Fe_2O_3$/CNT microsphere catalyst delivered excellent performance in terms of high-power density with high battery stability. The battery offered a power density of 253 mW $cm^{-2}$ and a specific capacity of 802 mA h $g^{-1}$. This performance of the battery is attributed to synergetic effect of oxide material and CNT, which lead to very less polarization losses (Hong et al., 2021). Zhang et al. (2020a) also prepared a hierarchically porous $Fe_3C$/Co/N-doped CNT catalyst for Zn-air batteries. A multi-step approach involving metal-organic framework (MOF)@Fe-Phen nanotube-derived composite, which contained nitrogen-rich doped CNTs. In $Fe_3C$/Co/N-doped CNTs, various kinds of nitrogen groups such as N-C, Fe-Nx, and Co-Nx, along with Co, $Fe_3C$, are uniformly distributed on the CNT. The composite was then used in Zn-air batteries, which offered a specific capacity of 670 mA h $kg^{-1}$ with an energy density of 800 Wh $kg^{-1}$. A Zn-air battery consisting of a hybrid catalyst made up of mixed oxide substrate with NiCo alloy and CNT showed a power density of 115 mW $cm^{-2}$(Li et al., 2020). Like Zn-air batteries, CNT and CNT-based composite electrodes have also been used for Li-air batteries cathode. Li et al. (2019) developed nitrogen-doped CNT as a cathode catalyst in Li-air batteries and showed a high discharge capacity of the order of 2203 mAh $g^{-1}$. Co-CoO nanoparticles embedded in the nitrogen-doped carbon nanorods catalyst were prepared by electrospinning method. This catalyst was a bifunctional catalyst which can improve the oxygen reduction and evolution reaction in metal-air batteries. It was demonstrated in Li-air battery that the cathode had good stability over 86 cycles with a specific capacity of 10,555 mAh $g^{-1}$ at 100 mA $g^{-1}$ (Hyun & Shanmugam, 2017). Material aspects and material potential for ORR and OER of some metal-oxide and CNT composite have been reported without the battery description. These reports again show the potential of these composites to be used in metal-air battery cathodes. Cheng et al. (2024) reported δ-$MnO_2$ nanosheets encapsulating single cobalt atoms-anchored CNT composite for OER. Material study and DFT calculations concluded that the composite material has great potential for OER due to superior conductivity, availability of additional reaction, and fast reaction kinetics. Cobalt oxide nanocrystal coupled CNT composites were also reported for enhanced ORR catalysts (Al-Hakemy et al., 2017; Liang et al., 2012). Xue et al. (2019) reported that a hollow mesoporous $Fe_2O_3$ Nanospindles/CNTs composite was prepared using a hydrolysis process combined with calcination. The composite was further used in Li-air batteries and offered a high discharge capacity of 7,730 mAh $g^{-1}$ with excellent stability· Reports showed that performance improvements are the result of combined effect of oxide and CNT presence (Levy et al., 2021; Zhang et al., 2014). Thus, the discussion concluded that synergetic effect of oxide with CNT can be used for the preparation of highly efficient bifunctional catalysts for metal-air batteries.

Metal-oxide/CNT-based anode for metal-air batteries have still not been explored and published. It can be expected from the metal-oxide/graphene composite anode discussion that CNT can also help in improving the anode performance.

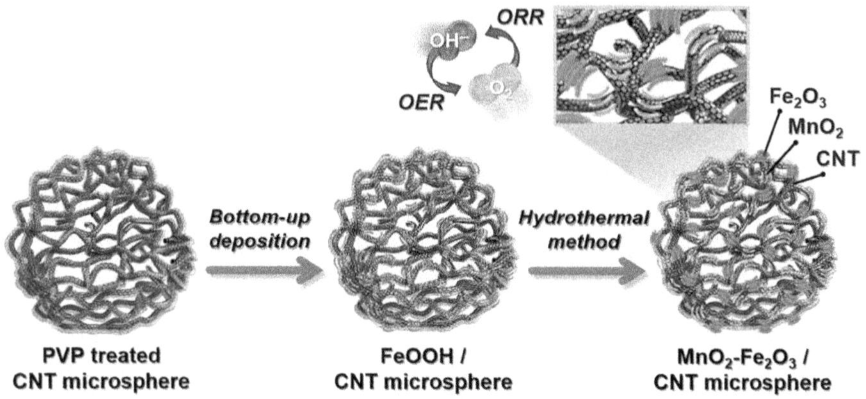


*Figure 5.6* Schematic of synthesis procedure for synthesizing the $MnO_2$-$Fe_2O_3$/CNT microsphere composite (Hong et al., 2021).

## 5.3.3 Metal-oxide/carbon composite electrodes

Other forms of carbon like carbon black, graphene quantum dots, activated carbon, carbon nanofibers (CNF), etc. also have been utilized for metal-oxide-carbon composite electrodes for metal-air batteries. The following is the discussion on some of the important reports on this composite formation and application in various kinds of metal-air batteries.

### *5.3.3.1 Cathode*

Slow ORR and OER are contributing to the low performance degradation of metal-air batteries, so scientists have developed efficient electrocatalysts for improving the reaction kinetics. Mahato et al. (2021) synthesized S, N co-doped graphene quantum dots decorated $TiO_2$ and supported with carbon catalyst, which has a great affinity toward ORR. A schematic of process used for synthesizing S, N co-doped graphene quantum dots decorated $TiO_2$ and supported with carbon catalyst is given in Figure 5.7. In the first step, $TiO_2$ nanoparticles were synthesized by doing hydrothermal treatment of titanium tetraisopropoxide with ethanol and water. Following the previous step, S, N co-doped graphene quantum dots decorated $TiO_2$ were prepared by hydrothermal treatment of last step obtained $TiO_2$ nanoparticles with citric acid and thiourea. Then, in the final step, S, N co-doped graphene quantum dots decorated $TiO_2$ were converted into S, N co-doped graphene quantum dots decorated $TiO_2$ supported with carbon again using hydrothermal treatment using 1:1 of graphene quantum dot decorated with $TiO_2$ and ketjen black. Experimental conditions like furnace temperature and reaction duration is given in schematic Figure 5.7.

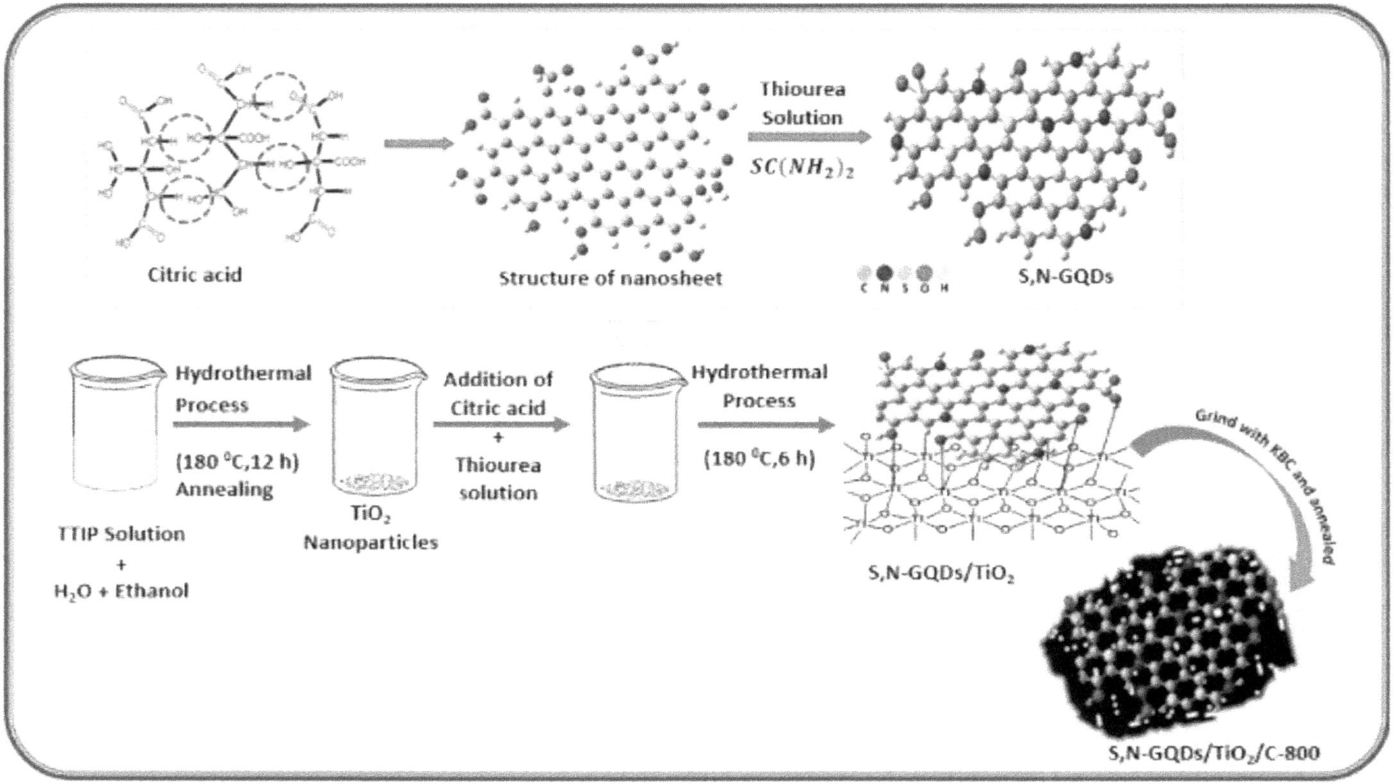


*Figure 5.7* Schematic of preparation strategy of S, N co-doped graphene quantum dots decorated $TiO_2$ and supported with carbon (Mahato et al., 2021).

A catalytic study of this material showed good ORR activity with an onset potential of 0.91 V with 80% retention in current density (Mahato et al., 2021). Chen et al. (2019) reported a bifunctional catalyst for Zn-air batteries made of porous carbon doped with N and S and inserted with hollow cobalt oxide nanoparticles. A one-step co-precipitation method was employed for the synthesizing the bifunctional catalysts and was further used for cathode in Zn-air batteries. Zn-air batteries offered a specific capacity of 745 mA h $g^{-1}$ with stable battery operation over 100 h. B, N-co doped carbon quantum dots also showed very good oxidation-reduction reactions with an onset potential of 0.92V (Syahputra et al., 2024). Lu et al. (2019) demonstrated $Fe_{0.1}Ni_{0.9}Co_2O_4$/activated carbon composites for Zn-air batteries cathode. The composites were prepared by a microwave-assisted hydrothermal method and displayed a power density of 99 mW $cm^{-2}$ at 100 Ma $cm^{-2}$ with a stable performance. Carbon and metal-oxide composites have also been explored for Li-air batteries. A catalyst of porous carbon networks decorated with cobalt on $CoFe_2O_4$ was used in Li-air batteries. Li-air batteries with a high current density of 100 mA $g^{-1}$ and a discharge capacity of 4320 mAh $g^{-1}$ were created using porous carbon networks embellished with cobalt on a CoFe2O4 catalyst (Athika & Elumalai, 2020). Core-shell-structured $NiCo_2O_4$ composites with GO/C were reported as electrocatalysts for Zn-air batteries by Fu et al. (2024). A composite of $Co_3O_4$ nanoparticles enclosed in a 3D N-doped porous carbon (Co-NpCs) oxygen electrocatalyst was prepared using a straightforward freeze-drying and in-situ pyrolysis technique. Oxygen catalysts showed better stability as compared to Pt/C and gave bifunctional activity $\Delta E$ = 0.849 V (Wang et al., 2020). $\Gamma$–$MnO_2$ (Flegler et al., 2016), lanthanum manganese cobalt oxide (García-Rodríguez, Flores-Lasluisa, Cazorla-Amorós, & Morallón, 2024), and Ni/NiO oxides (Athika, Junais, & Elumalai, 2023) were used with carbon for composite formation and further were used for Zn-air and Li-air batteries. Xi et al. (2023) prepared N-doped carbon nanosheets supported by $Co_3O_4$ nanoparticles ($Co_3O_4$@NCNs) using a one-step pyrolysis process. Zn-air batteries have a high-power density of 173.8 mW $cm^{-2}$ due to their 3D structure and significant nitrogen doping concentration. Like graphene and CNTs, metal-oxide-carbon composites were mostly explored for Zn-air batteries, Li-air batteries, and other kinds of metal-air batteries are under developing stage (Kurian et al., 2024; Ren, Yuan, Weng, & Yuan, 2018; Worku, Ayele, & Habtu, 2021).

#### *5.3.3.2 Anode*

Only one or two reports are available on the metal-oxide-carbon composites for metal-air anode applications. Hang et al. (2005b) studied the effect of various carbon materials in Fe/C composites for metal-air batteries.

Performance comparisons of vapor-grown carbon fiber, acetylene black, tubular CNF and platelet CNF, natural graphite, and iron have been done for metal-air battery anodes. It was calculated that acetylene black, tubular CNF, and platelet CNF exhibited a high surface area as compared to other kinds of carbon. Natural graphite and vapor-grown carbon fiber have surface area of the order of 8, and 13 $m^2 g^{-1}$, respectively. But in comparison to these surface area values, acetylene black, tubular CNF, and platelet CNF have surface areas of the order of 68, 92, and 91 $m^2 g^{-1}$, respectively. More surface area means that more active sites are available for electrochemical reactions. Acetylene black, tubular CNF, and platelet CNF displayed higher redox currents because of enhanced conductivity of these forms of carbon due to improved surface areas. It was observed that the performance was also influenced due to better dispersion of Fe in some special structures of carbon material, especially in case of tubular CNF (Hang et al., 2005a). The studies showed a potential of carbon material in composite formation for metal-air battery electrodes. $Fe_2O_3$-nanoloaded carbon and $Fe_2O_3$-nano/carbon mixed composites anode were reported for iron-air batteries by Hang et al. (2005b). $Fe_2O_3$-nanoloaded carbon in ratio of 1:8 showed the best performance as compared to $Fe_2O_3$-nano/carbon mixed composites. The performance was good due to better contact of $Fe_2O_3$ and nanocarbon as chemical incorporation method was used for this composite electrode preparation (Hang et al., 2005b). These efforts to improve the performance of electrodes can result in efficient metal-air batteries.

## 5.3.4 Metal-oxide/MXene composite electrodes

MXene is a new class of 2D materials as compared to graphene and other graphite derivatives. Like carbon nanomaterials, MXene also has exceptional characteristics like large surface area due to 2D layer structures, high conductivity, and easy structural modification because of the presence of different functional groups. Metal-oxide/MXene composite electrode have excellent potential for energy storage areas, but mainly Li-ion batteries and supercapacitors are the explored areas in this context. Only few reports are present on metal-oxide/MXene composite electrode for metal-air batteries.

### 5.3.4.1 *Cathode*

Metal-oxide/MXene composite electrocatalysts have been explored for making cathode more efficient. Lei et al. (2020) developed $NiCo_2O_4$/MXene hybrid for improving the sluggish OER and ORR reactions in Zn-air batteries. The hybrid was synthesized through a hydrothermal route and a strong bonding Ni/Co-F has been developed. A schematic of the synthesis procedure is shown in the Figure 5.8. Composite synthesis procedure was a two

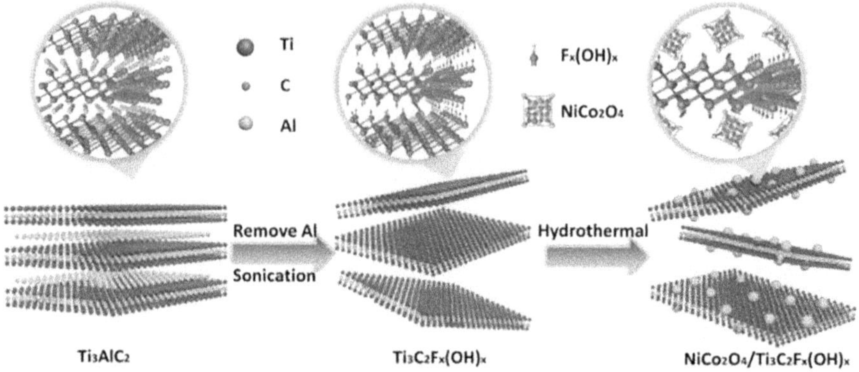


*Figure 5.8* Schematic of the process for forming $NiCo_2O_4/Ti_3C_2F_x(OH)_x$ hybrid composites (Lei et al., 2020).

steps process that was preparation of MXene $Ti_3C_2T_x$ multilayer nanosheets followed by hydrothermal synthesis of $NiCo_2O_4$/MXene hybrid. MXene $Ti_3C_2T_x$ multilayer nanosheets were prepared by reacting $Ti_3AlC_2$ powder with hydrofluoric acid. In the hydrothermal procedure, ethanol was used as solvent to form the solution of $Co(OAc)_2.4H_2O$ and $Ni(OAc)_2.2H_2O$. Then above solution was mixed with the MXene $Ti_3C_2T_x$ solution obtained by dissolving $Ti_3C_2T_x$ in distilled water. The reaction was carried out in the presence of ammonia for 10 h at 80°C, and it then underwent hydrothermal treatment for 3 h at 150°C. The resulting product was centrifuged and freeze dried. Zn-air battery consisting of the $NiCo_2O_4$/MXene composite electrode offered a peak power density of 55.1 mW $cm^{-2}$ and an energy efficiency of 66.1% at 1.0 mA $cm^{-2}$ (Lei et al., 2020).

Catalytic activity can be significantly improved by creating cationic vacancy, and similar approach has been employed for developing efficient Li-$O_2$ batteries. Xu et al. (2022) developed vacancy-enriched $V_{2-x}O_5$ on $V_2C$ MXene bifunctional electrocatalyst for Li-$O_2$ batteries. Battery operated with vacancy-enriched electrocatalyst gave energy efficiency of 83.4% at 100 mA $g^{-1}$ with excellent cycling performance of more than 500 cycles. Hybrid of $NiFe_2O_4/TiO_2$ also has been reported for Zn-air battery electrocatalysts. Composites had a very low potential gap of 0.75V for OER/ORR and battery could operate up to 500h at 10 mA $cm^{-2}$(Palanisamy et al., 2024).

To the best of our knowledge, no report has been published till date on the use of metal-oxide/MXene composites anode for the metal-air batteries (Table 5.1).

*Table 5.1* Summary of some key results of metal-oxide/carbon composites for electrode materials for metal-air batteries

| *Sr. No.* | *Type of electrode* | *Materials* | *Synthesis method* | *Battery type* | *Performance* | *References* |
|---|---|---|---|---|---|---|
| 1. | Cathode | $MnO_2$-$Fe_2O_3$/CNT | Hydrothermal | Zn-air batteries | Power density of 253 mW $cm^{-2}$, specific capacity of 802 mA h $g^{-1}$ | Hong et al. (2021) |
| 2. | Cathode | Laser-induced graphene (LIG)-CoNiFe | Laser writing | Zn-air batteries | Peak power density of 98.9 mW $cm^{-2}$ and energy density of 842 Wh $kgZn^{-1}$ | Lu et al. (2019) |
| 3. | Cathode | CoO/$Ti_3C_2T_x$ | Hydrothermal | Li-air batteries | 16,220 mAh $g^{-1}$ at 100 mA $g^{-1}$ | Al-Ogaili et al. (2021) |
| 4. | Cathode | $Mn_3O_4$/$Ti_3C_2$ MXene | Hydrothermal | Zn-air batteries | Power density of 150 mW $cm^{-2}$ | Xue et al. (2017) |
| 5. | Cathode | NiO/$Ti_3C_2$ | Hydrothermal | Li–$O_2$ battery | Capacity of 13350 mAh $g^{-1}$ | Athika and Elumalai (2020) |
| 5. | Cathode | Co/$CoFe_2O_4$-decorated porous carbon | Combustion synthesis | Li-air batteries | 4320 mAh $g^{-1}$ at a high current density of 100 mA $g^{-1}$ | Athika and Elumalai (2020) |
| 6. | Cathode | $MnO_2$@MXene | Hydrothermal | Zn-air batteries | Power density of 198.3 mW $cm^{-2}$ | Sun et al. (2024) |
| 7. | Cathode | FeCo-nitrogen-doped carbon (FeCo-NC) | Hydrothermal | Al-air batteries | Large specific power of 188 mW $cm^{-2}$ | Guo et al. (2020) |
| 8. | Anode | $Fe_2O_3$-nanoloaded carbon | Impregnation | Fe-air batteries | ------ | Hang et al. (2005b) |
| 9. | Anode | $Fe_3O_4$/rGO composites | Chemical precipitation | FeOx-air batteries | Stable discharge capacity of approximately 420 mA h $g^{-1}$ | Tan et al. (2020) |
| 10. | Anode | Zn-rGo | Layer-by-layer film | Zn-air batteries | ~20 mV at 1 mA $cm^{-2}$ and excellent long-life cyclability | Xia et al. (2019) |

## 5.4 CHALLENGES AND ENVIRONMENTAL IMPACT OF METAL-AIR BATTERIES

Metal-air batteries have high energy densities as compared to conventional batteries, so metal-air batteries are catching researchers' attention. But there are still many limiting factors that need to be addressed to achieve the full benefits of metal-air batteries. Use of aqueous electrolytes limits the choice of anode metals; as discussed, some metals like Na, K, and Li are unstable in aqueous electrolytes. Also, metals such as Fe, Al, and Zn have been utilized as anodes in aqueous electrolytes but corrosion is a common problem due to the aqueous nature of the electrolyte. So, the challenge is to find a non-aqueous efficient electrolyte so that the long-term stability of the anode can be assured irrespective of the type of metal used for anode. Li-air batteries are being explored extensively, but use of metals like Fe, Mg, and Al should also be studied because these can provide cheaper anodes as compared to the Li anodes. It was observed that use of non-aqueous electrolyte not only mitigates the electrode corrosion but also has a positive effect on cell voltage. A high charge/discharge overvoltage and a low power capacity are also performance concerning factors for metal-air batteries. These limitations are mainly related to the sluggish reaction rates of air electrodes, so development of efficient cathode catalysts can improve these constraints. Cycle life of metal-air batteries is also not good, which is again due to undesirable formation of dendrites on electrodes, slow ORR and OER reactions and unavailability of non-efficient cathode catalyst. All these factors need significant improvements and only then can the full potential of the metal-air batteries be achieved. In environmental concerns, metal-air batteries are less toxic and safe as compared to the conventional Li-ion, Na-ion, etc. batteries. But still, there are factors of concern like electrolyte management, air exposure, and managing manufacturing processes. Manufacturing material management and electrolyte disposal should be done in a proper way; otherwise, it can cause serious environmental issues. Also, air exposure requirements can degrade the battery components, which can cause environmental problems. Introducing efficient recycling processes and good management of by-products and disposal items can provide a safer way to utilize metal-air batteries.

## 5.5 CONCLUSIONS

Metal-air batteries recently got much attention from the scientists as theoretical energy densities are very high compared to conventional Li-ion batteries. There are also additional benefits associated with metal-air batteries like long cycle life, stability, non-toxic, and safe operations. A typical metal-air battery has three main components, which are a metal anode, an air cathode, and a separator to separate anode from the cathode. Researchers have developed different types of metal-air batteries depending upon the type of anode such as Li-air batteries, Zn-air batteries, Al-air batteries, Fe-air batteries, etc. Battery stability depends upon the kind of electrolyte as some metals are suitable for aqueous electrolytes and some are stable in non-aqueous electrolytes. However, incorporating an additional oxide or hydroxide layer or a conducting thin film can make them suitable for various kinds of electrolytes. Out of various types of metal-air batteries, Zn-air batteries and Li-air batteries are the most developed ones. Other classes of metal-air batteries are still under developing stage, although batteries like Fe-air batteries and Na-ions batteries are cost-effective as compared to the well-developed metal-air batteries. Excessive volume expansion and low catalytic properties of catalysts result in degradation of battery performance. It has been concluded that composite formation of metal-oxide with highly conducting carbon material can mitigate this problem. Several works have shown that composite electrolytes not only give the excellent performance but are also cost effective as compared to conventionally used Pt/C catalysts. Composite formation of metal-oxide and carbon materials also has a positive effect on anode electrode but a very few reports are available. These observations suggest that metal-oxide/carbon composites have the potential to resolve the performance degradation problem of metal-air batteries to some extent.

## REFERENCES


Ahuja, D., Kalpna, V., & Varshney, P. K. (2021). Metal air battery: A sustainable and low cost material for energy storage. *Paper presented at the Journal of Physics: Conference Series*. https://doi.org/10.1088/1742-6596/1913/1/012065

Al-Hakemy, A. Z., Nassr, A. B. A. A., Naggar, A. H., Elnouby, M. S., Soliman, H. M. A. E.-F., & Taher, M. A. (2017). Electrodeposited cobalt oxide nanoparticles modified carbon nanotubes as a non-precious catalyst electrode for oxygen reduction reaction. *Journal of Applied Electrochemistry*, *47*(2), 183–195. https://doi.org/10.1007/s10800-016-1027-0

Al-Ogaili, A. W. M., Cetinkaya, T., Pakseresht, S., & Akbulut, H. (2021). Graphene-based nanocomposite cathodes architecture with palladium and α-$MnO_2$ for high cycle life lithium-oxygen batteries. *Journal of Alloys and Compounds*, *854*, 157293. https://doi.org/10.1016/j.jallcom.2020.157293

Athika, M., & Elumalai, P. (2020). Porous carbon networks decorated with cobalt on $CoFe_2O_4$ as an air-breathing electrode for high-capacity rechargeable lithium-air batteries: Role of metallic cobalt nanoparticles. *ChemElectroChem*, *7*(20), 4188–4200. https://doi.org/10.1002/celc.202000908

Athika, M., Junais, P. M., & Elumalai, P. (2023). Boosting the bifunctional oxygen electrode activity of nitrogen-decorated carbon networks via Ni/NiO addition for Li-Air and Zn-air batteries. *Journal of The Electrochemical Society*, *170*(3), 030544. https://doi.org/10.1149/1945-7111/acc691

Bi, X., Amine, K., & Lu, J. (2020). The importance of anode protection towards lithium oxygen batteries. *Journal of Materials Chemistry A*, *8*(7), 3563–3573. https://doi.org/10.1039/C9TA12414D

Blurton, K. F., & Sammells, A. F. (1979). Metal/air batteries: Their status and potential — a review. *Journal of Power Sources*, *4*(4), 263–279. https://doi.org/10.1016/0378-7753(79)80001-4

Capellán-Pérez, I., Mediavilla, M., de Castro, C., Carpintero, Ó., & Miguel, L. J. (2014). Fossil fuel depletion and socio-economic scenarios: An integrated approach. *Energy*, *77*, 641–666. https://doi.org/10.1016/j.energy.2014.09.063

Chen, K., Yang, D.-Y., Huang, G., & Zhang, X.-B. (2021). Lithium–air batteries: Air-electrochemistry and anode stabilization. *Accounts of Chemical Research*, *54*(3), 632–641. https://doi.org/10.1021/acs.accounts.0c00772

Chen, S., Chen, S., Zhang, B., & Zhang, J. (2019). Bifunctional oxygen electrocatalysis of N, S-codoped porous carbon with interspersed hollow CoO nanoparticles for rechargeable Zn–air batteries. *ACS Applied Materials & Interfaces*, *11*(18), 16720–16728. https://doi.org/10.1021/acsami.9b02819

Cheng, F., & Chen, J. (2012). Metal–air batteries: From oxygen reduction electrochemistry to cathode catalysts. *Chemical Society Reviews*, *41*(6), 2172–2192. https://doi.org/10.1039/C1CS15228A

Cheng, Y., Cao, B., Xu, X., Peng, L., Liu, B., He, J., & Zhang, J. (2024). Oxygen vacancy rich δ-$MnO_2$ nanosheets encapsulating single cobalt atoms-anchored carbon nanotubes for efficient oxygen evolution. *Materials Today Energy*, *40*, 101515. https://doi.org/10.1016/j.mtener.2024.101515

Dobley, A. (2013). Catalytic batteries. In S. L. Suib (Ed.), *New and future developments in catalysis* (pp. 1–16). Elsevier. https://doi.org/10.1016/B978-0-444-53880-2.00001-6

Elalfy, D. A., Gouda, E., Kotb, M. F., Bureš, V., & Sedhom, B. E. (2024). Comprehensive review of energy storage systems technologies, objectives, challenges, and future trends. *Energy Strategy Reviews*, *54*, 101482. https://doi.org/10.1016/j.esr.2024.101482

Figueredo-Rodríguez, H. A., McKerracher, R. D., Insausti, M., Luis, A. G., de León, C. P., Alegre, C., … Walsh, F. C. (2017). A rechargeable, aqueous iron air battery with nanostructured electrodes capable of high energy density operation. *Journal of The Electrochemical Society*, *164*(6), A1148. https://doi.org/10.1149/2.0711706jes

Flegler, A., Hartmann, S., Weinrich, H., Kapuschinski, M., Settelein, J., Lorrmann, H., & Sextl, G. (2016). Manganese oxide coated carbon materials as hybrid catalysts for the application in primary aqueous metal-air batteries. *C – Journal of Carbon Research*, *2*(1), 4. https://doi.org/10.3390/c2010004

Fu, L., Ma, J., Zhang, Z., Wang, G., Yao, Y., & Wei, W. (2024). Core-shell-structured $NiCo_2O_4$ composite with GO/C to enhance its electrocatalytic performance as a cathode catalyst in zinc-air batteries. *Vacuum*, *224*, 113163. https://doi.org/10.1016/j.vacuum.2024.113163

García-Rodríguez, M., Flores-Lasluisa, J. X., Cazorla-Amorós, D., & Morallón, E. (2024). Enhancing interaction between Lanthanum manganese cobalt oxide and carbon black through different approaches for primary Zn–air batteries. *Materials*, *17*(10). https://doi.org/10.3390/ma17102309

Gnana Kumar, G., Christy, M., Jang, H., & Nahm, K. S. (2015). Cobaltite oxide nanosheets anchored graphene nanocomposite as an efficient oxygen reduction reaction (ORR) catalyst for the application of lithium-air batteries. *Journal of Power Sources*, *288*, 451–460. https://doi.org/10.1016/j.jpowsour.2015.04.029

Goel, P., Dobhal, D., & Sharma, R. C. (2020). Aluminum–air batteries: A viability review. *Journal of Energy Storage*, *28*, 101287. https://doi.org/10.1016/j.est.2020.101287

Guo, M., Zhang, X., Yang, T., Dang, Q., Li, X., Wang, Y., & Zhang, G. (2020). Binary FeCo-N-doped carbon/carbon nanotube composites for efficient oxygen reduction and high-performance aluminum-air battery. *Journal of Power Sources*, *456*, 227933. https://doi.org/10.1016/j.jpowsour.2020.227933

Gurushankar, K., Karuppaiya, C., Tari, V., & Kannan, K. (2024). Exploring MXenes and MBenes in metal sulfur and selenium battery technologies. In K. Kannan & V. Tari (Eds.), *Novel energy storage and conversion technologies for two-dimensional MXenes and MBenes*. IGI Global. https://doi.org/10.4018/979-8-3693-9546-2.ch005

Hang, B. T., Eashira, M., Watanabe, I., Okada, S., Yamaki, J.-I., Yoon, S.-H., & Mochida, I. (2005a). The effect of carbon species on the properties of Fe/C composite for metal–air battery anode. *Journal of Power Sources*, *143*(1), 256–264. https://doi.org/10.1016/j.jpowsour.2004.11.044

Hang, B. T., Watanabe, T., Eashira, M., Watanabe, I., Okada, S., & Yamaki, J.-I. (2005b). Comparative study of $Fe_2O_3$-nanoloaded carbon and $Fe_2O_3$-nano/carbon mixed composites for iron-air battery anodes. *Electrochemical and Solid-State Letters*, *8*(9), A476. https://doi.org/10.1149/1.1999914

He, J., Cao, L., Cui, J., Fu, G., Jiang, R., Xu, X., & Guan, C. (2024). Flexible energy storage devices to power the future. *Advanced Materials*, *36*(4), 2306090. https://doi.org/10.1002/adma.202306090

Hong, J. H., Kim, J. H., Park, G. D., Lee, J. Y., Lee, J.-K., & Kang, Y. C. (2021). A strategy for fabricating three-dimensional porous architecture comprising metal oxides/CNT as highly active and durable bifunctional oxygen electrocatalysts and their application in rechargeable Zn-air batteries. *Chemical Engineering Journal*, *414*, 128815. https://doi.org/10.1016/j.cej.2021.128815

Hyun, S., & Shanmugam, S. (2017). Mesoporous Co-CoO/N-CNR nanostructures as high-performance air cathode for lithium-oxygen batteries. *Journal of Power Sources*, *354*, 48–56. https://doi.org/10.1016/j.jpowsour.2017.04.029

Jiang, M., He, H., Yi, W.-J., Huang, W., Pan, X., Wang, M.-Y., & Chao, Z.-S. (2017). ZIF-67 derived Ag-$Co_3O_4$@N-doped carbon/carbon nanotubes composite and its application in Mg-air fuel cell. *Electrochemistry Communications*, *77*, 5–9. https://doi.org/10.1016/j.elecom.2017.01.023

Jung, K.-N., Hwang, S. M., Park, M.-S., Kim, K. J., Kim, J.-G., Dou, S. X., ... Lee, J.-W. (2015). One-dimensional manganese-cobalt oxide nanofibres as bi-functional cathode catalysts for rechargeable metal-air batteries. *Scientific Reports*, *5*(1), 7665. https://doi.org/10.1038/srep07665

Kalair, A., Abas, N., Saleem, M. S., Kalair, A. R., & Khan, N. (2021). Role of energy storage systems in energy transition from fossil fuels to renewables. *Energy Storage*, *3*(1), e135. https://doi.org/10.1002/est2.135

Kannan, K., & Tari, V. (Eds.). (2025). *Novel energy storage and conversion technologies for two-dimensional MXenes and MBenes*. https://www.amazon.com/dp/9798369395462

Kurian, M., Ranganath, S. P., Vijayakumar, V., Sivadasan, S., Kurian, R. M., Badiger, M. V., & Kurungot, S. (2024). F-doped nickel cobalt oxide-carbon composite electrocatalysts paired with mechanically robust anion-conducting chitosan membranes for flexible and rechargeable Zinc–air batteries. *ACS Applied Energy Materials*, *7*(16), 7037–7054. https://doi.org/10.1021/acsaem.4c01261

Lei, H., Tan, S., Ma, L., Liu, Y., Liang, Y., Javed, M. S., ... Mai, W. (2020). Strongly coupled $NiCo_2O_4$ nanocrystal/MXene hybrid through in situ Ni/Co–F bonds for efficient wearable Zn–air batteries. *ACS Applied Materials & Interfaces*, *12*(40), 44639–44647. https://doi.org/10.1021/acsami.0c11185

Levy, N. R., Tereshchuk, P., Natan, A., Haas, R., Schröder, D., Janek, J., … Ein-Eli, Y. (2021). Hybridization of carbon nanotube tissue and $MnO_2$ as a generic advanced air cathode in metal–air batteries. *Journal of Power Sources*, *514*, 230597. https://doi.org/10.1016/j.jpowsour.2021.230597

Li, F., Yin, Y., Zhang, C., Li, W., Maliutina, K., Zhang, Q., … Fan, L. (2020). Enhancing oxygen reduction performance of oxide-CNT through in-situ generated nanoalloy bridging. *Applied Catalysis B: Environmental*, *263*, 118297. https://doi.org/10.1016/j.apcatb.2019.118297

Li, Y., & Lu, J. (2017). Metal–Air batteries: Will they be the future electrochemical energy storage device of choice? *ACS Energy Letters*, *2*(6), 1370–1377. https://doi.org/10.1021/acsenergylett.7b00119

Li, Y., Zhang, Z., Duan, D., Han, Y., Wang, K., Hao, X., … Wu, F. (2019). An integrated structural air electrode based on parallel porous nitrogen-doped carbon nanotube arrays for rechargeable Li–air batteries. *Nanomaterials*, *9*(10), 1412. https://doi.org/10.3390/nano9101412

Liang, Y., Wang, H., Diao, P., Chang, W., Hong, G., Li, Y., … Dai, H. (2012). Oxygen reduction electrocatalyst based on strongly coupled cobalt oxide nanocrystals and carbon nanotubes. *Journal of the American Chemical Society*, *134*(38), 15849–15857. https://doi.org/10.1021/ja305623m

Liu, Y., Sun, Q., Li, W., Adair, K. R., Li, J., & Sun, X. (2017). A comprehensive review on recent progress in aluminum–air batteries. *Green Energy & Environment*, *2*(3), 246–277. https://doi.org/10.1016/j.gee.2017.06.006

Lu, Y.-T., Wu, J., Lin, Z.-X., You, T.-H., Lin, S.-C., Tiffany Chen, H.-Y., … Hu, C.-C. (2019). Enhanced oxygen evolution performance of spinel $Fe_{0.1}Ni_{0.9}Co_2O_4$/Activated carbon composites. *Electrochimica Acta*, *326*, 134986. https://doi.org/10.1016/j.electacta.2019.134986

Mahato, D., Kharwar, Y. P., Ramanujam, K., Haridoss, P., & Thomas, T. (2021). S, N co-doped graphene quantum dots decorated $TiO_2$ and supported with carbon for oxygen reduction reaction catalysis. *International Journal of Hydrogen Energy*, *46*(41), 21549–21565. https://doi.org/10.1016/j.ijhydene.2021.04.013

Martin, J. J., Neburchilov, V., Wang, H., & Qu, W. (2009, 22–23 October). Air cathodes for metal-air batteries and fuel cells. *Paper presented at the 2009 IEEE Electrical Power & Energy Conference (EPEC)*. https://doi.org/10.1109/EPEC.2009.5420955

Mei, D., Yuan, X., Ma, Z., Wei, P., Yu, X., Yang, J., & Ma, Z.-F. (2016). A $SnO_2$-based cathode catalyst for lithium-air batteries. *ACS Applied Materials & Interfaces*, *8*(20), 12804–12811. https://doi.org/10.1021/acsami.6b02402

Palanisamy, S., Gopalakrishnan, M., Ingavale, S., Etesami, M., Limphirat, W., Liu, W.-R., … Kheawhom, S. (2024). Sandwiched heterostructure of $NiFe_2O_4/TiO_2$ nanocrystals and MXene nanosheets for highly active oxygen electrocatalyst in rechargeable zinc-air batteries. *Journal of Energy Storage*, *98*, 113139. https://doi.org/10.1016/j.est.2024.113139

Pan, J., Xu, Y. Y., Yang, H., Dong, Z., Liu, H., & Xia, B. Y. (2018). Advanced architectures and relatives of air electrodes in Zn–air batteries. *Advanced Science*, *5*(4), 1700691. https://doi.org/10.1002/advs.201700691

Patra, S., & Tiwari, A. (2023). Chapter 19 - Graphene-based nanocomposites as electrode materials for Zn-air batteries. In A. Khan, M. Nazim, & A. Asiri

(Eds.), *Advances in electronic materials for clean energy conversion and storage applications* (pp. 395–412). Woodhead Publishing. https://doi.org/10.1016/B978-0-323-91206-8.00004-2

Pendashteh, A., Sanchez, J. S., Palma, J., Anderson, M., & Marcilla, R. (2019). Anchored $NiCoMnS_4$ nanoparticles on N-doped rGO: High-performance bifunctional electrocatalysts for rechargeable Zn-Air batteries. *Energy Storage Materials*, *20*, 216–224. https://doi.org/10.1016/j.ensm.2019.04.018

Peng, S.-H., Chen, T.-H., Lee, C.-H., Lu, H.-C., & Lue, S. J. (2020). Optimal cobalt oxide ($Co_3O_4$): Graphene (GR) ratio in $Co_3O_4$/GR as air cathode catalyst for air-breathing hybrid electrolyte lithium-air battery. *Journal of Power Sources*, *471*, 228373. https://doi.org/10.1016/j.jpowsour.2020.228373

Pozo-Gonzalo, C., & Ortiz-Vitoriano, N. (2022). Recent progress, advances, and future prospects in Na–O2 batteries. *Current Opinion in Electrochemistry*, *36*, 101120. https://doi.org/10.1016/j.coelec.2022.101120

Rahman, M. A., Wang, X., & Wen, C. (2013). High energy density metal-air batteries: A review. *Journal of The Electrochemical Society*, *160*(10), A1759. https://doi.org/10.1149/2.062310jes

Ren, J.-T., Yuan, G.-G., Weng, C.-C., & Yuan, Z.-Y. (2018). Rationally designed $Co_3O_4$–C nanowire arrays on Ni foam derived from metal organic framework as reversible oxygen evolution electrodes with enhanced performance for Zn–air batteries. *ACS Sustainable Chemistry & Engineering*, *6*(1), 707–718. https://doi.org/10.1021/acssuschemeng.7b03034

Ren, M., Zhang, J., & Tour, J. M. (2019). Laser-induced graphene hybrid catalysts for rechargeable Zn-air batteries. *ACS Applied Energy Materials*, *2*(2), 1460–1468. https://doi.org/10.1021/acsaem.8b02011

Ruan, Y., Lei, H., Xue, W., Wang, T., Song, S., Xu, H., … Mei, D. (2023). Cu-N-C assisted MnO nanorods as bifunctional electrocatalysts for superior long-cycle Zn-air batteries. *Journal of Alloys and Compounds*, *934*, 167781. https://doi.org/10.1016/j.jallcom.2022.167781

Sun, Q., Guo, Z., Shu, T., Li, Y., Li, K., Zhang, Y., … Yao, K. X. (2024). Lithium-induced oxygen vacancies in $MnO_2$@MXene for high-performance Zinc–air batteries. *ACS Applied Materials & Interfaces*, *16*(10), 12781–12792. https://doi.org/10.1021/acsami.3c18248

Sun, Q., Yang, Y., & Fu, Z.-W. (2012). Electrochemical properties of room temperature sodium–air batteries with non-aqueous electrolyte. *Electrochemistry Communications*, *16*(1), 22–25. https://doi.org/10.1016/j.elecom.2011.12.019

Syahputra, S., Sgreccia, E., Nallayagari, A. R., Vacandio, F., Kaciulis, S., Di Vona, M. L., & Knauth, P. (2024). Influence of nitrogen position on the electrocatalytic performance of B,N-codoped carbon Quantum dots for the oxygen reduction reaction. *Journal of The Electrochemical Society*, *171*(6), 066510. https://doi.org/10.1149/1945-7111/ad5872

Tan, W. K., Asami, K., Maegawa, K., Kumar, R., Kawamura, G., Muto, H., & Matsuda, A. (2020). $Fe_3O_4$-embedded rGO composites as anode for rechargeable $FeO_x$-air batteries. *Materials Today Communications*, *25*, 101540. https://doi.org/10.1016/j.mtcomm.2020.101540

Tan, W. K., Kawamura, G., Muto, H., & Matsuda, A. (2021). Current progress in the development of Fe-air batteries and their prospects for next-generation batteries. In K. Y. Cheong & L.-C. Chen (Eds.), *Sustainable materials for next generation energy devices* (pp. 59–83). Elsevier. https://doi.org/10.1016/j.mtcomm.2020.101540

Tari, V., Gupta, V., Joshi, A., Kurve, P., & Chandra, N. (2024). Biodiversity loss due to forest diversion. In A. Rathore (Ed.), *Impact of societal development and infrastructure on biodiversity decline* (pp. 137–165). IGI Global. https://doi.org/10.4018/979-8-3693-6950-0.ch010

Tari, V., Kannan, K., Koliar, J., Gurushankar, K., & Wahyuni, D. K. (2025). Influence of zinc oxide nanomaterials on crop plants: A prospective analysis. In D. Jabbrova, D. Sarkar, S. Rakesh, S. Singh, & R. Datta (Eds.), *Smart nanomaterials in agroforestry systems*. Springer. https://doi.org/10.1007/978-981-96-1337-3_8

Tari, V., & Patil, P. (2017a). Transfer factor of heavy metal in soil to plant from pesticide-contaminated area (mango orchards), Ratnagiri district, Maharashtra. *Research Journal of Chemistry and Environment*, *21*(1), 26–32.

Tari, V., & Patil, P. (2017b). Challenge of contamination of pesticide for Alphonso in Ratnagiri district, Maharashtra, India. *Research Journal of Chemistry and Environment*, *21*(12), 57–67

Tong, F., Wei, S., Chen, X., & Gao, W. (2021). Magnesium alloys as anodes for neutral aqueous magnesium-air batteries. *Journal of Magnesium and Alloys*, *9*(6), 1861–1883. https://doi.org/10.1016/j.jma.2021.04.011

Wang, H., Tan, R., Yang, Z., Feng, Y., Duan, X., & Ma, J. (2021). Stabilization perspective on metal anodes for aqueous batteries. *Advanced Energy Materials*, *11*(2), 2000962. https://doi.org/10.1002/aenm.202000962

Wang, N., Ning, S., Yu, X., Chen, D., Li, Z., Xu, J., … Chen, S. (2022a). Graphene composites with Ru-$RuO_2$ heterostructures: Highly efficient Mott–Schottky-type electrocatalysts for pH-universal water splitting and flexible zinc–air batteries. *Applied Catalysis B: Environmental*, *302*, 120838. https://doi.org/10.1016/j.apcatb.2021.120838

Wang, Y., Pan, W., Luo, S., Zhao, X., Kwok, H. Y. H., Xu, X., & Leung, D. Y. C. (2022b). High-performance solid-state metal-air batteries with an innovative dual-gel electrolyte. *International Journal of Hydrogen Energy*, *47*(33), 15024–15034. https://doi.org/10.1016/j.ijhydene.2022.03.011

Wang, Z.-L., Xu, D., Xu, J.-J., & Zhang, X.-B. (2014). Oxygen electrocatalysts in metal–air batteries: From aqueous to nonaqueous electrolytes. *Chemical Society Reviews*, *43*(22), 7746–7786. https://doi.org/10.1039/C3CS60248F

Wang, Z.-Y., Jiang, S.-D., Duan, C.-Q., Wang, D., Luo, S.-H., & Liu, Y.-G. (2020). In situ synthesis of $Co_3O_4$ nanoparticles confined in 3D nitrogen-doped porous carbon as an efficient bifunctional oxygen electrocatalyst. *Rare Metals*, *39*(12), 1383–1394. https://doi.org/10.1007/s12598-020-01581-4

Worku, A. K., Ayele, D. W., & Habtu, N. G. (2021). Recent advances and future perspectives in engineering of bifunctional electrocatalysts for rechargeable zinc–air batteries. *Materials Today Advances*, *9*, 100116. https://doi.org/10.1016/j.mtadv.2020.100116

Xi, W., Shen, M., Yin, X., Gao, B., He, L., Chen, Y., & Lin, B. (2023). Molten-salt confined synthesis of nitrogen-doped carbon nanosheets supported $Co_3O_4$ nanoparticles as a superior oxygen electrocatalyst for rechargeable Zn-air battery. *Journal of Power Sources*, *560*, 232692. https://doi.org/10.1016/j.jpowsour.2023.232692

Xia, A., Pu, X., Tao, Y., Liu, H., & Wang, Y. (2019). Graphene oxide spontaneous reduction and self-assembly on the zinc metal surface enabling a dendrite-free anode for long-life zinc rechargeable aqueous batteries. *Applied Surface Science*, *481*, 852–859. https://doi.org/10.1016/j.apsusc.2019.03.197

Xiang, W., Li, Y., Wu, M., Ma, J., & Sheng, Z. (2022). Co-doped $CeO_2$/N–C nanorods as a bifunctional oxygen electrocatalyst and its application in rechargeable Zn-air batteries. *Nanotechnology*, *33*(41), 415404. https://doi.org/10.1088/1361-6528/ac7ed1

Xu, H., Zheng, R., Du, D., Ren, L., Li, R., Wen, X., Shu, C. (2022). Cationic vanadium vacancy-enriched $V_{2-x}O_5$ on V2C MXene as superior bifunctional electrocatalysts for Li-$O_2$ batteries. *Science China Materials*, *65*(7), 1761–1770. https://doi.org/10.1007/s40843-021-1959-1

Xue, H., Ma, Y., Wang, T., Gong, H., Gao, B., Fan, X., He, J. (2019). Hollow mesoporous $Fe_2O_3$ nanospindles/CNTs composite: An efficient catalyst for high-performance Li-$O_2$ batteries. *Frontiers in Chemistry*, *7*. https://doi.org/10.3389/fchem.2019.00511

Xue, Q., Pei, Z., Huang, Y., Zhu, M., Tang, Z., Li, H., Zhi, C. (2017). $Mn_3O_4$ nanoparticles on layer-structured $Ti_3C_2$ MXene towards the oxygen reduction reaction and zinc–air batteries. *Journal of Materials Chemistry A*, *5*(39), 20818–20823. https://doi.org/10.1039/C7TA04532H

Yaqoob, L., Noor, T., & Iqbal, N. (2022). An overview of metal-air batteries, current progress, and future perspectives. *Journal of Energy Storage*, *56*, 106075. https://doi.org/10.1016/j.est.2022.106075

Zhang, C.-L., Liu, J.-T., Li, H., Qin, L., Cao, F.-H., & Zhang, W. (2020a). The controlled synthesis of $Fe_3C$/Co/N-doped hierarchically structured carbon nanotubes for enhanced electrocatalysis. *Applied Catalysis B: Environmental*, *261*, 118224. https://doi.org/10.1016/j.apcatb.2019.118224

Zhang, H., Qiao, H., Wang, H., Zhou, N., Chen, J., Tang, Y., Huang, C. (2014). Nickel cobalt oxide/carbon nanotubes hybrid as a high-performance electrocatalyst for metal/air battery. *Nanoscale*, *6*(17), 10235–10242. https://doi.org/10.1039/C4NR02125H

Zhang, L., Huang, Q.-A., Yan, W., Shao, Q., & Zhang, J. (2019). Design and fabrication of non-noble metal catalyst-based air-cathodes for metal-air battery. *The Canadian Journal of Chemical Engineering*, *97*(12), 2984–2993. https://doi.org/10.1002/cjce.23616

Zhang, P., Liu, X., Xue, J., & Jiang, K. (2020b). The role of microstructural evolution in improving energy conversion of Al-based anodes for metal-air batteries. *Journal of Power Sources*, *451*, 227806. https://doi.org/10.1016/j.jpowsour.2020.227806

Chapter 6

# Advanced electro catalysts based on hybrid composite materials for alkaline hydrogen evolution reaction

*Manikandan Krishnamurthy, Deepa G. Kaladevi, Valarmathi Narayanan, Mahadevi Pichandi, A Shalini, and Lakshmana Reddy Nagappagari*

## 6.1 INTRODUCTION TO HYDROGEN EVOLUTION REACTION (HER)

The global energy crisis and environmental challenges can be effectively tackled through the generation of solar fuels, particularly hydrogen (Kannan & Tari, 2025). Therefore, the development of non-toxic, cost-effective, and highly efficient photocatalysts is essential. Spinel ferrites, with the general formula $MFe_2O_4$ (where M = $Mg^{2+}$, $Co^{2+}$, $Ni^{2+}$, $Zn^{2+}$, etc.), have attracted significant attention in the field of solar energy conversion and storage, especially for hydrogen production via photocatalytic water splitting. Spinel ferrites are considered ideal candidates for solar fuel (hydrogen) generation due to their favorable magnetic properties for easy recovery, suitable band gap energy, low toxicity, excellent thermal stability, affordability, straightforward synthesis methods, and magnetic separability. Nevertheless, the solar-to-hydrogen (STH) conversion efficiency of spinel ferrite materials remains limited due to the rapid recombination of photogenerated electron-hole pairs, which hampers their photocatalytic performance. This review focuses on detailed preparation methods for spinel ferrites. Additionally, it highlights the advancements in solar hydrogen production achieved using spinel ferrites as the primary photocatalyst rather than as a cocatalyst. Ultimately, this review aims to inspire researchers to expand the application of spinel ferrite-based materials for economical and eco-friendly solar fuel (hydrogen) generation. Since its discovery, photocatalysis has proven to be a valuable and rapidly evolving field of research. In the early stages, semiconductors such as $TiO_2$, ZnO, $PtO_2$, and others were employed to perform various photocatalytic processes, including hydrogen production, photodegradation, water purification, $CO_2$ conversion into energy fuels, microbial disinfection, and organic transformations (Gurushankar et al., 2024; Wang et al., 2017; Lim et al., 2018; Ying et al., 2017; Zhang et al., 2017). The growing global population, coupled with rapid social and economic advancements, has led to a significant rise in energy demand. Fossil fuels like coal, oil, and natural gas have remained the primary energy sources; however, they pose serious challenges due to their finite, non-renewable nature, which contributes

 DOI: 10.1201/9781003584742-6

to environmental degradation and increasing pollution levels (Tari, Gupta, et al., 2024; Siddiqui et al., 2022; Tari, Siddiqui, et al., 2024; Tari et al., 2025). Consequently, there has been a rapid expansion in the adoption of clean and renewable energy sources such as solar, wind, and tidal energy, which are now regarded as viable alternatives. Nevertheless, fluctuations in environmental conditions affect the consistency and stability of these energy supplies, highlighting the need for advancements in the development and scaling up of energy production from these renewable sources. Water splitting using solar energy to produce hydrogen is a highly promising strategy for sustainable energy generation. Hydrogen is viewed as an ideal fuel for the future due to its zero emissions, high combustion efficiency, and much higher specific energy density compared to traditional energy sources. For instance, 1 kg of hydrogen can produce 140 MJ of energy, which is nearly four times the energy output of methane (33 kJ/g) (Chu & Majumdar, 2012; Cheng & Yi, 2017; Binz et al., 2017; Schlapbach & Züttel, 2001; McCrory et al., 2015; Carmo et al., 2013; Roger et al., 2017; Crabtree et al., 2004; Dresselhaus & Thomas, 2001; Chu & Majumdar, 2012; Turner, 2004; Shi & Zhang, 2016). Several techniques have been explored for hydrogen production, including photocatalytic, photobiological, and photoelectrochemical methods. Among these, photocatalytic (PC) water splitting powered by solar energy is considered the most viable approach to addressing the global energy crisis. This technique is particularly effective for hydrogen generation because it is simple, cost-efficient, eco-friendly, and supports sustainable development. Both organic and inorganic materials have been developed as photocatalysts for solar hydrogen production. Some of these materials can absorb ultraviolet (UV) light, while others are optimized for visible light absorption, which represents 4% and 43% of the solar spectrum, respectively. Currently, significant research efforts are focused on developing photocatalysts with enhanced visible light absorption, high efficiency, and superior photostability. The hydrogen evolution reaction (HER) is one of the most fundamental and widely studied electrochemical processes in modern science and technology. It plays a pivotal role in the production of hydrogen gas ($H_2$), a clean and sustainable energy carrier that is central to the global transition toward renewable energy systems. HER is the cathodic half-reaction in water electrolysis, where water is split into hydrogen and oxygen gases using electrical energy. This process is not only essential for hydrogen production but also finds applications in fuel cells, energy storage, and various industrial processes (Tee et al., 2017). Hydrogen is often hailed as the "fuel of the future" due to its high energy density, zero carbon emissions when burned, and versatility in energy applications. However, the efficient and sustainable production of hydrogen remains a significant challenge. HER is a key step in addressing this challenge, as it directly influences the efficiency, cost, and scalability of hydrogen production technologies. Understanding the mechanisms, kinetics, and materials involved in HER is therefore critical for advancing clean energy technologies (Konieczny et al., 2008; Turner et al., 2008; Dresselhaus & Thomas, 2001; Liu et al., 2022; Bongaarts & O'Neill, 2018; Huhta, 2022; Chu & Majumdar, 2012; Yu et al., 2021). Recent research on the HER has focused on improving the efficiency, scalability, and sustainability of hydrogen production through advancements in catalyst design, reaction mechanisms, and system integration. Below is a summary of key trends and breakthroughs in HER research, along with their implications for the future of clean energy.

## 6.2 HISTORICAL CONTEXT AND SIGNIFICANCE

The study of HER dates back to the early 19th century when scientists first began exploring the electrolysis of water. The discovery that hydrogen gas could be produced by passing an electric current through water laid the foundation for modern electrochemistry. Over the years, HER has been extensively studied in the context of both fundamental science and industrial applications. In recent decades, the growing demand for clean energy has renewed interest in HER. The global push to reduce greenhouse gas emissions and combat climate change has highlighted the importance of hydrogen as a renewable energy source. Unlike fossil fuels, hydrogen produces only water as a byproduct when used in fuel cells or combustion processes. This makes it an attractive alternative for decarbonizing sectors such as transportation, industry, and power generation. However, the majority of hydrogen produced today is derived from fossil fuels through processes like steam methane reforming, which emits significant amounts of carbon dioxide. To realize the full potential of hydrogen as a clean energy carrier, it is essential to produce it sustainably through water electrolysis powered by renewable energy sources such as solar, wind, or hydropower. HER is at the heart of this process, making it a cornerstone of the hydrogen economy. HER is an electrochemical reduction reaction that occurs at the cathode during water electrolysis. The overall reaction involves the reduction of protons ($H^+$) or water molecules ($H_2O$) to produce hydrogen gas ($H_2$). The efficiency of HER is influenced by several factors, including the electrode material, electrolyte composition, temperature, and applied voltage. The reaction typically proceeds through a series of steps involving the adsorption, desorption, and recombination of hydrogen atoms on the electrode surface (Abdalla et al., 2018; Zhang et al., 2016).

### 6.2.1 Catalysts for HER

The efficiency of HER is highly dependent on the choice of catalyst. An ideal catalyst should exhibit high activity, stability, and cost-effectiveness. The most widely used and effective catalyst for HER is platinum (Pt), which has a low overpotential and high exchange current density. However, the high cost and scarcity of Pt have driven research into alternative catalysts, including: transition metal-based catalysts: Researchers have made significant progress in developing catalysts based on earth-abundant transition metals such as nickel (Ni), cobalt (Co), iron (Fe), and molybdenum (Mo). These materials offer a cost-effective alternative to platinum (Pt) while maintaining reasonable catalytic activity. Two-dimensional (2D) materials such as graphene, molybdenum disulfide ($MoS_2$), and MXenes have gained attention for their high surface area, tunable electronic properties, and excellent conductivity. Recent studies have deepened our understanding of how HER mechanisms vary with pH. In acidic media, protons ($H^+$) are directly reduced, while in alkaline media, water molecules ($H_2O$) are the primary reactants. b. Role of Adsorbed Intermediate. The adsorption strength of hydrogen atoms ($H_ad_s$) on the catalyst surface is a critical factor in HER efficiency. Recent work has focused on optimizing this parameter to balance the Volmer, Heyrovsky, and Tafel steps. a. Membrane Electrode Assemblies (MEAs) Advances in MEA design have improved the efficiency and durability of HER in proton exchange membrane (PEM) electrolyzers. Transition metal-based materials such as nickel (Ni), cobalt (Co), iron (Fe), and molybdenum (Mo) have shown promise as cost-effective alternatives to Pt. Carbon materials such as graphene, carbon nanotubes, and doped carbons have been explored for their high surface area and conductivity. Heteroatom doping (e.g., nitrogen, sulfur, or phosphorus) can enhance the catalytic activity of carbon-based materials. Metal-organic frameworks (MOFs) and single-atom catalysts offer unique opportunities for optimizing active sites and improving catalytic performance. Researchers are increasingly focusing on catalysts derived from abundant and inexpensive materials to enable large-scale hydrogen production (Sazali, 2020; Chen et al., 2024).

### 6.2.2 Applications of HER

HER is a key process in several important technologies, including:

1. **Water Electrolysis:**
    - HER is the cathodic reaction in water electrolyzers, which produce hydrogen gas from water using electricity.
    - Electrolysis is a clean and sustainable method for hydrogen production when powered by renewable energy.
2. **Fuel Cells:**
    - In proton exchange membrane fuel cells (PEMFCs), HER occurs in reverse at the anode, where hydrogen is oxidized to produce electricity.
3. **Energy Storage:**
    - Hydrogen produced via HER can be stored and used as a renewable energy carrier, addressing the intermittency of solar and wind power.

4. **Industrial Processes:**
    - HER is used in various industrial applications, including ammonia production, petroleum refining, and metal processing.

### 6.2.3 Challenges and future directions

Despite significant progress, several challenges remain in optimizing HER for practical applications:

1. **Catalyst Cost and Availability:**
    - The reliance on expensive and scarce materials like platinum limits the scalability of HER technologies.
2. **Overpotential and Efficiency:**
    - High overpotentials and energy losses reduce the overall efficiency of HER.
3. **Durability and Stability:**
    - Catalysts must withstand harsh operating conditions, including acidic or alkaline environments and high current densities.
4. **Integration with Renewable Energy:**
    - Developing systems that efficiently integrate HER with renewable energy sources is critical for sustainable hydrogen production.

This book chapter is mainly reviewing the future research efforts focused on addressing these challenges through the development of advanced catalysts, novel electrode materials, and innovative system designs. The ultimate goal is to enable cost-effective, efficient, and scalable hydrogen production to support the global transition to a clean energy future. The HER is a cornerstone of modern electrochemistry and clean energy technologies. Its role in hydrogen production, energy storage, and industrial processes makes it a critical area of research and development. As the world seeks to reduce its reliance on fossil fuels and mitigate climate change, HER will play an increasingly important role in enabling a sustainable hydrogen economy. By advancing our understanding of HER mechanisms, developing innovative catalysts, and optimizing system designs, we can unlock the full potential of hydrogen as a clean and renewable energy source. This introduction provides a comprehensive overview of HER, its significance, mechanisms, applications, and challenges. Let me know if you'd like to dive deeper into any specific aspect.

- Hydrogen evolution reaction
- Challenges and future directions
- Scalability and cost-effectiveness
- Integration with renewable energy sources
- Industrial applications of hydrogen evolution

Hydrogen energy is a clean, carbon-free, and potential future energy source. Hydrogen generation by electrochemical water splitting has become a promising approach, which is highly dependent on the electroactivity of HER electrocatalysts. The transition metal phosphides (TMPs) have become promising electrocatalysts due to low overpotential, good electrochemical stability, and high electrocatalytic activity. The electrochemical properties of five typical TMPs ($Ni_xP_y$, $Co_xP_y$, $Mo_xP_y$, $W_xP_y$, and $Fe_xP_y$) under different phosphating degrees are intensively compared. Since the electrocatalytic performance of current TMPs is insufficient for industrial applications, new design and preparation methods for nanostructure-controlled TMPs electrocatalysts are proposed (Mehta et al., 2024).

In particular, the development of element-doped, self-supporting nanomaterials is the future development direction of TMPs electrocatalysts that can be applied across the whole pH range. With interest in renewable energy sources and in the decarbonization of industry rapidly accelerating, alkaline water electrolysis can now be regarded as a key technology enabling efficient energy conversion and storage. Since an alkaline environment is suitable for a vast range of materials with satisfactory chemical stability under operating conditions, the topic of catalysts for the alkaline route could give rise to confusion in the community (Khalil et al., 2024).

The focus of this research is on analyzing the current situation in the electrocatalysis of the HER in an alkaline and a neutral environment, presenting the main group of materials studied, and discussing their potential to achieve industrial relevance. It addresses the main limitations of common parameters used for evaluating and comparing the catalytic activity and stability of selected catalysts.

Furthermore, the provides a comprehensive comparison of catalyst activities with respect to the individual groups depending on their composition as well as with respect to the most used cations and cation-based materials. For the sake of clarity, the comparison is also presented in graphical form. Finally, based on the literature data, fundamental material characteristics to be evaluated in the development of new catalysts for electrochemical water splitting are proposed (Mehta et al., 2024).

## 6.3 HYDROGEN EVOLUTION REACTION

The generally accepted mechanism of the HER in alkaline media and neutral media consists of three elementary reaction steps (Yin et al., 2024; Durovič et al., 2021; Ferriday et al., 2021; Koper, 2011; Marković & Ross, 2002; Conway & Tilak, 2002)

$$H_2O + e^- \rightleftarrows H_{ads} + OH^- \text{(Volmer step)} \tag{6.1}$$

$$H_{ads} + H_2O + e^- \rightleftarrows H_2 + OH^- \text{(Heyrovsky step)} \quad (6.2)$$

$$2H_{ads} \rightleftarrows H_2 \text{(Tafel step)} \quad (6.3)$$

The Volmer step (6.1) describes the splitting of a water molecule and adsorption of hydrogen on a free site of the electrode/catalyst surface. This is then followed by hydrogen production via an electrochemical process (Heyrovsky step (6.2)) and/or a chemical process (Tafel step (6.3)). Several authors have reported that, at low overpotentials, the HER mechanism consists of the Volmer step followed by a parallel Heyrovsky and Tafel step, while at high overpotentials the Tafel step is negligible and the reaction proceeds via the Volmer–Heyrovsky mechanism (Wang et al., 2016; Zou and Zhang, 2015; Bockris & Potter, 1952. Sapountzi et al., 2017). The electrocatalyst activity toward the HER is strongly influenced by the strength of the Me–H bond (Jaksic et al., 2000). Figure 6.1 showed two ways of the HER.

### 6.3.1 Scalability and cost-effectiveness

Scalability of hydrogen evolution has three categories: modular design, mass production, and large-scale electrolysis

1. **Modular design** is depending on the demand for hydrogen, electrolyzers may be easily scaled up or down thanks to their modular architecture.
    - Standardized Components: Modular electrolyzers consist of standardized components, making it easier to scale up or down.
    - Flexible Configuration: Modular designs allow for flexible configuration, enabling customization to meet specific hydrogen demands.

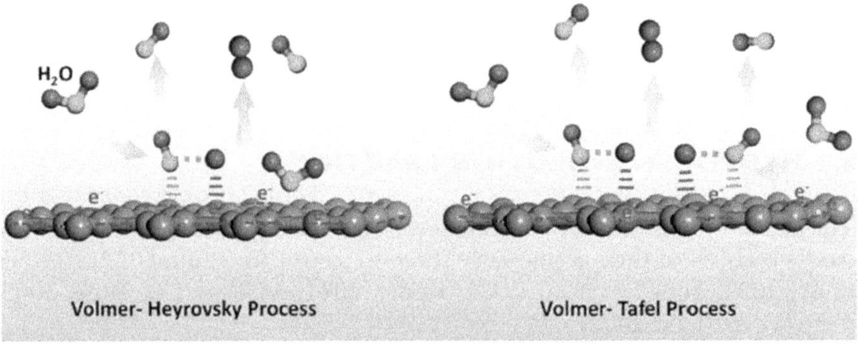


*Figure 6.1* The two ways the hydrogen evolution reaction.

**(Reproduced with permission from Ref Ferriday et al. (2021). Copyright @ MDPI 2021.)**

- Easy Maintenance: Modular components can be easily replaced or maintained, reducing downtime and increasing overall efficiency.

2. **Mass Production** of electrolyzers can reduce costs and increase scalability.
   - Economies of Scale: Mass production reduces costs per unit, making electrolyzers more competitive.
   - Improved Manufacturing Efficiency: Mass production enables manufacturers to optimize their processes, reducing waste and increasing efficiency.
   - Increased Availability: Mass-produced electrolyzers can be widely available, accelerating the transition to hydrogen-based energy.
3. **Large-scale electrolysis** systems can produce significant amounts of hydrogen, making them suitable for industrial applications.
   - High-Volume Hydrogen Production: Large-scale electrolysis enables the production of significant amounts of hydrogen, making it suitable for industrial applications.
   - Improved Energy Efficiency: Large-scale electrolysis systems can achieve higher energy efficiency due to economies of scale.
   - Reduced Costs: Large-scale electrolysis reduces costs per unit of hydrogen produced, making it more competitive with traditional energy sources.

**Benefits**

- Increased Efficiency: Modular design, mass production, and large-scale electrolysis can increase overall efficiency and reduce costs.
- Improved Scalability: These approaches enable the scalability of hydrogen production, making it suitable for various applications.
- Accelerated Transition: By reducing costs and increasing efficiency, these approaches can accelerate the transition to a hydrogen-based energy economy.

**Cost-Effectiveness**

- Reduced Energy Consumption: Improving energy efficiency reduces the cost of hydrogen production.
- Economies of Scale: Large-scale production reduces costs per unit.
- Competitive with Fossil Fuels: Hydrogen production costs are decreasing, making it more competitive with fossil fuels.

**Challenges**

- High Initial Investment: Setting up a large-scale electrolyzer facility requires significant initial investment.
- Energy Consumption: Hydrogen production through electrolysis requires significant amounts of energy.
- Material Costs: The cost of materials used in electrolyzers, such as platinum or iridium, can be high.

### 6.3.2 Future directions

- Advanced Materials: Researching alternative materials with lower costs and improved performance.
- Efficiency Improvements: Continuing to improve energy efficiency and reduce energy consumption.
- Integration with Renewable Energy: Integrating hydrogen production with renewable energy sources to reduce costs and increase sustainability.

### 6.3.3 Integration with renewable energy sources

Wind power, solar power, and water power are technologies that can be used as the main sources of renewable energy so that the target of decarbonization in the energy sector can be achieved. However, when compared with conventional power plants, they have a significant difference. The share of renewable energy has made a difference and posed various challenges, especially in the power generation system. The reliability of the power system can achieve the decarbonization target but this objective often collides with several challenges and failures, such that the achievement of the target is very vulnerable; even so, the challenges and technological solutions are still very rarely discussed in the literature. This study carried out specific investigations on various technological solutions and challenges, especially in the power system domain. The results of the review of the solution matrix and the interrelated technological challenges are the most important parts to be developed in the future. Developing a matrix with various renewable technology solutions can help solve RE challenges. The potential of the developed technological solutions is expected to be able to help and prioritize them especially cost-effective energy. In addition, technology solutions that are identified in groups can help reduce certain challenges. The categories developed in this study are used to assist in determining the specific needs and increasing transparency of the renewable energy integration process in the future.

Specifically, the review in this research is to study the integration of variable renewable energy (VRE) systems that are connected with modern power systems and technology to overcome challenges. Besides, the need for power system technology in increasing VRE market share with complex integration is also discussed. The collection of challenges undertaken in this study was drawn from a variety of literature relating to technology solutions in integrating VRE. The challenges developed can consistently integrate VRE, which is the root problem of this analysis. The results of this analysis are supplemented by data from interviews with experts who have helped in investigations related to technology solutions and their challenges.

The list of challenges as explained earlier will be refined with literature and reviews relating to challenges according to their level and challenges related to overall causality. The relationship between the challenges and the technological solutions analyzed shows that the two are mutually exclusive. Therefore, the analysis methodology applied in this study aims to find out the causes, management tools, and the standard tools. Besides, the purpose of applying this method is to identify the main causes of certain problems and events as the root causes (Kibria et al., 1995; Conway & Bai, 1986; Wendt et al., 2005). Categories with failure modes on micro-networks that can be used to find various errors and resolutions are found in the method (Andersen & Fagerhaug, 2002; Douglas-Smith et al., 2020; Shinsel et al., 2020; Dileep, 2020). The method is applied to identify the increasing symptoms of penetration of VRE collected from various literature. The symptoms analyzed represent various effects that have adverse effects on performance characteristics for the power system. The identification of challenges found in the literature is then mapped based on the symptoms of each specific VRE characteristic that is the root of the problem.

### 6.3.4 Major challenges

In the context of the dual carbon goals, we must promote the strategic significance of hydrogen energy, analyze the technological progress in the production, storage, and application of hydrogen energy, examine the associated challenges and opportunities, and actively promote the realization of the Hydrogen China strategy.

At present, the development of the global hydrogen industry is in its infancy, and hydrogen still accounts for a very low percentage of end-use energy consumption. Although the countries with plans for using hydrogen energy possess 75% of the aggregate global economy, the hydrogen industry has not formed a complete chain as a joint force to facilitate the production and routine activities, due to several factors.

1) The key materials and equipment components for hydrogen production are technically demanding, complex, and costly, and the technology gap is wide between countries and sectors. Some key technologies are still monopolized by foreign countries. For example, for the core component of hydrogen production by PEM, the Nafionperfluorosulfonic acid membrane produced by DuPont has more than 90% of the global market share. Therefore, it is difficult to make breakthroughs in technology.
2) Hydrogen production by water electrolysis is the most promising way to realize the mass production of green hydrogen, but its cost is still too high, mainly because of the price of electricity. Therefore, it still cannot completely replace hydrogen production from fossil fuels in the short term. At present, hydrogen production using water electrolysis accounts for less than 2% of the total hydrogen production in China, and we still need to rely on hydrogen production from coal to guarantee the supply of hydrogen energy in the near future.

3) Owing to constraints in the distribution of renewable energy resources in China, there is a large mismatch between hydrogen production and hydrogen use in both time and space, and a connected hydrogen storage and transportation network has not yet been formed. The northwest region of China has abundant wind and sunlight resources, but the demand for hydrogen is mainly in the economically developed and densely populated southeast region.
4) Compared with the petrochemical energy industry, the hydrogen industry is still emerging, and there is no overall plan for the required infrastructure and facilities. For example, there is a severe shortage of supporting facilities such as urban hydrogen refueling stations, hydrogen transmission pipelines, and purification systems for industrial by-product hydrogen. Therefore, it is difficult to effectively integrate the incomplete upstream and downstream components needed for hydrogen energy to form an industrial chain.
5) The current demand for hydrogen is low and dominated by its use in hydrogen fuel cells and energy carrier vehicles, which are currently immature technologies and small in scale. Although hydrogen is an energy carrier, the demand for it has not been fully developed in traditional energy-intensive industries or in new hydrogen energy application scenarios.
6) The technical standards related to hydrogen energy have not been fully developed; few guidelines regarding hydrogen quality, storage, and transportation, hydrogen refueling stations, and safety are available. To regulate the development of the hydrogen energy market, it is urgent to formulate sound international, national, and industrial standards regarding the technical processes, devices, equipment, production, and operation conditions of new hydrogen energy areas, such as hydrogen production from renewable energy, liquid hydrogen storage, and industrial green hydrogen.

### 6.3.5 Future prospects

Prospects for hydrogen energy industrial chain development: As with the current development of the natural gas industry, China is building a hydrogen energy industrial system that will consist of hydrogen production, storage, transportation, refueling, and application (Figure 6.2) (Caineng et al., 2022). Below, we describe the prospects of the future development of hydrogen in view of the technological challenges, economic efficiency, and the layout and planning of the hydrogen industry in relation to the various links of the industry chain.

1) Hydrogen is a secondary energy source. In terms of hydrogen production, to achieve zero carbon emissions, the development of hydrogen will inevitably rely on technological breakthroughs in renewable

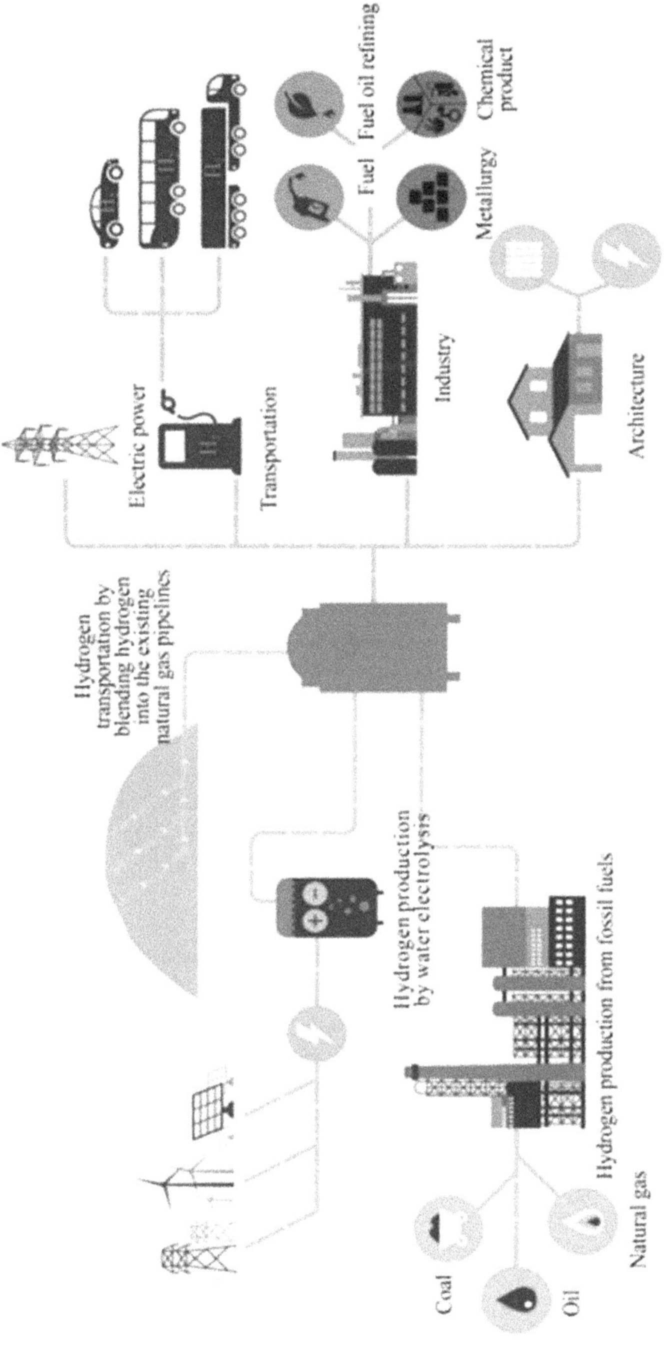


*Figure 6.2* Diagram of the hydrogen industry chain with green hydrogen as the core.

**(Reproduced with permission from Caineng et al. (2022). Copyright @ Elsevier 2022.)**

energy sources such as solar and wind power. The economic advantages of green hydrogen can be realized only by reducing the costs of both electricity and equipment. Compared with Japan, Korea, and other countries, China has a larger territory, including the vast deserts, grasslands, and offshore resources, which can provide abundant renewable energy resources such as solar, wind, and tidal energy. Therefore, we have inherent advantages in developing green hydrogen, which can accelerate the realization of the Hydrogen China strategy.

2) For long-distance transportation, hydrogen will be mainly transported together with natural gas through existing gas pipelines or by dedicated hydrogen pipelines to be built. For short and medium distances, various storage and transportation technologies will be combined and developed according to local conditions. With technological breakthroughs in hydrogen production, the concept of a green hydrogen economy will be realized through the large-scale use of hydrogen transmission networks for downstream applications such as industry, transportation, and construction activities.
3) In terms of hydrogen application, with the gradual increase in industrial and technological development, hydrogen fuel cells are highly expected to be the primary change in the transportation field. In the chemical industry that uses hydrogen for oil refining, ammonia synthesis, methanol production, and steelmaking, green hydrogen will gradually replace gray hydrogen. In other traditional energy-intensive industries, hydrogen energy will also replace fossil fuels and act as an energy carrier for the energy supply. In building construction, the distributed combined supply of green hydrogen energy for cooling, heating, and power will be an important way to conserve energy and reduce emissions. Meanwhile, additional hydrogen energy application scenarios will be gradually developed.

## 6.4 APPLICATION OF TRANSITION METAL PHOSPHIDES (TMPs) IN PHOTOCATALYTIC $H_2$ EVOLUTION

In order to boost the activity of photocatalytic $H_2$ generation for industrial and large-scale applications, a large number of researchers have been devoted to developing new photocatalysts and enhancing existing photocatalytic systems. Some effective measures have been taken such as: anion and cation doping, (Hlalele et al., 2019; Erdiwansyah et al., 2021; Zou et al., 2021; Kumaravel et al., 2019) constructing heterojunctions, (Reda et al., 2020; Du et al., 2020; Wang et al., 2019; Li et al., 2019) dye sensitization, (Gao et al., 2019) and loading co-catalysts. ((Xiao & Jiang, 2019. and Low et al., 2017) Noble metal co-catalysts, such as Pt, Au, Pd, and Ru, (Yu et al., 2019; Guo et al., 2020) have been intensively researched in photocatalytic $H_2$ evolution. However, the high-price and scarcity limit the large-scale application of rare precious metals. TMPs, as co-catalysts, can cooperate with the host catalyst to enhance the efficiency of the photocatalytic system because of their excellent metallic conductivity and the suitable Fermi level position. In addition, the low cost and easy access also endow TMP an outstanding co-catalyst. (Chen et al., 2021) TMPs, including $Ni_2P$, WP, FeP, $Co_2P$, etc., are often employed in combination with metal sulfides, metal oxides, g-$C_3N_4$ (Liu et al., 2020), and other photocatalytic materials to achieve better photocatalytic effect.

## 6.5 APPLICATIONS OF TMPS IN ELECTROCHEMICAL HER

Strategies for enhancing electrochemical performance: Although the intrinsic structures of TMPs meet the criteria of outstanding electrocatalysts, there still exist many methods that could further improve their HER performance by structural modification.

1. **Fine surface design:** With the electrolysis going on, gas bubbles will gradually generate from the surface of the electrode. This phenomenon should be positive. However, plenty of bubbles adsorbed on the catalytic surface would impede the contact between electrocatalysts and the electrolyte, which would greatly decrease the electrochemical active area, hence lowering the activity of the electrocatalysts. To avoid such a negative effect, it is of great significance to develop electrocatalysts with hydrophilic and aerophobic surfaces from which bubbles are detached easily. In this regard, an MoS2 nanosheet array-like film with a "superaerophobic" surface is reported. The "superaerophobic" MoS2 film can greatly promote the HER electrocatalytic performance, which even surpasses Pt/C at high current density (Wang et al., 2020). Inspired by this report, our group synthesized vertical Ni2P nanosheets grown on Ni foam as high-performance HER electrocatalysts. The vertical Ni2P nanosheets have been proven to be aerophobic by providing a rapid renewal of small gas bubbles and a fixed working area, thus increasing the efficiency of HER (Zhao et al., 2021). Therefore, the design of the electrode with an aerophobic surface is very important for HER and other gas evolution reactions.
2. **Coupled with carbon materials:** Electrocatalytic efficiency is also affected by the electrical conductivity and dispersity of the catalyst. An electrocatalyst with poor electrical conductivity results in a voltage drop across the electrode, producing an extra overpotential to lower the apparent activity of the catalysts and consume more energy. On the other hand, excellent dispersity of electrocatalysts allows full use of active sites on catalysts to participate in electrode reactions to improve the electrocatalytic efficiency. Compounding TMPs with carbon materials like carbon nanotubes (CNTs), graphite, graphene,

and porous carbon can not only increase the dispersity of the active components due to their huge surface area, but also improve the conductivity of the hybrid catalysts (Shang et al., 2009). Not only so, the coupled carbon materials can also modulate the electron density and the distribution of electronic potential in the hybrid materials, thus enhancing their HER activity and stability (Zhou et al., 2014). Taking CoP as an example, a nanohybrid which consists of CNTs decorated with CoP nanocrystals exhibits greatly improved HER activity than with CoP nanocrystals only (Zheng et al., 2015). In a word, compositing catalysts with carbon materials can make full use of the excellent conductivity and favorable dispersity of carbon materials, and benefit improving the electrocatalytic activity.

3. **Doping with other elements:** Doping TMPs with other elements is another way to improve their catalytic activity. Jaramillo and coworkers successfully doped S into MoP to form molybdenum phosphosulfide (MoP|S) on the surface of MoP by a post-sulfidation process. (Zhao et al., 2004). At the same catalyst loading, reaching a hydrogen-generating current density of 10 mA cm2 required the overpotentials of 86 and 117 mV for MoP|S and MoP, respectively, indicating that MoP with a phosphosulfide surface substantially improved the activity over that of pure MoP. The reason might be that S and P can tune each other's electronic properties to produce an active catalyst phase. Jin's group investigated the structure and HER activity of a sequence of pyrite-phase nickel phosphoselenide nanomaterials through thermal conversion of $Ni(OH)_2$ nanoflakes. By simply adjusting the raw ratios of phosphorus and selenium, they acquired four different nanomaterials, respectively, i.e., $NiP_2$, Se-doped $NiP_2$, P-doped $NiSe_2$, and $NiSe_2$. The results showed that Se-doped $NiP_2$ performed the highest HER activity, providing another example of improving the HER catalytic activity by doping. Cation doping is also accessible. Tungsten-doped nickel phosphide microspheres were proved to be high-activity HER catalysts. It might be conjectured that by doping tungsten into NixP, more lattice defects would form, and many electrons would be offered to P atoms for better activity. 36 To date, exact reasons for the high activity of the doped materials are not entirely clear yet, inspiring researchers to elucidate their origins by means of further characterization and DFT calculations.

## 6.6 NEW TRENDS IN ELECTROCHEMICAL HYDROGEN GENERATION

Trends in electrochemical HER undergoing rapid development are briefly discussed in this section. Such trends are not only suitable for TMPs, but are also instructive for designing many other non-noble metal electrocatalysts.

### 6.6.1 From coating nanocrystal dispersion on planar electrodes to 3D self-supported electrodes

In recent years. Three-dimensional (3D) self-supported electrodes are becoming more and more popular in electrochemical research studies. Compared with traditional electrodes coated on 2D substrates by spin-coating, dip-coating, and sputtering, 3D self-supported electrodes can always exhibit much better performance than the planar ones. (Yang et al., 2019). Common 3D current collectors like carbon cloth (CC), Cu foam and Ni foam possess excellent conductivity, robust skeleton, and large surface area, and thus effectively increase the contact area between catalyst and electrolyte. To combine electrocatalysts and substrates, polymer binders are always necessary. However, the polymer binders will increase the series resistance, block active sites and inhibit diffusion, resulting in a weakened catalytic activity. If the electrocatalysts grow directly on the 3D structures to form 3D self-supported electrodes, polymer binders are free of use to avoid the above disadvantages. For example, Sun and coworkers have successfully developed a series of well-designed TMPs (FeP, WP, and CoP) with different structures on CC as HER electrocatalysts with extremely enhanced HER activity. Generally, the use of 3D self-supported electrodes can greatly increase the current density by nearly 5–6 times at the same overpotential.

### 6.6.2 Combining experiment and theory

With the rapid development of computer technology, theoretical calculation is becoming more and more powerful to predict promising catalysts, ascertain reactive sites and get insight into the reaction mechanism. Extraordinary progress in DFT has created infinite possibilities in the computer-aided design of catalysts with the most favorable activity (Sun et al., 2021). As mentioned above, Liu and Rodriguez predicted that $Ni_2P(001)$ had comparable HER activity with the (NiFe) hydrogenase by DFT calculations, which was proven by Schaak and coworkers later. However, Hu et al. recently found that if the HER completely occurred on Ni2P(001), the theoretical overpotential should be at least as high as 0.31 V, which was inconsistent with the latest reports, implying the existence of other widely available active sites. To verify this conjecture, they synthesized uniform hexagonal $Ni_2P$ nanowires without exposing (0001) facets as calculated models, and found that either (1%120) or (11 % 21) % facets were the active sites as well (Kaladevi et al., 2020). Even so, various active sites or facets of $Ni_2P$ are still to be studied. Wang's group also employed DFT calculations to elucidate the "hydrogen deliverer" behavior of the P-terminated surface on (001)-MoP as mentioned. Not only that, when Mo3P is calculated in the same way, the corresponding hydrogen binding energy was positive even with the initial two hydrogen atoms adsorbed, substantially explaining the origin of the poor HER activity of $Mo_3P$. From acid to all pH. The water splitting reaction can be considered as a combination of two half reactions: HER and oxygen evolution reaction (OER). In consideration of thermodynamic convenience and potential in proton exchange membrane fuel cells, most HER catalysts are studied and functioned well in acidic media. However, almost all the best OER catalysts work well only in neutral or basic media (Jin et al., 2014). In order to achieve overall water splitting, it is of great importance to find HER catalysts that work well in a wide pH range so that they can have the chance to cooperate with OER catalysts. The first attempt to use the TMPs as HER catalysts in alkaline solution was made by Schaak et al. However, their $Ni_2P$ hollow nanoparticles decomposed to metal nickel quickly in 1.0 M KOH with the HER activity declining rapidly (Wang et al., 2015). The reason is not clear yet. Later reports have shown that under both strong acidic and strong alkaline conditions, not only $Ni_2P$, but also other TMPs like CoP, MoP, $Ni_5P_4$, and $WP_2$ can offer excellent catalytic performance and maintain the high performance for dozens of hours. This pH-tolerable ability makes TMPs one of the most attractive electrocatalysts for overall water splitting.

### 6.6.3 From unfunctional HER to bifunctional overall water splitting

Another idea for cost-effective overall water splitting is to design earth-abundant bifunctional electrocatalysts for both HER and OER. This concept can be accomplished by either tactfully integrating highly active HER and OER electrocatalysts, or developing electrocatalysts which can be highly active and stable for both HER and OER under the same pH conditions. In particular, TMPs can be partially oxidized under oxidizing potentials, and these oxidized intermediates (oxides and phosphates) are always highly capable of OER. In this way, the highly active HER catalysts (TMPs) and highly active OER catalysts (the oxidized intermediates) are integrated. Shalom's group explored the OER performance of Ni5P4 in alkaline media, which was even superior to metal Ni. They attributed the OER activity of Ni5P4 to the fast electrochemical formation of highly OER-active NiOOH (Zheng et al., 2015) Yoo et al. found that operating CoP nanoparticles upon anodic potential under alkaline conditions would cause unique metamorphosis, leading to efficient but irreversible performance of CoP toward OER.42 Based on XPS, X-ray absorption near edge spectroscopy (XANES), and EXAFS (X-ray absorption fine structure), they expected that the local structure of the post-OER sample was composed mainly of bis-oxo/hydroxo-bridged Co ions in discrete molecular units arranged into a phosphate-enriched amorphous network. Besides the good OER activity of cobalt oxide, the assistance of the surface phosphate group and the synergetic effect between cobalt oxides and phosphates were suggested to make a contribution to the high OER activity. It needs to be emphasized that irreversible changes occur on TMPs after OER experiments, and the post-OER TMPs are not capable of HER anymore. It is also allowed to directly combine TMPs with OER-active materials during the synthetic process. By reacting porous anodized Co oxide films with phosphorus vapor, Tour's group directly fabricated porous Co phosphide/phosphate thin films, which performed superior activity for both HER and OER with onset overpotentials of 35 mV and 220 mV for HER and OER, respectively (Xiao et al., 2014). The development of bifunctional electrocatalysts for both HER and OER is still an ongoing challenge. It deserves more efforts to design and fabricate new and efficient bifunctional electrocatalysts.

### 6.6.4 Applications in photocatalytic and photoelectrocatalytic hydrogen evolution

#### *6.6.4.1 As cocatalysts in photocatalytic and photoelectrocatalytic hydrogen evolution*

In photocatalytic water splitting process, semiconductors always have large surface overpotentials for water oxidation or reduction, which need to be compensated with extra photovoltage or the application of external bias. Electrocatalysts with high HER or OER activity have been used as cocatalysts to lower the overpotential on the surface of semiconductors (Popczun et al., 2013). The high activity of TMPs toward HER makes them potential cocatalysts for photocatalytic water splitting. Besides, TMPs are in favor of charge separation at the cocatalyst/semiconductor interface which is advantageous for the high photocatalytic efficiency, especially when a "Z-scheme" heterogeneous structure is formed. Schaak, Lewis, and co-workers have successfully synthesized FeP nanoparticles with a uniform, hollow morphology (Ledendecker et al., 2015). When immobilized on TiO2, FeP nanoparticles could act as cocatalysts to enhance the photocatalytic generation of hydrogen under UV irradiation in a neutral aqueous solution. Fu's group has also found that both Ni2P and Co2P can be used as efficient cocatalysts for hydrogen evolution in a system with CdS nanorods as photosensitizers (Ryu et al., 2015). TMPs can be used as cocatalysts in photoelectrochemical hydrogen evolution as well. Huang and coworkers demonstrated that Ni12P5 nanoparticles had efficient and stable catalytic activity for HER. Furthermore, introducing Ni12P5 nanoparticles to silicon nanowires displayed efficient hydrogen generation in the voltage range of 0.1 V to 0.4 V, and the solar power conversion efficiency of the resulting composite was even larger than that of silicon nanowires decorated with Pt particles under the illumination of a 350 W Xe lamp, whose light intensity was adjusted to 100 mW $cm^2$ (Yang et al., 2015). These wonderful studies make TMPs competent earth-abundant cocatalysts for sustainable photocatalytic and photoelectrochemical hydrogen production.

#### *6.6.4.2 As photocatalysts in hydrogen evolution*

It is also worth noting that since the P content can adjust the band gap of metal phosphides by limiting the electron delocalization, choosing metal phosphides with appropriate P content may have possibilities to achieve photocatalytic hydrogen evolution by phosphides themselves. Wu et al. demonstrated that with the assistance of Pt and triethanolamine as a cocatalyst and electron donor, respectively, semimetallic $MoP_2$ could perform actively in water reduction to generate hydrogen under visible light (Yang et al., 2015; Ran et al., 2014; Ryu et al., 2015; Callejas et al., 2014). This attempt may motivate researchers to find and design appropriate TMPs as novel photocatalysts for effective water splitting under visible light. Figure 6.3 represents industrial applications of hydrogen (Vergara et al., 2024).

### 6.6.5 Electrode materials and surface modifications

Surface modification of electrode materials is a crucial technique for enhancing their electrochemical performance, stability, and durability. By altering the surface properties of electrode materials, researchers can improve conductivity, corrosion resistance, and catalytic activity, making them suitable for applications in energy storage, sensing, and catalysis. This chapter explores various electrode materials used for surface modification and their impact on performance.

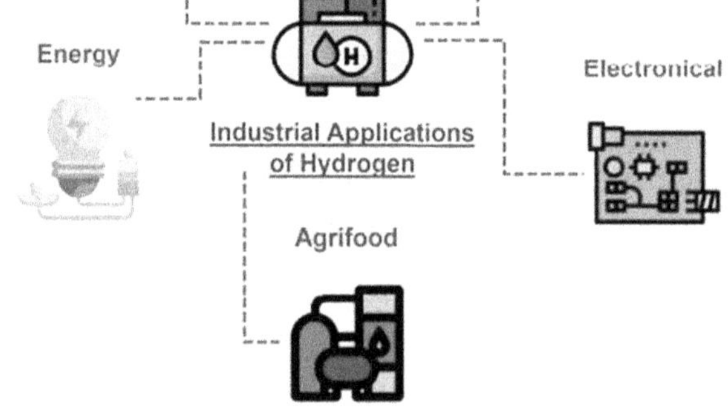


*Figure 6.3* Industrial applications of hydrogen.

**(Reproduced with permission from Ref Vergara et al. (2024). Copyright @ MDPI 2024.)**

## 6.7 TYPES OF ELECTRODE MATERIALS FOR SURFACE MODIFICATION

1. **Metal-Based Electrodes**
    - Platinum (Pt) and its alloys: Commonly used in fuel cells due to high catalytic activity and stability.
    - Nickel (Ni) and Cobalt (Co) alloys: Employed in supercapacitors and batteries for their excellent redox properties.
    - Titanium (Ti) and Stainless Steel: Used in biomedical applications due to corrosion resistance and biocompatibility.
2. **Carbon-Based Electrodes**
    - Graphene and Carbon Nanotubes (CNTs): Offer high surface area and excellent conductivity, making them suitable for energy storage devices.
    - Activated Carbon: Frequently used in supercapacitors for its porous structure and ion accessibility.
3. **Conductive Polymer Electrodes**
    - Polyaniline (PANI) and Polypyrrole (PPy): Improve charge storage capacity and enhance electrode stability.
    - Poly (3,4-ethylenedioxythiophene) (PEDOT): Provides high conductivity and mechanical flexibility.
4. **Composite Electrodes**
    - Metal Oxide Composites (e.g., $MnO_2$, $RuO_2$): Enhance capacitance and catalytic performance in energy storage applications.
    - Hybrid Materials (e.g., Ni-Co sulfides on graphene): Combine advantages of different materials for improved efficiency.

### 6.7.1 Surface modification techniques

1. Electrodeposition: Used for coating metals and alloys with nanoscale materials to enhance conductivity and stability.
2. Chemical Vapor Deposition (CVD): Applied for graphene and carbon-based modifications to increase surface area.
3. Plasma Treatment: Enhances adhesion and functionalization of surface layers for improved performance.
4. Electrochemical Etching: Increases surface roughness and active sites for better electrochemical interaction.
5. Self-Assembled Monolayers (SAMs): Used for precise molecular-level modifications in sensing applications.

### 6.7.2 Metal and alloy electrodes

Metal and alloy electrodes have gained significant attention in various fields, including energy storage, catalysis, and sensing, due to their excellent electrical conductivity, catalytic properties, and mechanical stability.

Understanding their performance and characterizing their properties are crucial for optimizing their applications in devices such as batteries, supercapacitors, fuel cells, and electrochemical sensors. This chapter explores the methodologies for evaluating the performance of metal and alloy electrodes and the characterization techniques used to analyze their structural, chemical, and electrochemical properties (Vergara et al., 2024).

### 6.7.3 Performance evaluation of metal and alloy electrodes

1. **Electrochemical Performance**
   The electrochemical performance of metal and alloy electrodes is evaluated using techniques such as:
   - **Cyclic Voltammetry (CV):** CV measures the electrochemical activity and stability of electrodes by recording current as a function of applied voltage. The peak current and potential provide insights into the redox reactions, charge transfer kinetics, and capacitance.
   - **Galvanostatic Charge-Discharge (GCD):** GCD evaluates the energy storage capacity and cycling stability of electrodes. Parameters such as specific capacitance, energy density, and power density are calculated from the charge-discharge curves.
   - **Electrochemical Impedance Spectroscopy (EIS):** EIS analyzes the resistance components of an electrode, including charge transfer resistance ($R_1\Omega$) and solution resistance ($R_1\Omega$). It provides information about ion diffusion and electrode-electrolyte interface properties.
2. **Catalytic Activity**
   Metal and alloy electrodes often serve as catalysts in electrochemical reactions, such as the OER, HER, and oxygen reduction reaction (ORR). Performance is evaluated using (Wu et al., 2015; Bayer, 2004; Davis, 2001; Holmberg & Matthews, 2009; Aliofkhazraei, 2014; Musil, 2000; Smith, 2020):
   - **Overpotential Measurements:** The overpotential required to drive a reaction at a specific current density indicates the catalytic efficiency.
   - **Turnover Frequency (TOF):** TOF measures the number of reaction events per active site per second, providing insights into the intrinsic activity of the material.
   - **Tafel Slope Analysis:** The Tafel slope reveals the reaction mechanism and the rate-determining step in the catalytic process.
3. **Mechanical and Thermal Stability**
   The mechanical and thermal stability of metal and alloy electrodes is critical for long-term operation in harsh environments. Stress-strain testing, thermal cycling, and microstructural analysis are employed to evaluate these properties.

### 6.7.4 Characterization techniques for metal and alloy electrodes

1. **Structural Characterization**
   - **X-ray Diffraction (XRD):** XRD identifies the crystalline structure, phase composition, and crystallite size of the electrodes. It is particularly useful for detecting phase transitions during electrochemical cycling.
   - **Scanning Electron Microscopy (SEM):** SEM provides high-resolution images of the electrode surface morphology, revealing features such as grain boundaries, cracks, and pores.
   - **Transmission Electron Microscopy (TEM):** TEM offers insights into the nanostructure and lattice defects in electrodes. Advanced techniques like high-resolution TEM (HRTEM) can resolve atomic-scale features.
2. **Chemical Composition Analysis**
   - **Energy Dispersive X-ray Spectroscopy (EDS):** EDS is often coupled with SEM or TEM to analyze the elemental composition of electrodes.
   - **X-ray Photoelectron Spectroscopy (XPS):** XPS determines the oxidation states and chemical environment of elements on the electrode surface, providing insights into the surface chemistry.
   - **Inductively Coupled Plasma Optical Emission Spectroscopy (ICP-OES):** ICP-OES quantitatively analyzes the elemental composition of dissolved samples, often used after electrochemical testing.
3. **Surface and Interface Analysis**
   - **Atomic Force Microscopy (AFM):** AFM measures surface roughness and topography with nanometer precision, enabling the study of changes in surface structure after electrochemical cycling.
   - **Contact Angle Measurements:** Contact angle analysis evaluates the wettability of the electrode surface, which influences electrolyte interaction and ion transport.
   - **Fourier Transform Infrared Spectroscopy (FTIR):** FTIR identifies functional groups and chemical bonds on the electrode surface, providing insights into surface modifications during operation.
4. **Electrochemical Active Surface Area (ECSA) Analysis**
   ECSA quantifies the active surface area participating in electrochemical reactions. It is calculated using CV by measuring the double-layer capacitance or redox peak currents.

### 6.7.5 Case studies in metal and alloy electrodes

1. Nickel-Cobalt Alloys for Supercapacitors: Nickel-cobalt alloys exhibit excellent redox activity and high specific capacitance, making them ideal for supercapacitor electrodes. Studies have shown

that incorporating graphene or conductive polymers enhances the electrochemical performance by providing additional active sites and improving conductivity.

2. Platinum-Based Electrodes for Fuel Cells: Platinum and its alloys are widely used as catalysts in proton exchange membrane fuel cells (PEMFCs). Characterization techniques such as XPS and TEM have demonstrated the role of alloying elements in enhancing catalytic activity and durability.
3. Titanium-Based Alloys for Biomedical Sensors: Titanium alloys are employed in biosensors due to their biocompatibility and corrosion resistance. Electrochemical and surface analyses have revealed their suitability for detecting biomolecules with high sensitivity.

### 6.7.6 Challenges and future directions

Despite their potential, metal and alloy electrodes face challenges such as limited availability of resources, surface degradation, and scalability issues. Future research should focus on:

1. Developing cost-effective synthesis methods.
2. Enhancing stability through surface engineering and protective coatings.
3. Exploring novel alloy compositions and nanostructures.

The performance evaluation and characterization of metal and alloy electrodes are crucial for advancing their applications in energy storage, catalysis, and sensing. By employing a combination of electrochemical, structural, and chemical analysis techniques, researchers can gain a comprehensive understanding of these materials. Innovations in material design and characterization will pave the way for more efficient and sustainable technologies.

### 6.7.7 Coatings and surface treatments performance evaluation and characterization

Coatings and surface treatments are indispensable in enhancing the performance, durability, and functionality of materials across various industries. These processes modify the surface properties of substrates to improve resistance to corrosion, wear, and environmental degradation while imparting specific functionalities such as hydrophobicity or antimicrobial activity. To ensure optimal performance, evaluating and characterizing these coatings and treatments is essential, involving a range of mechanical, chemical, and functional analyses.

### 6.7.8 Performance evaluation of coatings and surface treatments

1. Mechanical Performance
   I. **Adhesion Testing:** Adhesion strength between a coating and its substrate is a critical parameter, often measured using pull-off tests or scratch tests. These methods evaluate the bond integrity under mechanical stress.
   II. **Wear Resistance:** Wear resistance is assessed through tests like pin-on-disk, where the coating is subjected to frictional forces to determine its durability under operational conditions.
   III. **Hardness Tests:** Techniques such as Vickers or nanoindentation are used to measure the hardness of coatings, providing insights into their ability to resist deformation.

2. Corrosion Resistance
   I. **Salt Spray Testing:** This accelerated test exposes coated materials to a saline mist, simulating a corrosive environment to evaluate the protective capabilities of the coating.
   II. **Electrochemical Testing:** Techniques like potentiodynamic polarization and electrochemical impedance spectroscopy (EIS) provide a quantitative understanding of the coating's ability to prevent corrosion by analyzing parameters such as corrosion potential and resistance.

3. Functional Performance
   I. **Thermal Stability:** Thermogravimetric analysis (TGA) and differential scanning calorimetry (DSC) are used to determine the thermal resistance of coatings, particularly for high-temperature applications.
   II. **Hydrophobicity and Surface Energy:** Contact angle measurements assess surface wettability, which is critical for applications requiring water or oil resistance.
   III. **Antimicrobial Activity:** Biological assays evaluate the effectiveness of coatings in inhibiting microbial growth, particularly in biomedical and food industries.

### 6.7.9 Characterization techniques for coatings and surface treatments

1. Structural Characterization
   I. **X-ray Diffraction (XRD):** XRD identifies the crystalline phases and structural integrity of coatings, offering insights into their composition and stability.

II. **Scanning Electron Microscopy (SEM):** SEM provides detailed images of surface morphology, revealing defects, cracks, and overall quality of the coating.
III. **Transmission Electron Microscopy (TEM):** TEM offers atomic-level insights into the microstructure and nanostructure of coatings, aiding in understanding their properties.

2. **Chemical Analysis**
   I. **X-ray Photoelectron Spectroscopy (XPS):** XPS determines the chemical composition and oxidation states of elements within the coating, particularly for surface layers.
   II. **Energy Dispersive X-ray Spectroscopy (EDS):** Coupled with SEM, EDS provides elemental analysis to confirm the uniformity and composition of the coating material.
   III. **Fourier Transform Infrared Spectroscopy (FTIR):** FTIR identifies chemical bonds and functional groups, particularly useful for polymer-based or organic coatings.

3. **Surface and Interface Analysis**
   I. **Atomic Force Microscopy (AFM):** AFM measures surface roughness and topography at the nanoscale, revealing changes after surface treatments.
   II. **Nanoindentation:** This technique evaluates the mechanical properties like hardness and elastic modulus of coatings at very small scales.

4. **Electrochemical Characterization**
   I. **Cyclic Voltammetry (CV):** CV provides insights into the electrochemical activity of functional coatings, such as those used in batteries or sensors.
   II. **Electrochemical Impedance Spectroscopy (EIS):** EIS evaluates the resistance and capacitance of coated materials, critical for assessing their protective and functional properties.

### 6.7.10 Applications of coatings and surface treatments

I. **Automotive Industry:** Coatings protect against corrosion and wear, while surface treatments improve fuel efficiency and performance.
II. **Aerospace Sector:** Thermal barrier coatings enhance the durability of components exposed to high temperatures and extreme conditions.
III. **Biomedical Field:** Surface treatments improve biocompatibility, reduce friction, and provide antimicrobial properties for implants and medical devices.
IV. **Renewable Energy:** Anti-reflective coatings for solar panels and protective layers for wind turbines enhance efficiency and longevity.

### 6.7.11 Challenges and future directions

While coatings and surface treatments offer immense benefits, challenges remain in ensuring long-term performance, eco-friendliness, and cost-effectiveness. Innovations such as nanostructured coatings, self-healing layers, and sustainable materials are paving the way for next-generation applications. Advanced characterization techniques, including in situ monitoring, will further enhance the understanding and optimization of these technologies (Brown, 2021; Patel, 2022; Wang, 2019; Zhang, 2021).

The study successfully demonstrates the hydrothermal synthesis of zinc cobalt sulfide (ZCS) and reduced graphene oxide/ZCS (rZCS) nanoparticles for energy storage and hydrogen evolution applications. Comprehensive characterization using XRD, BET, XPS, FESEM, and HR-TEM confirmed the structural and morphological properties of the synthesized materials. Electrochemical evaluations revealed that the rZCS electrode exhibits exceptional catalytic activity for the HER, requiring only 132 mV overpotential to achieve a current density of 10 mA $cm^{-2}$. Additionally, the rZCS-based supercapacitor electrode displayed a high specific capacitance of 891 F $g^{-1}$ at 1 A $g^{-1}$, along with remarkable cyclic stability (99% retention after 5000 cycles). When assembled into an asymmetric supercapacitor (rZCS//rGO), the device delivered a specific capacitance of 181 F $g^{-1}$, 92% retention after 10,000 cycles, and high energy and power densities of 29 Wh $kg^{-1}$ and 2215 W $kg^{-1}$, respectively. These outstanding electrochemical performances highlight the potential of rZCS-based materials for advanced energy storage and conversion technologies (Manikandan et al., 2023). This study presents a novel method for synthesizing amorphous metal–organic frameworks (aMOFs) by amorphizing coordinately saturated crystalline MOFs (cMOFs) under physical force and alkaline conditions. The resulting aMOFs retained the structural and compositional integrity of their crystalline precursors (Ni-cMOF and NiCo-cMOFn) while exhibiting nanocrystalline features at the nanodomain scale. These materials feature a 2D sheet-like architecture with a $Ni_3(\mu_3\text{-}O)$ secondary building unit, contributing to their superior electrocatalytic performance. When tested for overall seawater splitting, the aMOFs demonstrated exceptional oxygen selectivity and stability, maintaining performance for over 24 hours at 340 mA—a record for aMOF-based systems and the first such application in seawater. The study suggests that this direct amorphization strategy could pave the way for developing advanced aMOFs for electrocatalytic applications (Deepa et al., 2024; Krishnamurthy et al., 2024).

In a recent study published in *Physica Scripta*, researchers developed a novel bifunctional electrode material composed of zinc manganese cobalt sulfide (ZMCS) and its reduced graphene oxide composite (rZMCS) for efficient energy storage and conversion applications. The team employed a hydrothermal synthesis route to create these materials, which were thoroughly characterized using advanced techniques including XRD, XPS,

FESEM, TEM, elemental mapping, and BET analysis. The composite material demonstrated exceptional electrocatalytic performance for overall water splitting, with rZMCS exhibiting remarkably low overpotentials of 170 mV for the HER and 230 mV for the OER at a current density of 10 mA $cm^{-2}$ in alkaline medium. Moreover, the electrode maintained stable operation for 15 hours during continuous water splitting. For supercapacitor applications, the rZMCS electrode showed a high specific capacitance of 1167 F $g^{-1}$ at 1 A $g^{-1}$ and outstanding cycling stability, retaining 98% of its initial capacitance after 5000 cycles. When configured into an asymmetric supercapacitor device with rGO, the system delivered impressive energy and power densities of 47.6 Wh $kg^{-1}$ and 925 W $kg^{-1}$, respectively, while maintaining 86% capacity retention over 10,000 cycles. These findings highlight the great potential of the rZMCS/rGO composite as a multifunctional material for next-generation energy storage and conversion technologies, offering both high performance and durability for practical applications (Manikandan et al., 2024).

MOFs have emerged as highly promising materials for photocatalytic hydrogen ($H_2$) production due to their unique structural and functional properties. Composed of metal nodes connected by organic linkers, MOFs possess a highly porous and tunable architecture, which allows for extensive functionalization to enhance their photocatalytic performance. Their large surface area, ability to absorb visible light, and efficient charge separation mechanisms make them ideal candidates for solar-driven $H_2$ generation. This chapter explores the fundamental mechanisms of MOF-based photocatalysis, detailing the various synthetic approaches used to tailor their properties for optimal activity. Additionally, it highlights recent advancements in MOF design, including modifications to improve light absorption and minimize charge recombination. The discussion also addresses the key challenges in this field, such as stability and efficiency limitations, while presenting strategies like heterostructure formation, doping, and co-catalyst integration to overcome these barriers and enhance photocatalytic $H_2$ production. Overall, MOFs demonstrate significant potential as next-generation photocatalysts for sustainable energy applications.

## 6.8 CONCLUSION

In conclusion, alkaline water electrolysis stands as a well-established and promising technology for large-scale green hydrogen production, playing a pivotal role in the global transition toward sustainable energy systems. As hydrogen emerges as a key energy carrier for the future, the development of efficient and cost-effective electrocatalysts for the alkaline HER is critical. This chapter highlights significant advancements in non-precious materials, metals, alloys, and high-surface-area materials, which have demonstrated enhanced catalytic performance. Key factors influencing catalyst effectiveness, such as composition, metal ions, and derived materials, have been thoroughly analyzed, supported by visual aids to facilitate understanding. Moving forward, it is essential to focus on fundamental material attributes when designing new catalysts, ensuring they meet the demands of scalability, efficiency, and sustainability. By addressing these considerations, alkaline water electrolysis can further solidify its position as a cornerstone technology for renewable energy conversion and storage, driving the transition to a greener and more sustainable industrial and societal framework.

## REFERENCES


Abdalla, A. M., Hossain, S., Nisfindy, O. B., Azad, A. T., Dawood, M., & Azad, A. K. (2018). Hydrogen production, storage, transportation and key challenges with applications: A review. *Energy Conversion and Management*, *165*, 602–627. https://doi.org/10.1016/j.enconman.2018.03.088

Aliofkhazraei, M. (2014). *Anti-corrosion nanocoatings: A handbook*. CRC Press.

Andersen, B., & Fagerhaug, T. (2002). Root cause analysis: Simplified tools and techniques. *The Journal for Healthcare Quality (JHQ)*, *24*(3), 46–47. https://doi.org/10.1097/01445442-200205000-00012

Bayer, R. G. (2004). *Mechanical wear fundamentals and testing*. ASTM International.

Binz, C., Gosens, J., Hansen, T., & Hansen, U. E. (2017). The politics of energy system transformation—Explaining coal phase-out in the UK, Germany and the Netherlands. *World Development*, *96*, 418–429. https://doi.org/10.1016/j.worlddev.2017.03.021

Bockris, J. O. M., & Potter, E. C. (1952). The mechanism of hydrogen evolution at nickel cathodes in aqueous solutions. *The Journal of Chemical Physics*, *20*(4), 614–628. https://doi.org/10.1063/1.1700493

Bongaarts, J., & O'Neill, B. C. (2018). Global warming policy: Is population left out in the cold? *Science*, *361*(6403), 650–652. https://doi.org/10.1126/science.aat8680

Brown, A. (2021). Nanostructured electrode materials for energy storage and catalysis. *Materials Science Review*, *30*(2), 245–260.

Caineng, Z., Li, J., Zhang, X., Jin, X., Xiong, B., Yu, H., Liu, X., Wang, S., Li, Y., Zhang, L., Miao, S., Zheng, D., Zhou, H., Song, J., & Pan, S. (2022). Industrial status, technological progress, challenges, and prospects of hydrogen energy. *Natural Gas Industry B*, *9*, 427–447. https://doi.org/10.1016/j.ngib.2022.04.006

Callejas, J. F., McEnaney, J. M., Read, C. G., Crompton, J. C., Biacchi, A. J., Popczun, E. J., Gordon, T. R., Lewis, N. S., & Schaak, R. E. (2014). Electrocatalytic and photocatalytic hydrogen production from acidic and neutral-pH aqueous solutions using iron phosphide nanoparticles. *ACS Nano*, *8*(11), 11101–11107. https://doi.org/10.1021/nn5048553

Carmo, M., Fritz, D. L., Mergel, J., & Stolten, D. (2013). A comprehensive review on PEM water electrolysis. *International Journal of Hydrogen Energy*, *38*(12), 4901–4934. https://doi.org/10.1016/j.ijhydene.2013.01.151

Chen, C., Zhou, X., Guo, R., & Li, Y. (2021). Tuning the interfacial charge transfer of S-scheme heterojunctions for efficient photocatalysis. *Chemical Engineering Journal*, *418*, 129431. https://doi.org/10.1016/j.cej.2021.129431

Chen, J., Ni, J., Xu, H., He, G., & Chen, H. (2024). Advanced materials for hydrogen storage. *Journal of Alloys and Compounds*, *1004*, 175883.

Cheng, Q., & Yi, H. (2017). Comprehensive review of hydrogen energy and storage systems: Current status and future prospects. *Renewable and Sustainable Energy Reviews*, *67*, 683–691. https://doi.org/10.1016/j.rser.2016.09.043

Chu, S., & Majumdar, A. (2012). Opportunities and challenges for a sustainable energy future. *Nature*, *488*, 294–303. https://doi.org/10.1038/nature11475

Conway, B. E., & Bai, L. (1986). Determination of adsorption of OPD H species in the cathodic hydrogen evolution reaction at Pt in relation to electrocatalysis. *Journal of Electroanalytical Chemistry and Interfacial Electrochemistry*, *198*(1), 149–175. https://doi.org/10.1016/0022-0728(86)90033-1

Conway, B. E., & Tilak, B. V. (2002). Interfacial processes involving electrocatalytic evolution and oxidation of $H_2$, and the role of chemisorbed H. *Electrochimica Acta*, *47*(22), 3571–3594. https://doi.org/10.1016/S0013-4686(02)00329-8

Crabtree, G. W., Dresselhaus, M. S., & Buchanan, M. V. (2004). The hydrogen economy. *Physics Today*, *57*(12), 39–44. https://doi.org/10.1063/1.1878333

Davis, J. R. (2001). *Surface engineering for corrosion and wear resistance*. ASM International.

Deepa, K., Shalini, A., Jaisankar, V., Dixit, S., Devanesan, S., Selvankumar, T., Song, K. S., & Lo, H.-M. (2024). Development of biodegradable polymer composites for sustainable packaging solutions. *Polymer for Advanced Technologies*, *36*(1).

Dileep, G. J. R. E. (2020). A survey on smart grid technologies and applications. *Renewable Energy*, *146*, 2589–2625. https://doi.org/10.1016/j.renene.2019.08.092

Douglas-Smith, D., Iwanaga, T., Croke, B. F. W., & Jakeman, A. J. (2020). Certain trends in uncertainty and sensitivity analysis: An overview of software tools and techniques. *Environmental Modelling & Software*, *124*, 104588. https://doi.org/10.1016/j.envsoft.2019.104588

Dresselhaus, M. S., & Thomas, I. L. (2001). Alternative energy technologies. *Nature*, *414*(6861), 332–337. https://doi.org/10.1038/35104599

Du, C., Yan, B., Lin, Z., & Yang, G. (2020). Enhanced carrier separation and increased electron density in 2D heavily N-doped $ZnIn_2S_4$ for photocatalytic hydrogen production. *Journal of Materials Chemistry A*, *8*(1), 207–217. https://doi.org/10.1039/C9TA11318E

Durovič, M., Hnát, J., & Bouzek, K. (2021). Electrocatalysts for the hydrogen evolution reaction in alkaline and neutral media: A comparative review. *Journal of Power Sources*, *493*, 229708. https://doi.org/10.1016/j.jpowsour.2021.229708

Erdiwansyah, Mahidin, Husin, H., Nasaruddin, Zaki, M., & Muhibbuddin. (2021). A critical review of the integration of renewable energy sources with various technologies. *Protection and Control of Modern Power Systems*, *6*(1), 3. https://doi.org/10.1186/s41601-021-00181-3

Ferriday, T. B., Middleton, P. H., & Kolhe, M. L. (2021). Review of the hydrogen evolution reaction—A basic approach. *Energies*, *14*(24), 8535. https://doi.org/10.3390/en14248535

Gao, H., Cao, R., Xu, X., Zhang, S., Yongshun, H., Yang, H., Deng, X., & Li, J. (2019). Construction of dual defect mediated Z-scheme photocatalysts for enhanced photocatalytic hydrogen evolution. *Applied Catalysis B: Environmental*, *245*, 399–409. https://doi.org/10.1016/j.apcatb.2019.01.004

Guo, F., Peng, L., Yang, Y., Li, Y., Wang, Y., & Yu, H. (2020). A review on sustainable photocatalytic-based hydrogen production from water splitting: Challenges and strategies toward high efficiency. *Applied Energy*, *264*, 114752. https://doi.org/10.1016/j.apenergy.2020.114752

Gurushankar, K., Karuppaiya, C., Tari, V., & Kannan, K. (2024). Exploring MXenes and MBenes in metal sulfur and selenium battery technologies. In K. Kannan & V. Tari (Eds.), *Novel energy storage and conversion technologies for two-dimensional MXenes and MBenes*. IGI Global. https://doi.org/10.4018/979-8-3693-9546-2.ch005

Hlalele, T. S., Sun, Y., & Wang, Z. (2019). Faults classification and identification on smart grid: Part-A status review. *Procedia Manufacturing*, *35*, 601–606. https://doi.org/10.1016/j.promfg.2019.05.085

Holmberg, K., & Matthews, A. (2009). *Coatings tribology* (2nd ed.). Elsevier.

Huhta, K. (2022). The role of hydrogen in the global energy transition. *Global Environmental Change*, *73*, 102454.

Jaksic, J. M., Vojnović, M. V., & Krstajić, N. V. (2000). Kinetic analysis of hydrogen evolution at Ni–Mo alloy electrodes. *Electrochimica Acta*, *45*(25–26), 4151–4158. https://doi.org/10.1016/S0013-4686(00)00549-1

Jin, Z., Li, P., Huang, X., Zeng, G., Jin, Y., Zheng, B., & Xiao, D. (2014). Three-dimensional amorphous tungsten-doped nickel phosphide microsphere as an efficient electrocatalyst for hydrogen evolution. *Journal of Materials Chemistry A*, 2(43), 18593–18599.

Kaladevi, G., Wilson, P., & Pandian, K. (2020). Simultaneous and selective electrochemical detection of sulfite and nitrite in water sources using homogeneously dispersed Ag nanoparticles over PANI/rGO nanocomposite. *Journal of the Electrochemical Society*, *167*(2), 027514. https://doi.org/10.1149/1945-7111/ab6973

Kannan, K., & Tari, V. (Eds.). (2025). *Novel energy storage and conversion technologies for two-dimensional MXenes and MBenes*. IGI Global.

Khalil, M., Kadja, G. T. M., Nugroho, F. A. A., Sutanto, L. G., Jiwanti, P. K., Abdi, F. F., Hussin, F., & Aroua, M. K. (2024). A comprehensive review on hydrogen production from renewable resources. *Renewable and Sustainable Energy Reviews*, *206*, 114869. https://doi.org/10.1016/j.rser.2024.114869

Kibria, M. F., Mridha, M. S., & Khan, A. H. (1995). Electrochemical studies of a nickel electrode for the hydrogen evolution reaction. *International Journal of Hydrogen Energy*, *20*(6), 435–440.

Konieczny, A., Mondal, K., Wiltowski, T., & Dydo, P. (2008). Catalyst development for thermocatalytic decomposition of methane to hydrogen. *International Journal of Hydrogen Energy*, *33*(1), 264–272. https://doi.org/10.1016/j.ijhydene.2007.07.054

Koper, M. T. M. (2011). Thermodynamic theory of multi-electron transfer reactions: Implications for electrocatalysis. *Journal of Electroanalytical Chemistry*, *660*(2), 254–260. https://doi.org/10.1016/j.jelechem.2010.12.017

Krishnamurthy, M., Kumar, N. D., & Swaminathan, M. (2024). Review on hierarchically porous BEA and ZSM-5 zeolites and its catalytic applications. *ES Materials & Manufacturing*, *24*, 1151. https://doi.org/10.30919/esmm1151

Kumaravel, V., Mathew, S., Bartlett, J., & Pillai, S. C. (2019). Photocatalytic hydrogen production using metal doped $TiO_2$: A review of recent advances. *Applied Catalysis B: Environmental*, *244*, 1021–1061. https://doi.org/10.1016/j.apcatb.2018.11.080

Ledendecker, M., Krick Calderon, S., Papp, C., Steinrück, H.-P., Antonietti, M., & Shalom, M. (2015). *Angewandte Chemie International Edition*, *54*(44), 12361–12365.

Li, Y., Yin, Z., Ji, G., Liang, Z., Xue, Y., Guo, Y., Tian, J., Wang, X., & Cui, H. (2019). 2D/2D/2D heterojunction of $Ti_3C_2$ MXene/$MoS_2$ nanosheets/$TiO_2$ nanosheets with exposed (001) facets toward enhanced photocatalytic hydrogen production activity. *Applied Catalysis B: Environmental*, *246*, 12–20. https://doi.org/10.1016/j.apcatb.2019.01.051

Lim, J., Park, D., Jeon, S. S., Roh, C.-W., Choi, J., Yoon, D., Park, M., Jung, H., & Lee, H. (2018). Hierarchical PtNi nanostructures with high-performance oxygen reduction activity and durability. *Advanced Functional Materials*, *28*, 1704796. https://doi.org/10.1002/adfm.201704796

Liu, J., Huang, Z. Y., Fan, M., Yang, J. H., Xiao, J., & Wang, Y. (2022). Coordination structures design of single-atom catalysts for efficient electrochemical energy conversion. *Nano Energy*, *104*, 107915. https://doi.org/10.1016/j.nanoen.2022.107915

Liu, Y., Chen, M., Li, C., Huang, W., Zhao, C., & Liu, S. (2020). Review on S-scheme photocatalysts. *Chinese Journal of Catalysis*, *41*(10), 1440–1455.

Low, J., Jiang, C., Cheng, B., & Yu, J. (2017). A review of direct Z-scheme photocatalysts. *Small Methods*, *1*(5), 1700080. https://doi.org/10.1002/smtd.201700080

Manikandan, M., Karthigeyan, K. A., Papanasam, E., & Manikandan, E. (2024). Investigation of nanostructured materials for enhanced photocatalytic activity. *Physica Scripta*, *99*, 125987.

Manikandan, M., Kulurumotlakatla, D., Manikandan, E., Karthigeyan, K. A., & Al-Kahtani, A. A. (2023). Design and performance of a novel hybrid energy storage system for renewable energy applications. *Journal of Energy Storage*, *72*, 108769. https://doi.org/10.1016/j.est.2023.108769

Marković, N. M., & Ross, P. N. (2002). Surface science studies of model fuel cell electrocatalysts. *Surface Science Reports*, *45*(4), 117–229. https://doi.org/10.1016/S0167-5729(01)00022-X

McCrory, C. C., Jung, S., Ferrer, I. M., Chatman, S. M., Peters, J. C., & Jaramillo, T. F. (2015). *Journal of the American Chemical Society*, *137*, 4347.

Mehta, S., Thakur, R., Rani, S., Nagaraja, B. M., Mehla, S., & Kainthla, I. (2024). Hydrogen economy: Present status and future prospects. *International Journal of Hydrogen Energy*, *82*, 1061–1080. https://doi.org/10.1016/j.ijhydene.2024.01.045

Musil, J. (2000). Hard and superhard nanocomposite coatings. *Surface and Coatings Technology*, *125*(1–3), 322–330. https://doi.org/10.1016/S0257-8972(99)00586-1

Patel, R. (2022). Enhancing electrode performance via surface engineering techniques. *Electrochemical Advances*, *18*(4), 335–350.

Popczun, E. J., McKone, J. R., Read, C. G., Biacchi, A. J., Wiltrout, A. M., Lewis, N. S., & Schaak, R. E. (2013). Nanostructured nickel phosphide as an electrocatalyst for the hydrogen evolution reaction. *Journal of the American Chemical Society*, *135*(25), 9267–9270. https://doi.org/10.1021/ja403440e

Ran, J., Zhang, J., Yu, J., Jaroniec, M., & Qiao, S. Z. (2014). Earth-abundant cocatalysts for semiconductor-based photocatalytic water splitting. *Chemical Society Reviews*, *43*(24), 7787–7812. https://doi.org/10.1039/C3CS60425J

Reda, S. M., Khairy, M., & Mousa, M. A. (2020). Photocatalytic activity of nitrogen and copper doped $TiO_2$ nanoparticles prepared by microwave-assisted sol-gel process. *Arabian Journal of Chemistry*, *13*(1), 86–95. https://doi.org/10.1016/j.arabjc.2017.05.002

Roger, M. A., Shipman, M., & Symes, M. D. (2017). Earth-abundant catalysts for electrochemical and photoelectrochemical water splitting. *Nature Reviews Chemistry*, *1*, 0003. https://doi.org/10.1038/s41570-016-0003

Ryu, J., Jung, N., Jang, J. H., Kim, H. J., & Yoo, S. J. (2015). In situ transformation of hydrogen-evolving CoP nanoparticles: Toward efficient oxygen evolution catalysts bearing dispersed morphologies with Co-oxo/hydroxo molecular units. *ACS Catalysis*, *5*(7), 4066–4074. https://doi.org/10.1021/acscatal.5b00349

Sapountzi, F. M., Gracia, J. M., Weststrate, C. J. J., Fredriksson, H. O. A., & Niemantsverdriet, J. W. (2017). Electrocatalysts for the generation of hydrogen, oxygen and synthesis gas. *Progress in Energy and Combustion Science*, *58*, 1–35. https://doi.org/10.1016/j.pecs.2016.09.001

Sazali, N. (2020). Hydrogen energy: Challenges and perspectives. *International Journal of Hydrogen Energy*, *45*(38), 18753–18765. https://doi.org/10.1016/j.ijhydene.2020.05.195

Schlapbach, L., & Züttel, A. (2001). Hydrogen-storage materials for mobile applications. *Nature*, *414*, 353–358. https://doi.org/10.1038/35104634

Shang, M., Wang, W., Zhang, L., Xu, Y., & Wu, L. (2009). Synthesis of flower-like $Bi_2WO_6$ superstructures and their high visible-light photocatalytic performances. *Journal of Hazardous Materials*, *172*(1), 338–344. https://doi.org/10.1016/j.jhazmat.2009.07.013

Shi, Y., & Zhang, B. (2016). Recent advances in transition metal phosphide nanomaterials: Synthesis and applications in hydrogen evolution reaction. *Chemical Society Reviews*, *45*(6), 1529–1541. https://doi.org/10.1039/C5CS00434A

Shinsel, S. R., Riemke, R. L., & Hoffmann, V. H. (2020). Challenges and solution technologies for the integration of variable renewable energy sources—A review. *Renewable Energy*, *145*, 2271–2285. https://doi.org/10.1016/j.renene.2019.06.147

Siddiqui, N., Faiyaz, T., & Tari, V. (2022). Ganga and Yamuna: Through the lens of National Green Tribunal. *Nature, Environment and Pollution Technology*, *21*(4), 1657–1664. https://doi.org/10.46488/NEPT.2022.v21i04.018

Smith, J. (2020). Advancements in surface modification of electrodes. *Electrochemical Journal*, *25*(3), 112–125.

Sun, J., Qiu, Y., Zhang, J., Zhang, M., & Zeng, D. (2021). A novel $Cu_2O$-based Z-scheme photocatalyst with enhanced photocatalytic performance. *Applied Surface Science*, *538*, 148042. https://doi.org/10.1016/j.apsusc.2020.148042

Tari, V., Gupta, V., Joshi, A., Kurve, P., & Chandra, N. (2024). Biodiversity loss due to forest diversion. In A. Rathore (Ed.), *Impact of societal development and infrastructure on biodiversity decline* (pp. 137–165). IGI Global. https://doi.org/10.4018/979-8-3693-6950-0.ch010

Tari, V., Kannan, K., Koliar, J., Gurushankar, K., & Wahyuni, D. K. (2025). Influence of zinc oxide nanomaterials on crop plants: A prospective analysis. In D. Jabbrova, D. Sarkar, S. Rakesh, S. Singh, & R. Datta (Eds.), *Smart nanomaterials in agroforestry systems*. Springer.

Tari, V., Siddiqui, N., Rathi, D., Siddiqui, N. N., & Wahyuni, D. K. (2024). Need for an evolved groundwater justice in rural area Uttar Pradesh, India. *Nature, Environment and Pollution Technology*, *23*(1), 125–137. https://doi.org/10.46488/NEPT.2024.v23i01.009

Tee, S. Y., Win, K. Y., Teo, W. S., Koh, L. D., Liu, S., Teng, C. P., & Han, M. Y. (2017). Recent progress in energy-driven water splitting. *Advanced Science*, *4*(5), 1600337. https://doi.org/10.1002/advs.201600337

Turner, J., Sverdrup, G., Mann, M. K., Maness, P. C., Kroposki, B., Ghirardi, M., Evans, R. J., & Blake, D. (2008). Renewable hydrogen production. *International Journal of Energy Research*, *32*(5), 379–407. https://doi.org/10.1002/er.1372

Turner, J. A. (2004). Sustainable hydrogen production. *Science*, *305*(5686), 972–974. https://doi.org/10.1126/science.1103197

Vergara, D., Fernández-Arias, P., Lampropoulos, G., & Antón-Sancho, Á. (2024). Hydrogen revolution in Europe: Bibliometric review of industrial hydrogen applications for a sustainable future. *Energies*, *17*(8), 3658. https://doi.org/10.3390/en17153658

Wang, H., Zang, S., Zhang, Y., Liu, Q., Dong, H., Zhang, Q., Yu, M., & Zheng, Y. (2020). Recent progress on S-scheme photocatalysts for environmental purification. *Catalysts*, *10*(9), 1064. https://doi.org/10.3390/catal10091064

Wang, J., Wang, G., Wang, X., Wu, Y., Su, Y., & Tang, H. (2019). 3D/2D direct Z-scheme heterojunctions of hierarchical $TiO_2$ microflowers/g-$C_3N_4$ nanosheets with enhanced charge carrier separation for photocatalytic $H_2$ evolution. *Carbon*, *149*, 618–626. https://doi.org/10.1016/j.carbon.2019.04.046

Wang, J., et al. (2016). Recent progress in cobalt-based heterogeneous catalysts for electrochemical water splitting. *Advanced Materials*, *28*(2), 215–230. https://doi.org/10.1002/adma.201502696

Wang, P., Zhang, X., Zhang, J., Wan, S., Guo, S., Lu, G., Yao, J., & Huang, X. (2017). Precise tuning in platinum–nickel/nickel sulfide interface nanowires for synergistic hydrogen evolution catalysis. *Nature Communications*, *8*, 14580. https://doi.org/10.1038/ncomms14580

Wang, X. (2019). Surface engineering of metal and carbon-based electrodes for high-performance supercapacitors. *Journal of Energy Chemistry*, *42*, 67–79. https://doi.org/10.1016/j.jechem.2019.04.006

Wang, X., Kolen'ko, Y. V., Bao, X.-Q., Kovnir, K., & Liu, L. (2015). One-step synthesis of self-supported nickel phosphide nanosheet array cathodes for efficient electrocatalytic hydrogen generation. *Angewandte Chemie International Edition*, *54*(28), 8306–8310. https://doi.org/10.1002/anie.201502577

Wendt, H., et al. (2005). Electrocatalysis and electrocatalysts for low temperature fuel cells: Fundamentals, state of the art, research and development. *Química Nova*, *28*(6), 1066–1075. https://doi.org/10.1590/S0100-40422005000600024

Wu, T., Chen, S., Zhang, D., & Hou, J. (2015). First principles study of optoelectronic and photocatalytic properties of $BiVO_4$ polymorphs. *Journal of Materials Chemistry A*, *3*(20), 10360–10367.

Xiao, P., Sk, M. A., Thia, L., Ge, X., Lim, R. J., Wang, J.-Y., Lim, K. H., & Wang, X. (2014). Molybdenum phosphide as an efficient electrocatalyst for the hydrogen evolution reaction. *Energy & Environmental Science*, *7*(8), 2624–2629. https://doi.org/10.1039/C4EE00957F

Xiao, Y., & Jiang, Y. (2019). Strategies in constructing ternary heterojunctions for photocatalytic hydrogen evolution. *Chemistry – An Asian Journal*, *14*(21), 3863–3879.

Yang, Y., Fei, H., Ruan, G., & Tour, J. M. (2015). Porous cobalt-based thin film as a bifunctional catalyst for hydrogen generation and oxygen generation. *Advanced Materials*, *27*(20), 3175–3180. https://doi.org/10.1002/adma.201500894

Yang, Z., Wang, H., Li, Y., Cui, W., & Dong, F. (2019). Construction of $Cu_2O$/g-$C_3N_4$ heterojunction with enhanced visible light photocatalytic activity for hydrogen evolution. *ACS Sustainable Chemistry & Engineering*, 7(3), 2895–2903.

Yin, H., Rong, F., & Xie, Y. (2024). A review of typical transition metal phosphides electrocatalysts for hydrogen evolution reaction. *International Journal of Hydrogen Energy*, *52*, 350–375. https://doi.org/10.1016/j.ijhydene.2023.10.262

Ying, J., Jiang, G., Cano, Z. P., Han, L., Yang, X.-Y., & Chen, Z. (2017). Recent progress in low-cost and high-performance anode catalysts for direct alcohol fuel cells. *Nano Energy*, *40*, 88–115. https://doi.org/10.1016/j.nanoen.2017.07.031

Yu, J., Wang, K., Xiao, W., Cheng, B., & Fan, J. (2019). Enhanced photocatalytic $CO_2$-reduction activity of direct Z-scheme g-$C_3N_4$–$TiO_2$ photocatalysts. *Journal of Materials Chemistry A*, 7(20), 12022–12034.

Yu, Z. Y., Duan, Y., Feng, X. Y., Yu, X., Gao, M. R., & Yu, S. H. (2021). A universal design of single-atom catalysts with a tunable coordination environment for electrochemical energy conversion. *Advanced Materials*, *33*(6), 2007100. https://doi.org/10.1002/adma.202007100

Zhang, F., Zhao, P. C., Niu, M., & Maddy, J. (2016). The survey of key technologies in hydrogen energy storage. *International Journal of Hydrogen Energy*, *41*(33), 14535–14552. https://doi.org/10.1016/j.ijhydene.2016.05.303

Zhang, L., Han, L., Liu, H., Liu, X., & Luo, J. (2017). Metal-organic framework-derived CoP/reduced graphene oxide hybrid as an efficient electrocatalyst for hydrogen evolution. *Angewandte Chemie International Edition*, *56*, 13694–13698.

Zhang, Y. (2021). Recent developments in conductive polymer-based electrode modifications for energy applications. *Advanced Functional Materials*, *31*(7), 2102564.

Zhao, J., Wu, T., Wu, K., & Oikawa, K. (2004). Photodegradation of organic compounds with composite catalysts of $TiO_2$ and other semiconductors. *Journal of Molecular Catalysis A: Chemical*, *216*(1), 257–264.

Zhao, W., Bai, S., Jin, Y., Wang, X., Song, H., Li, X., & Zhang, L. (2021). Engineering S-scheme heterojunctions for photocatalytic water splitting. *Chemical Society Reviews*, *50*(6), 3763–3783.

Zheng, Y., Jiao, Y., Jaroniec, M., & Qiao, S. Z. (2015). Advancing the electrochemistry of the hydrogen evolution reaction through combining experiment and theory. *Angewandte Chemie International Edition*, *54*(1), 52–65. https://doi.org/10.1002/anie.201407031

Zhou, L., Ding, X., Liu, S., Zhao, Y., Ren, Y., Zhou, M., & Zhang, Y. (2014). A simple method to prepare $Bi_2WO_6$ with excellent visible-light-driven photocatalytic activity. *Applied Surface Science*, *301*, 499–503. https://doi.org/10.1016/j.apsusc.2014.02.051

Zou, C., Xue, H., Xiong, B., Zhang, G., Pan, S., Jia, C., Wang, Y., et al. (2021). Connotation, innovation and vision of "carbon neutrality". *Natural Gas Industry B*, *8*(5), 523–537. https://doi.org/10.1016/j.ngib.2021.08.009

Zou, X., & Zhang, Y. (2015). Noble metal-free hydrogen evolution catalysts for water splitting. *Chemical Society Reviews*, *44*(15), 5148–5180.

Chapter 7

# Nanostructured carbon-based heterogeneous electrocatalysts for oxygen evolution reaction in alkaline media

*P. K. Kannan, G. Bhavani, E. Praveen, and Mariappan Anandkumar*

## 7.1 OXYGEN EVOLUTION REACTION IN ELECTROCATALYSTS

Electricity-driven water splitting happens through two reactions at different electrodes, separated by a membrane. At the cathode, hydrogen ions team up with electrons to form hydrogen gas $\left(2H^+ + 2e^- \rightarrow H_2\right)$, which is called the hydrogen evolution reaction (HER). On the flip side, at the anode, water gets oxidized, producing oxygen gas, protons, and electrons $\left(2H_2O \rightarrow O_2 + 4H^+ + 4e^-\right)$ in the oxygen evolution reaction (OER). Since OER needs a transfer of four electrons, it's slower compared to HER, which only requires two (Kannan and Tari, 2025; Gurushankar et al., 2024). The OER is essential for the efficiency of electrochemical water splitting. Historically, $IrO_2$ and $RuO_2$ have been the top choices for OER catalysts due to their outstanding performance (Li et al., 2023). However, their dependence on precious metals renders them costly and less sustainable for widespread use (Tari et al., 2025, 2022; Gupta et al., 2024). This issue has led to an increasing need for affordable, high-efficiency alternatives that can improve OER kinetics and boost water-splitting performance. Identifying these materials is vital for making clean hydrogen production more attainable and economical.

The catalyst-driven water splitting process involves the four important steps, which involve injection of a proton into the electrolyte, which in turn will recombine with the electron transferred to the cathode.

Here's a simple rundown of the steps:

1. Hydroxyl adsorption: The hydroxide ion sticks to the catalyst surface, forming a hydroxyl group and releasing an electron. $OH^- +^* \rightarrow HO^* + e^-$
2. Hydroxyl oxidation to oxygen: The hydroxyl group is oxidized to an oxygen intermediate, releasing a proton and an electron. $HO^* \rightarrow O^* + e^- + H^+$

DOI: 10.1201/9781003584742-7

3. Formation of hydroperoxide: The oxygen intermediate reacts with another hydroxide ion to form a hydroperoxide species and releases an electron. $O^{*} + OH^{-} \rightarrow HOO^{*} + e^{-}$
4. Oxygen evolution: The hydroperoxide species breaks down, releasing oxygen gas and freeing up the catalyst surface again. $HOO^{*} \rightarrow^{*} + O_{2(g)} + e^{-} + H^{+}$

These steps are crucial for designing effective OER catalysts. The aim is to reduce the energy needed for these intermediate steps, especially for the formation of O* and HOO*, to boost the catalyst's performance.

## 7.2 METAL-FREE OER ELECTROCATALYSTS BASED ON NANOCARBONS

### 7.2.1 Metal doped carbon catalysts

Carbon is generally inert in most electrochemical reactions, including the HER. To enhance its catalytic activity, structural modifications at the surface or subsurface level are necessary to improve the binding energies of reactants and reaction intermediates. Many of the existing surface modification techniques used for this purpose have been adapted from strategies developed for the OER (Song et al., 2020).

Single heteroatom doping, such as nitrogen (N), sulfur (S), or phosphorus (P), into carbon matrices has been widely reported to modify the electronic properties of graphene or carbon nanosheets (Gao et al., 2024). For instance, nitrogen doping introduces lone pair electrons that can alter the local charge density and create positively polarized carbon sites, which serve as active centers for OER. Studies have demonstrated that N-doped carbon nanosheets synthesized via gas-foaming or pyrolysis methods exhibit a high density of edge sites and a large surface area, leading to enhanced catalytic activity and stability. The precise incorporation of N atoms disrupts the conjugated $\pi$-system of graphene, creating localized electronic states that favor the adsorption of reaction intermediates during OER.

Building on the success of single-element doping, multi-element doping strategies have been developed to further tune the electronic and structural properties of carbon electrocatalysts. By simultaneously introducing two or more heteroatoms (e.g., N and S or N and P), synergistic effects can be achieved that optimize both the binding energies of oxygen intermediates and the overall conductivity of the carbon framework. For example, dual-doped carbon materials have been shown to possess enhanced defect densities and tailored electronic structures, thereby lowering the overpotential required for OER. In the past decade, many studies have focused on designing such catalysts using artificial intelligence, which has grabbed the interest of the scientific community.

The advancement of computational methodologies, particularly the integration of big data and artificial intelligence, has significantly contributed to reducing the costs associated with catalyst development. Techniques such as density functional theory (DFT), molecular dynamics, and machine learning (ML) are extensively utilized for catalyst pre-screening. Based on these computational insights, targeted material synthesis is conducted for experimental validation, facilitating the rational design of high-performance OER electrocatalysts.

Wu's group has used ML to test 511 combinations of the transition-metal-supported double-atom catalysts out of 16,767 structures and studied their catalytic performance for OER (Li et al., 2023). Simulations suggested that catalysts having carbon surface with elements such as Cu, Pd, Cd, Pt, Ag, Au, Ni, and Zn show superior catalytic activity as shown in Figure 7.1. They also estimated the overpotential for OER, and this result suggested not only the presence of transition metal, but the transition metal's occupancy in the carbon lattice can also impact the performance of the catalyst.

Structural engineering of carbon materials to increase the density of edge and defect sites is another effective strategy to boost OER performance. Edge sites in graphene and carbon nanosheets exhibit higher chemical reactivity compared to the basal planes, as they expose unsaturated carbon atoms that can act as active catalytic centers. Techniques such as plasma treatment, chemical etching, and controlled oxidation have been employed to produce carbon structures rich in defects and edge sites. These methods not only enhance the active surface area but also facilitate the adsorption and desorption of oxygenated intermediates during the OER process (Muhammad et al., 2024). The introduction of vacancies and other

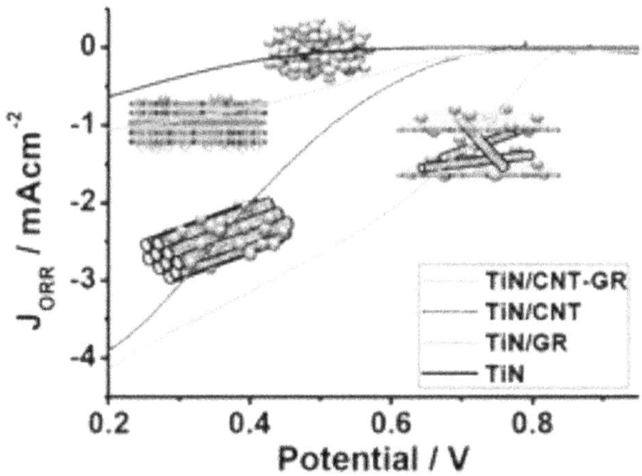


*Figure 7.1* Binding energies of various transition elements.

**(Reproduced with permission from (Wu et al., 2022).)**

topological defects disrupts the electronic symmetry of carbon, creating localized states that are conducive to electron transfer reactions.

Intrinsic defects in nanocarbon structures, such as vacancies, voids, Stone–Wales defects, dislocations, and grain boundaries (Zhang et al., 2017), typically arise from atomic vacancies and/or lattice reconstruction, leading to disruptions in electron–hole symmetry. These defective regions, which may feature dangling bonds, hydrogen saturation, or reconstructed configurations devoid of dangling bonds, have been extensively shown to modulate the local density of $\pi$-electrons and enhance chemical reactivity.

In-plane topological defects, including pentagonal and heptagonal rings replacing the conventional hexagonal lattice, are commonly observed in graphitic carbon frameworks. These structural irregularities locally induce Gaussian curvature, alter bond lengths, and rehybridize electron orbitals, thereby redistributing electron density and enhancing oxygen reduction reaction (ORR) activity. Moreover, the presence of Stone–Wales defects in N-doped graphene has been demonstrated to significantly improve ORR performance by reducing the HOMO–LUMO energy gap in nanoscale graphene domains and modifying reaction pathways (Tiwari et al., 2023).

Additionally, line defects such as edges, cracks, and voids in graphene sheets have been identified as efficient catalytic sites for electrochemical reactions, including ORR. Localized electronic states along graphene edges influence the band structure and alter the electronic properties of the material (Gong et al., 2024). Furthermore, pentagonal rings at zigzag edges, as well as pentagon–pentagon–octagon chain configurations, have been shown to induce local charge and spin polarization, thereby further enhancing ORR catalytic activity.

## 7.3 TRANSITION METAL AND HETEROATOM DOPED NANOCARBONS

These catalysts involve utilization of non-noble metals such as transition metals and non-metals, which are doped into various carbon structures. They exhibit comparable performance when compared to noble metals but also a substantial reduction in the cost of the catalysts.

### 7.3.1 Co-N-C catalysts

Cobalt coordinated with nitrogen in a carbon matrix (Co-N-C) represents one of the most studied systems in the realm of non-noble metal electrocatalysis. In these catalysts, cobalt centers are atomically dispersed and coordinated by nitrogen atoms, forming $Co–N_x$ motifs that mimic the active sites found in natural enzymes such as cytochrome c oxidase. The synergy between the metal center and the doped carbon framework enhances the

OER activity by providing optimal adsorption energies for intermediates such as *OH, *O, and *OOH. Recent studies report that optimizing the coordination environment and the local electronic structure of the Co centers can significantly reduce the overpotential and improve the reaction kinetics (Ding et al., 2022). Advanced characterization techniques, such as X-ray absorption spectroscopy (XAS) and Mössbauer spectroscopy, have been crucial in elucidating the nature of these active sites.

### 7.3.2 Ni-N-C catalysts

Nickel-based catalysts have also garnered attention due to their earth abundance and favorable electronic properties. In Ni-N-C systems, the incorporation of nitrogen not only stabilizes the nickel centers but also modifies the surrounding carbon matrix to facilitate electron transfer. The Ni–$N_x$ active sites exhibit a unique electronic configuration that enables them to interact with OER intermediates more effectively. Reports indicate that Ni-N-C catalysts, when engineered to have a high density of accessible Ni sites within a porous carbon network, achieve remarkable catalytic efficiency and durability under alkaline conditions (Zhuang et al., 2016). The interaction between the nickel sites and the nitrogen functionalities plays a pivotal role in lowering the activation energy for the OER, thus enhancing overall performance.

### 7.3.3 Fe-N-C catalysts

Iron-based doped carbon catalysts (Fe-N-C) have attracted significant research interest because of iron's redox flexibility and its potential to form highly active Fe–$N_x$ sites. In these catalysts, the iron centers embedded in a nitrogen-doped carbon framework exhibit high catalytic activity for OER as well as for complementary reactions such as the oxygen reduction reaction (ORR). The activity of Fe-N-C catalysts is often attributed to the optimized binding energies of key reaction intermediates and the efficient charge transfer facilitated by the conductive carbon matrix. Moreover, the presence of iron can induce additional defect sites within the carbon lattice, further boosting the catalytic performance. Recent works have highlighted that careful control of the synthesis parameters—such as pyrolysis temperature and precursor composition—is essential for achieving a high density of well-defined Fe–$N_x$ active centers (Ni et al., 2022).

### 7.3.4 Nanocarbon-based hybrid electrocatalysts with transition metal nitrides (TMN)

Transition metal nitrides (TMNx) exhibit exceptional electrical conductivity, catalytic activity, and chemical stability, making them prime candidates for hybrid electrocatalysts. When integrated with nanocarbon materials such as graphene (Ding et al., 2022), carbon nanotubes (CNTs) (Yang et al., 2024), or carbon dots, these hybrids leverage synergistic effects to achieve superior electrocatalytic performance. TMN ensures efficient electron transfer during catalytic processes due to its high conductivity, while nanocarbon materials provide a robust and conductive support matrix that prevents agglomeration of the active material and enhances structural integrity.

### 7.3.5 Synergistic properties of nanocarbon-TMN hybrids

Nanocarbon-based TMN hybrids capitalize on the complementary properties of their constituents. TMN contributes a high density of catalytic active sites and excellent stability under harsh conditions, while nanocarbon materials offer unique advantages, including high surface area, excellent electrical conductivity, and mechanical strength. This combination ensures uniform dispersion of TMN nanoparticles on the carbon matrix, resulting in increased exposure of active sites and enhanced interaction with reactants. Additionally, the unique surface chemistry of nanocarbon facilitates strong interfacial bonding with TMN, further boosting catalytic performance (Porter et al., 2024).

### 7.3.6 Applications in electrochemical reactions

#### 7.3.6.1 Oxygen reduction reactions (ORR)

One of the primary applications of TMN-carbon hybrids is in oxygen reduction reactions (ORR), a critical process in fuel cells. TMN's high catalytic activity for ORR, coupled with the excellent conductivity of nanocarbon materials, results in catalysts with significantly improved energy conversion efficiency. For example, studies have shown that molybdenum nitride ($Mo_2N$) integrated with graphene (TiN/CNT–GR) demonstrates exceptional ORR activity due to the synergy between $Mo_2N$'s active sites and graphene's conductive framework (Youn et al., 2013). The electrocatalytic performance of TiN/CNT–GR as shown in Figure 7.2 indicates its promise as a ORR electrocatalyst for PEMFCs and DMFCs.

##### *7.3.6.2 Hydrogen evolution reactions (HER)*

In HERs, TMN-carbon hybrids exhibit high catalytic efficiency under both acidic and alkaline conditions. The hybrid structure not only ensures efficient electron transport but also prevents the degradation of TMN under extreme electrochemical conditions. Notable examples include titanium nitride (TiN)-based hybrids with CNTs, which exhibit enhanced HER performance due to the effective charge transfer and robust structural integrity. Theerthagiri et al. (2020) has clearly summarized the HER performance of various TMNs and the results are summarized in Table 7.1.

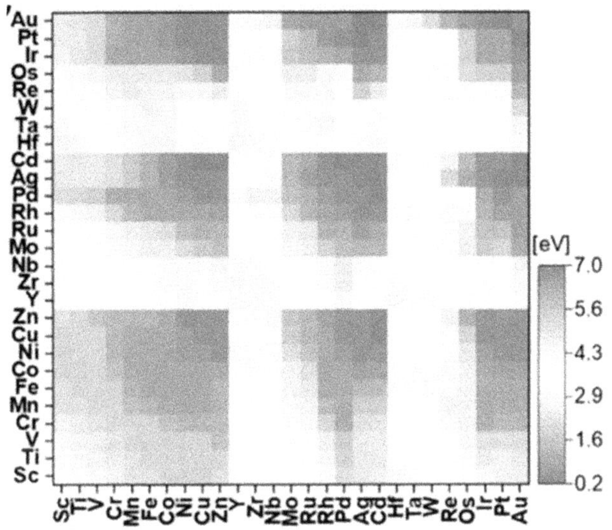


*Figure 7.2* Performance of TiN/CNT–GR as an oxygen ORR electrocatalyst.

**(Reprinted from (Youn et al., 2013) with permission from RSC.)**

Abghoui and Skúlason (2017) used DFT calculations to screen for stable and active catalyst material among TMNx for catalyzing the HER. It was concluded from the results shown in Figure 7.3 that TaN, HfN, MoN, and ScN are found to be catalyzing OER reactions with an overpotential comparable to the Pt.

#### *7.3.6.3 Durability and stability in harsh conditions*

The durability of TMN-carbon hybrids is a standout feature, particularly in applications like fuel cells and electrolyzers, where materials are exposed to harsh electrochemical environments. The carbon matrix's protective properties mitigate corrosion and degradation of TMN, ensuring long-term stability. Furthermore, the intrinsic thermal and chemical stability of TMN enables these hybrids to perform reliably across a wide range of temperatures and pH conditions (X. Li et al., 2024).

#### *7.3.6.4 Emerging examples and applications*

##### *7.3.6.4.1 $Mo_2N$-graphene hybrids*

One of the most studied hybrids is the $Mo_2N$-graphene system, which has demonstrated remarkable performance in HER. The $Mo_2N$ nanoparticles provide abundant active catalytic sites, while graphene enhances electron mobility and prevents aggregation. This synergy results in a catalyst with outstanding efficiency and longevity (Meng et al., 2022).

*Table 7.1* HER performance of TMN based Electrocatalysts

| *Electrocatalysts* | *Nitridation process* | *Electrolyte* | *Overpotential (mV) at 10 mA cm$^{-2}$* | *Tafel slope (mV/dec)* |
|---|---|---|---|---|
| $Ni_3N$/Ni | Nitridation with $NH_3$ | 1.0 M KOH | 64 mV at 100 mA cm$^{-2}$ | 29.3 |
| $Ni_2N_{1-x}$ | Plasma-induced nitridation | 1.0 M KOH | 55 | 54 |
| $Ni_3N$ | Reactive magnetron sputtering | 1.0 M KOH | 159 mV at 100 mA cm$^{-2}$ | 91 |
| $Ni_3N$ | Nitridation with $NH_3$ | 1.0 M KOH | 121 mV at 100 mA cm$^{-2}$ | 109 |
| $Ni_3N$ | Solid state growth and heat treated under $N_2$ atmosphere | 1.0 M KOH | 50 mV at 100 mA cm$^{-2}$ | 120 |
| NiCoN/C | Nitridation with $NH_3$ | 1.0 M KOH | 142 | - |
| $Ni_3N$-NiMoN | Nitridation with $NH_3$ | 1.0 M KOH | 31 | 64 |
| NiCoN | Nitridation with $NH_3$ | 1.0 M NaOH | 145 | 105.2 |
| IrNiN | Nitridation with $NH_3$ | 0.1 M $HClO_4$ | 100 mV at 0.613 mA cm$^{-2}$ | 36 |
| NiMoN,/C | Nitridation with $NH_3$ | 0.1 M $HClO_4$ | 78 mV at 100 mA cm$^{-2}$ | 35 |
| $Ni_3FeN$ | Nitridation with $NH_3$ | 1.0 M KOH | 238 | 46 |
| Pt-loaded $Ni_3N$/Ni@C | Urea-urea glass-route | 1.0 M KOH | 117 | 47.3 |
| P-Fe,N/RGO | Nitridation with $NH_3$ | 0.5 M $H_2SO_4$ | 64.8 | 48.7 |
| $Fe_3N$/N-doped rGO | Nitridation with $NH_3$ | 0.5 M $H_2SO_4$ | 94 | 49 |
| $Ni_3FeN$ | Nitridation with $NH_3$ | 1.0 M KOH | 45 | 75 |
| $FeNi_3N$ | Nitridation with $NH_3$ | 1.0 M KOH | 75 | 98 |
| $Fe_2Ni_2N$ | Nitridation with $NH_3$ | 1.0 M KOH | 110 | 101 |
| B/$Fe_3N$@C | Ethylenediamine nitridation | 1.0 M KOH | 216 | 133 |
| $Ni_3N$/VN-NF | Nitridation with $NH_3$ | 1.0 M KOH | 56 | 47 |
| $MoS_2$-VN | Nitridation with $NH_3$ | 0.5 M $H_2SO_4$ | 85 | 53.31 |
| VMoN | Magnetron sputtering deposition | 0.5 M $H_2SO_4$ | 108 | 60 |

| | | | | |
|---|---|---|---|---|
| VN-Co-P | Nitridation with $NH_3$ | 1.0 M KOH | 137 | 81 |
| Co/N-VN | Nitridation with $NH_3$ | 0.1 M KOH | 179 | 123 |
| $MoS_2$,/TiN | Nitridation with $NH_3$ | 0.5 M $H_2SO_4$ | 146 | 44.8 |
| TiN | Chemical vapor deposition | 1 M $HClO_4$ | 92 mV at of 1 mA cm | 54 |
| TiN - NFG | Urea glass route | 0.5 M $H_2SO_4$ | 161 | 66 |
| Ni-FeP/TIN/CC | Nitridation with NH, | 1.0 M KOH | 75 | 73 |
| WN/CC | Chemical vapor deposition | 1.0 M KOH | 130 | 57.1 |
| WN/CC | Chemical vapor deposition | 0.5 M $H_2SO_4$ | 134 | 59.6 |
| WN-$W_2C$ | WOx/aniline precursor route | 0.5 M $H_2SO_4$ | 242 | 85 |
| WN/CC | Nitridation with $NH_3$ | 0.5 M $H_2SO_4$ | 198 | 92 |
| Ni-Mo-N | Nitridation with $NH_3$ | 0.5 M $H_2SO_4$ | 72 mV at 20 mA $cm^{-2}$ | 39 |
| Ni-Mo-N | Nitridation with $NH_3$ | 1 M KOH | 57 mV at 20 | 40 |
| MoN@NPCNCs | Nano-$SiO_2$ template assisted electrospinning process | 0.5 M $H_2SO_4$ | 72 | 53.21 |
| MoN-NC | Nitridation with $NH_3$ | 0.5 M $H_2SO_4$ | 62 | 54 |
| g-$C_3N_4$@MoN | Salt-templated method and dicyandiamide precursor route | 1.0 M KOH | 110 | 57.8 |
| N-rich 2D $Mo_3No_6$ | Simple salt-templated route | 1.0 M KOH | 94 | 66 |
| $Mo_2C/Mo_2N$ | Hexamethylenetetramine precursor route | 0.5 M $H_2SO_4$ | 205 | 71.8 |
| γ-$Mo_2N$ | Urea glass method | 0.5 M H2SO4 | 381 | 100 |
| γ-$Mo_2N$ | Urea glass method | 1.0 M KOH | 353 | 108 |
| $MoN_3$ | Nitridation with $NH_3$ | 0.5 M $H_2SO_4$ | 148 | 114 |
| $Mo_2N$/NC | p-phenylenediamine precursor route | 0.5 M $H_2SO_4$ | 217 | 115.7 |
| MoN | Simple salt-templated route | 1.0 M KOH | 389 | 120 |
| $Mo_2N$ | g-$C_3N_4$ as nitrogen source | 0.5 M $H_2SO_4$ | 400 mV at 197 $mAcm^{-2}$ | 145 |

*Source*: Reprinted from (Theerthagiri et al., 2020) with permission from Elsevier.

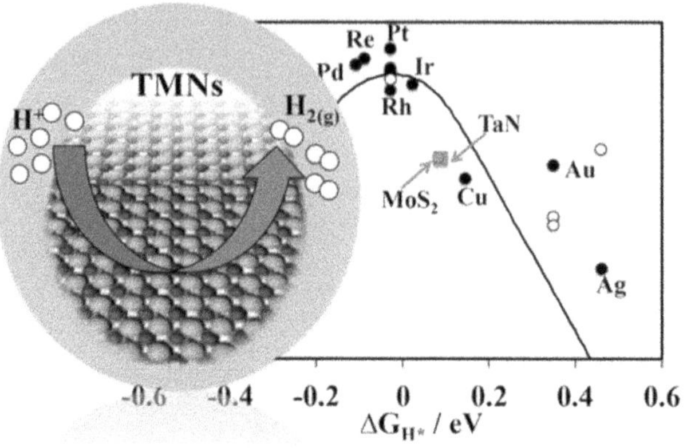


*Figure 7.3* Effective charge transfer reactions in transition metal oxides.

**(Reprinted with permission from (Abghoui & Skúlason, 2017. Copyright 2017 American Chemical Society.)**

#### *7.3.6.4.2 VN-MWCNT hybrids*

Another promising example is vanadium nitride (VN) integrated with multi-walled carbon nanotubes (MWCNTs), which has shown significant improvements in both HER and supercapacitor applications. The MWCNTs provide excellent electrical pathways and mechanical reinforcement, while VN offers high catalytic activity and durability (Adalati et al., 2022).

#### *7.3.6.4.3 NbN-carbon dots*

Niobium nitride (NbN) combined with carbon dots has recently emerged as a potential candidate for electrocatalysis (Zhong et al., 2016). The carbon dots' unique optical and electrical properties complement NbN's stability and activity, resulting in a hybrid material suitable for advanced fuel cell technologies.

#### *7.3.6.4.4 Potential for next-generation energy systems*

The integration of TMN with nanocarbon materials is paving the way for next-generation energy systems. Their remarkable performance in ORR and HER, combined with their durability and adaptability, makes them strong contenders for future advancements in sustainable energy technologies. Continued research into novel combinations of TMN and nanocarbon materials, along with advancements in synthesis techniques, promises to unlock even greater potential for these hybrids.

## 7.4 NANOCARBON-BASED HYBRID ELECTROCATALYSTS WITH TRANSITION METAL SULFIDES (TMS)

TMSx are widely recognized for their high catalytic activity, layered structure, and affordability, making them promising candidates for hybrid electrocatalysts. When integrated with nanocarbon materials such as graphene, CNTs (Hu et al., 2024), or nitrogen-doped carbon, these hybrids exhibit remarkable electrocatalytic properties and enhanced durability, addressing key challenges in energy applications. The hybridization process improves charge transport, structural integrity, and the exposure of active catalytic sites, enabling superior performance in HER, OER, and energy storage systems (Zhang et al., 2017).

### 7.4.1 Synergistic properties of nanocarbon-TMS hybrids

The layered structure of TMS materials, such as molybdenum disulfide ($MoS_2$) and tungsten disulfide ($WS_2$), provides a high density of active sites. However, their relatively low intrinsic conductivity can hinder catalytic performance (Li et al., 2024). This limitation is effectively addressed by hybridizing TMS with nanocarbon materials. Graphene, CNTs, and carbon nanodots serve as conductive supports, facilitating efficient electron transfer during catalytic reactions. Additionally, functionalization of nanocarbon surfaces promotes stronger interactions with TMS, ensuring uniform dispersion and enhanced activity (Zhu et al., 2021).

### 7.4.2 Applications in electrochemical reactions

#### *7.4.2.1 Hydrogen evolution reactions (HER)*

Nanocarbon-TMS hybrids have demonstrated exceptional HER performance under both acidic and alkaline conditions. For instance, $MoS_2$ nanosheets supported on graphene exhibit enhanced catalytic activity due to the strong electronic interaction between the two components (Y. Li et al., 2021). Similarly, $WS_2$-CNT hybrids show improved durability and activity in alkaline HER systems (Zhu et al., 2021). These hybrids overcome the conductivity limitations of TMSx and provide a robust framework that prevents catalyst degradation.

#### *7.4.2.2 Oxygen evolution reactions (OER)*

In water-splitting applications, TMS-carbon hybrids are effective catalysts for OER. The incorporation of nanocarbon materials reduces the overpotential required for OER, improving energy efficiency. For example, $Ni_3S_2$ embedded in nitrogen-doped graphene enhances OER performance due to the synergistic effects of the hybrid structure (Liu et al., 2024).

#### 7.4.2.3 *Applications in energy storage*

Beyond electrocatalysis, TMS-carbon hybrids are promising candidates for energy storage devices, particularly in lithium-sulfur (Li-S) batteries. The hybrids improve redox kinetics and suppress the shuttle effect of lithium polysulfides, leading to better cycling stability and energy efficiency (Yi et al., 2022). $CoS_2$ embedded in porous carbon has demonstrated excellent performance in Li-S batteries, benefiting from the hybrid's synergistic properties. Moreover, $MoS_2$-carbon hybrids have shown potential in supercapacitors, offering high capacitance and cycling stability. The performance of TMS as a cathode is clearly summarized in Table 7.2.

#### 7.4.2.4 *Emerging examples and applications*

##### *7.4.2.4.1 $MoS_2$-graphene hybrids*

$MoS_2$-graphene hybrids have been extensively studied for HER. The $MoS_2$ nanosheets provide abundant catalytic sites, while graphene enhances electron mobility and prevents aggregation. This combination results in catalysts with outstanding efficiency and stability (Yi et al., 2022).

*Table 7.2* Details of transition metal sulfides and their battery properties (Yi et al., 2022)

| *Cathode* | *Electrolyte* | *Current density/ mA $g^{-1}$* | *Cycle number-capacity (retention)/mAh $g^{-1}$* |
|---|---|---|---|
| a-$NbS_3$ | 1 M LiPF6 + EC/DMC (1:1 by volume) | | 10–281 |
| a-$TiS_3$/S/KB | 80 $Li_2S$ - 20$P_2S_5$ solid electrolyte | 0.064 mA $cm^{-2}$ | 50–650 |
| $MoS_3$/MWCNTs | 1 M LiTFSI + DOL/DME (1:1 by volume) + 0.1 M 1 $LiNO_3$ | 0.5 mA cm-$^{-2}$ | 200–291.7 |
| a-$FeS_4$/C | 1.2 M LiPF6 + EC/DMC (1:1 by volume) + 5% FEC | 100 | 500–644 |
| a-$TiS_4$ | 1 M LiPF6 + EC/DMC (1:1 by volume) | 400 | 2–500 |
| $P_{0.4}VS_{5.0}$ | 1 M LiPF6 + EC/DMC (1:1 by volume) | 0.1 C | 50–445 |

#### 7.4.2.4.2 $WS_2$-CNT hybrids

$WS_2$ integrated with MWCNTs exhibits significant improvements in both HER and energy storage applications. The CNTs act as conductive pathways and reinforce the mechanical strength of the hybrid structure (Zhu et al., 2021).

#### 7.4.2.4.3 $CoS_2$-porous carbon hybrids

Cobalt sulfide ($CoS_2$) embedded in porous carbon matrices has shown remarkable performance in lithium-sulfur batteries. The porous carbon provides a conductive network and prevents the aggregation of $CoS_2$ particles, enhancing both capacity and cycling stability (Yi et al., 2022).

#### 7.4.2.4.4 Potential for next-generation energy systems

Nanocarbon-TMSx hybrids hold immense potential for advancing next-generation energy systems. Their outstanding performance in HER, OER, and energy storage applications, combined with their durability and adaptability, makes them ideal for sustainable energy technologies. Future research focused on exploring novel combinations of TMSx and nanocarbon materials, as well as optimizing synthesis techniques, will further unlock their potential for diverse applications.

## 7.5 NANOCARBON-BASED HYBRID ELECTROCATALYSTS WITH TRANSITION METAL SELENIDES (TMSe)

Transition metal selenides (TMSe) have garnered significant attention for their exceptional electrocatalytic activity, robust chemical stability, and affordability. However, challenges such as limited conductivity and insufficient active site exposure often hinder their standalone performance. Hybridizing TMSe with nanocarbon materials, such as graphene, CNTs, and carbon dots, has emerged as an effective strategy to address these limitations. This synergistic combination improves catalytic performance by enhancing conductivity, increasing active sites, and facilitating charge transfer, making these hybrids ideal for various energy applications.

### 7.5.1 Synergistic advantages of nanocarbon-TMSe hybrids

The hybridization of nanocarbon materials with TMSe leverages the unique properties of both components. TMSes, such as cobalt selenide ($CoSe_2$), nickel selenide ($NiSe_2$), and molybdenum diselenide ($MoSe_2$), exhibit excellent catalytic activity due to their layered structures and transition metal centers, which provide abundant active sites. However, their relatively low electrical conductivity can be a bottleneck for high-performance electrocatalysis. Integrating nanocarbon materials resolves this issue by offering a highly conductive matrix that ensures rapid electron transfer and uniform dispersion of TMSex nanoparticles.

Nanocarbon materials also provide mechanical robustness and chemical stability, protecting the active material from degradation under harsh electrochemical conditions. The resulting hybrid materials exhibit improved durability and operational stability, enabling their use in demanding applications such as fuel cells and electrolyzers.

## 7.5.2 Applications in electrochemical reactions

### *7.5.2.1 Hydrogen evolution reaction (HER)*

Nanocarbon-TMSex hybrids have demonstrated remarkable performance in the HER. For example, $MoSe_2$ nanosheets supported on reduced graphene oxide (rGO) have shown excellent HER activity in both acidic and alkaline media, achieving a low overpotential of 200 mV at a current density of 10 mA $cm^{-2}$. The strong electronic interaction between $MoSe_2$ and rGO enhances charge transfer and stability, significantly outperforming pristine $MoSe_2$ (Namsheer et al., 2023).

### *7.5.2.2 Oxygen evolution reaction (OER) and oxygen reduction reaction (ORR)*

TMSe-carbon hybrids have also been extensively studied for OER and oxygen reduction reactions (ORR), which are critical for rechargeable metal-air batteries and fuel cells. A notable example is manganese selenide (MnSe) anchored on MWCNTs. As shown in Figure 7.4, MnSe@MWCNT composite achieved an impressive overpotential of 290 mV at 10 mA $cm^{-2}$ for OER, surpassing traditional catalysts like $RuO_2$, while displaying an onset potential of 0.94 V for ORR. This dual functionality makes it a promising candidate for bifunctional electrocatalysts in energy devices (Singh et al., 2022).

### *7.5.2.3 Strategies for enhancing performance*

Advancements in the synthesis and design of nanocarbon-TMSe hybrids have further improved their catalytic performance. It is done by creating TMSenanosheets, nanodots, or nanorods to increase surface area and active site density; by introducing defects in the TMSe lattice or doping the nanocarbon matrix with heteroatoms (e.g., N, S, or P) to modulate electronic properties; by enhancing the interaction between TMSe and nanocarbon

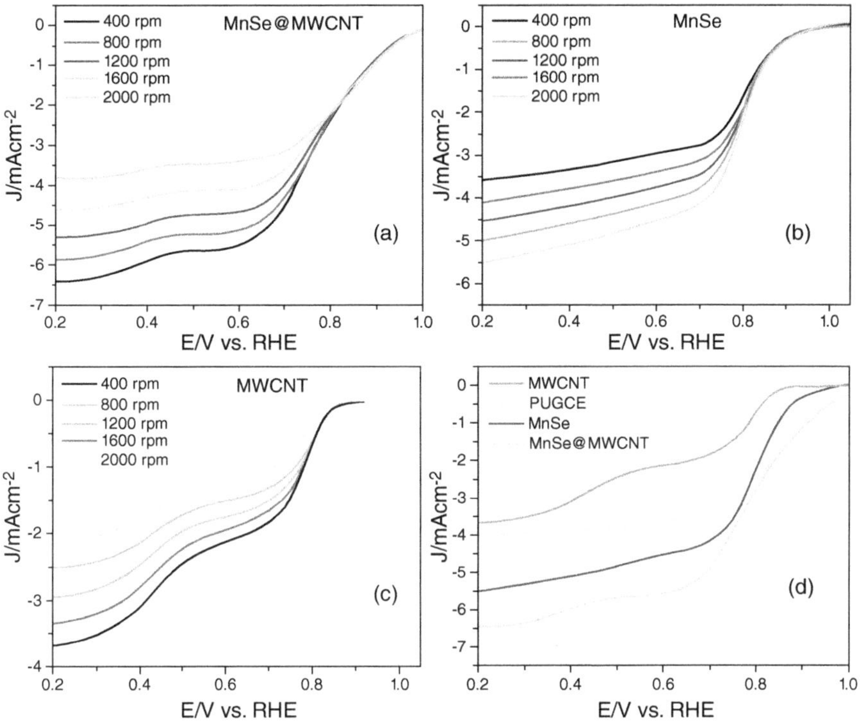


*Figure 7.4* Linear sweep voltammograms of (a) the MnSe@MWCNT modified electrode, (b) MnSe, and (c) MWCNT collected using the RDE set-up. (d) Comparison of the ORR activity of these catalysts at a rotating speed of 2000 rpm (Singh et al., 2022).

through covalent bonding or functionalization; and by employing advanced methods such as hydrothermal synthesis, chemical vapor deposition (CVD), or atomic layer deposition (ALD) to achieve uniform dispersion and robust interfaces.

### *7.5.2.4 Energy storage applications*

Beyond water splitting, TMSe-carbon hybrids have shown tremendous promise in energy storage devices, including lithium-ion batteries, sodium-ion batteries, and supercapacitors. For instance, $CoSe_2$-carbon hybrids have demonstrated high specific capacitance and exceptional cycling stability, with applications in supercapacitors achieving energy densities exceeding 40 Wh/kg. The strong interaction between $CoSe_2$ and the carbon matrix enhances ion diffusion and charge storage capability, resulting in superior electrochemical performance as shown in Figure 7.5 (Sajjad et al., 2021).

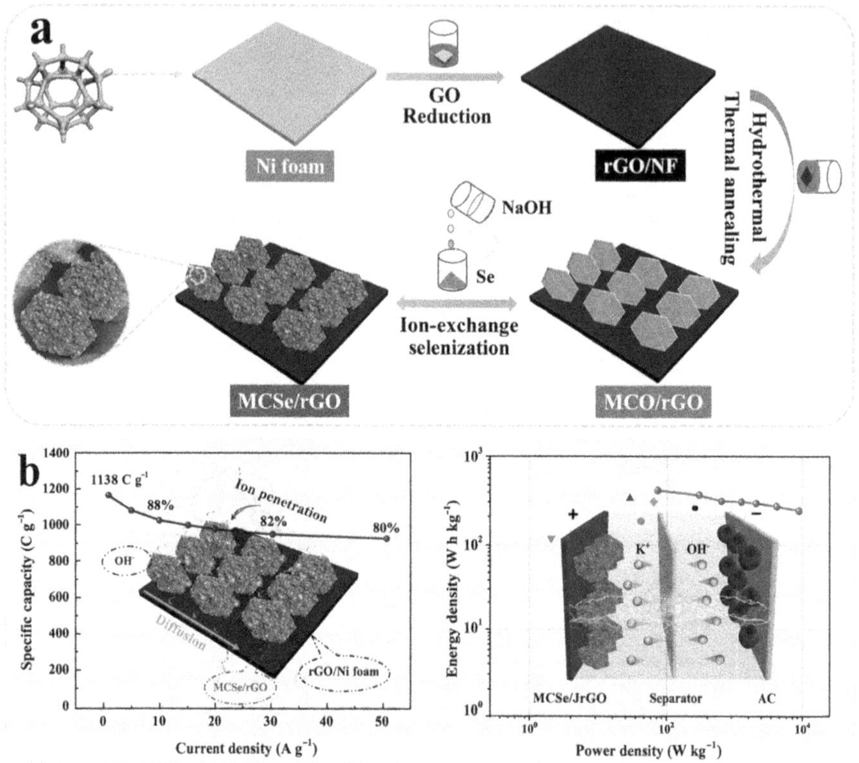


*Figure 7.5* Schematic illustration of the synthesis procedure of hierarchical MCSe/rGO composites, (b) capacitance vs. current density plot, and Ragone plot.

**(Reprinted from (Sajjad et al., 2021) with permission from Elsevier.)**

Similarly, $NiSe_2$-carbon hybrids have been employed in lithium-ion batteries, where their layered structures facilitate rapid lithium-ion intercalation and deintercalation. This structural advantage, combined with the high conductivity of nanocarbon materials, ensures excellent capacity retention and cycling stability.

Nanocarbon-based hybrids incorporating transition metal selenides represent a promising class of materials for next-generation energy systems. The enhanced conductivity, increased active sites, and improved charge transfer dynamics achieved through hybridization position these materials as leading candidates for sustainable energy conversion and storage technologies. With continued advancements in synthesis and design, nanocarbon-TMSex hybrids are poised to play a pivotal role in addressing the global energy challenge.

## 7.6 NANOCARBON-BASED HYBRID ELECTROCATALYSTS WITH TRANSITION METAL PHOSPHIDES (TMP)

Transition metal phosphides (TMPx), such as nickel phosphide ($Ni_2P$), cobalt phosphide (CoP), and molybdenum phosphide (MoP), have emerged as promising materials for electrocatalysis due to their unique electronic structures, high intrinsic catalytic activities, and robustness under diverse electrochemical conditions. However, challenges such as limited conductivity, aggregation, and susceptibility to degradation hinder their standalone efficiency. Hybridizing TMPx with nanocarbon materials, including graphene, CNTs, and carbon nanofibers (CNFs), effectively overcomes these challenges. The resulting nanocarbon-based TMP hybrids exhibit enhanced catalytic performance, conductivity, and long-term durability, making them highly suitable for reactions like the HER, OER, and ORR.

### 7.6.1 Synergistic properties of nanocarbon-TMP hybrids

Nanocarbon materials offer a high surface area, excellent electrical conductivity, and mechanical stability, which complement the catalytic properties of TMPx. TMPx serve as the active catalytic centers, while nanocarbon provides a conductive matrix that facilitates rapid charge transfer and prevents the agglomeration of TMP nanoparticles. This combination ensures the uniform distribution of TMPx, exposes more active sites, and enhances interaction with reactants.

The strong chemical interactions between TMPx and nanocarbon substrates further boost the performance of the hybrids. For instance, doping nanocarbon materials with heteroatoms such as nitrogen can modulate the electronic structure at the TMP-nanocarbon interface, improving catalytic efficiency. Studies have shown that CoP nanoparticles anchored on nitrogen-doped CNTs exhibit significantly lower overpotentials for HER due to the synergistic effects of the hybrid structure (G. Li et al., 2021).

### 7.6.2 Applications in electrochemical reactions

#### 7.6.2.1 *Hydrogen evolution reaction (HER)*

Nanocarbon-TMP hybrids have demonstrated exceptional HER performance in both acidic and alkaline media. For example, the $Ni_2P$/N-doped reduced graphene oxide (NRGO) hybrid demonstrates remarkable catalytic activity for HER, with an onset overpotential of 37 mV and a Tafel slope of 59 mV $dec^{-1}$. The enhanced conductivity of the rGO facilitates rapid electron transport, while the $Ni_2P$ nanoparticles provide abundant active sites for hydrogen production (Gómez et al., 2019).

Another notable example is the CoP-CNT hybrid, which exhibited superior catalytic activity for HER in alkaline media, demonstrated by a low overpotential and a small Tafel slope. The synergistic effect between CoP and CNTs facilitated efficient electron transfer and provided abundant active sites for hydrogen production. The hybrid catalyst showed remarkable stability over prolonged electrochemical cycling, maintaining its performance without significant degradation. The structural integrity of the catalyst was preserved, attributed to the robust interaction between CoP nanoparticles and the CNT support (Zhao et al., 2022).

#### *7.6.2.2 Oxygen evolution reaction (OER)*

For OER, TMP hybrids have shown remarkable activity and stability under oxidative conditions. The MoP@NC hybrid exhibited superior catalytic activity for HER across a wide pH range. Specifically, it achieved overpotentials ($\eta_{10}$) of 96 mV in acidic, 191 mV in neutral, and 149 mV in alkaline solutions, with corresponding Tafel slopes of 49.2, 95.0, and 61.7 mV $dec^{-1}$. The synergistic effect between MoP and the N-doped carbon layer provided abundant active sites and facilitated efficient electron transfer, contributing to the enhanced performance. The MoP@NC electrocatalyst demonstrated remarkable stability over prolonged electrochemical cycling, maintaining its performance without significant degradation. The robust interaction between the MoP nanowires and the N-doped carbon coating preserved the structural integrity of the catalyst (Pi et al., 2020).

#### *7.6.2.3 Bifunctional catalysis (HER and OER)*

Bifunctional electrocatalysts capable of efficiently catalyzing both HER and OER are critical for water-splitting applications. Cobalt phosphide (CoP) nanoparticles embedded in CNFs with CNT grafting, referred to as CNF@CoP-CNTs. This catalyst demonstrated remarkable performance, achieving an overpotential of 65 mV at 10 mA $cm^{-2}$ for HER and 367 mV at 50 mA $cm^{-2}$ for OER. The enhanced activity is attributed to the strong interaction between CoP nanoparticles and the CNF matrix, which facilitates efficient charge transfer and maintains structural stability during electrochemical reactions (Yu et al., 2022).

### 7.6.3 Recent advancements in nanocarbon-TMPx hybrid synthesis

Recent advancements in the synthesis of nanocarbon-TMP hybrids have focused on optimizing the size, morphology, and electronic properties of TMPx, as well as enhancing the interaction between TMPs and nanocarbon supports. Synthesis methods like hydrothermal techniques and solvothermal methods enable precise control over the size and shape of TMP nanoparticles, enhancing their catalytic efficiency (Downes et al., 2022). Functionalizing nanocarbon materials with active groups improves the binding of TMPs, leading to stronger interfacial interactions and enhanced charge transfer (Liang et al., 2013). Methods such as chemical vapor deposition (CVD) and electrodeposition have been employed to uniformly disperse TMPx on nanocarbon substrates (Zhang et al., 2020).

#### *7.6.3.1 Energy storage applications*

In addition to electrocatalysis, nanocarbon-TMP hybrids have shown significant potential in energy storage devices, such as lithium-ion batteries (LIBs) and supercapacitors. For instance:

- **$Ni_2P$-CNT Hybrids:** The solvent-induced structural modifications of $Ni_2P$/CNT hybrids lead to improved electrochemical performance, highlighting the importance of nanocarbon-TMP hybrids in energy storage. The structural tuning optimizes the electrochemical stability and charge storage capacity, making these hybrids highly efficient for supercapacitor applications (Zhang et al., 2023).
- **CoP-Graphene Hybrids:** CoP-based composites exhibit enhanced electrochemical performance, including high specific capacitance, excellent cycling stability, and good rate capability. This is due to the synergistic effects between CoP and conductive carbon materials, which improve charge transfer and enhance electrochemical performance. CoP shows potential for high-performance energy storage devices, especially in supercapacitors, due to its favorable electronic structure and active sites for charge storage (Xing et al., 2021).

### 7.6.4 Emerging examples and future directions

**MoP-Carbon Dot Hybrids:** The development of MoP quantum dots encapsulated in nitrogen/phosphorus-doped carbon (P/N-doped carbon) for efficient HER in both acidic and alkaline electrolytes. This hybrid material shows excellent catalytic activity and stability due to the synergistic effects of MoP quantum dots and the doped carbon, making it promising for hydrogen production in diverse environments (Y. Chen et al., 2023).

**CoP-Nanocarbon Composites:** he CoP/N-doped carbon composite demonstrates enhanced catalytic performance for both the OER and ORR, making it a promising material for energy storage applications. The hollow structure and nitrogen doping significantly improve conductivity and stability, contributing to its superior performance (Gan et al., 2024).

**NiP-CNT Hybrids:** This hybrid catalyst exhibits enhanced hydrogenation activity, demonstrating its potential in catalytic reactions. The CNT support plays a critical role in stabilizing the NiP alloy, improving its catalytic performance and efficiency (Ju & Li, 2006).

**Nanocarbon-based hybrid:** Electrocatalysts with transition metal phosphides represent a breakthrough in sustainable energy technologies. Their enhanced catalytic activity, conductivity, and stability make them ideal candidates for applications in water splitting, fuel cells, and energy storage systems. Continued advancements in synthesis techniques and interface engineering are expected to unlock the full potential of these hybrids, paving the way for their integration into next-generation energy solutions.

## 7.7 NANOCARBON-BASED ELECTROCATALYSTS WITH NON-HYBRID METALS

Nanocarbon materials, such as graphene, CNTs, and CNFs, play a pivotal role as supports for non-hybrid metals, significantly enhancing their catalytic activity, stability, and durability. Non-hybrid metals like platinum (Pt), palladium (Pd), and gold (Au) are highly efficient catalysts in various electrochemical reactions, including the ORR, carbon dioxide ($CO_2$) reduction, and HER. The synergy between these metals and nanocarbon substrates creates advanced electrocatalysts with superior performance and practical applicability.

### 7.7.1 Platinum-based nanocarbon hybrids

Platinum (Pt) is widely regarded as the benchmark catalyst for ORR and HER due to its high catalytic activity. However, the high cost and scarcity of Pt necessitate strategies to improve its utilization and stability. Nanocarbon materials, particularly nitrogen-doped graphene and CNTs, provide a conductive matrix that disperses Pt nanoparticles uniformly and prevents aggregation, thereby maximizing active site exposure.

**ORR Performance:** The Pt–N-rGO hybrids demonstrated outstanding performance for the oxygen reduction reaction (ORR), with an electron transfer number close to 4.0 and an onset potential of 0.90 eV. These hybrids also exhibited excellent stability and resistance to methanol in alkaline environments. The interactions between Pt–N-rGO and oxygen molecules suggest that the catalyst's performance may be enhanced by oxygen-modified growth during preparation. This high level of electrocatalytic activity could reduce the reliance on costly platinum, opening new avenues for applications in lithium-air batteries (Chen et al., 2021).

**HER Activity:** PtNPs/CNFs synthesized via electrospinning and carbonization process show high catalytic activity for both HER and electrochemical sensor, as well as long-term stability for HER, serving as a bifunctional catalyst (T. Yang et al., 2015).

### 7.7.2 Palladium-based nanocarbon hybrids

Palladium (Pd) is an attractive alternative to Pt due to its lower cost and comparable catalytic properties, particularly in $CO_2$ reduction reactions and fuel cell applications. The integration of Pd nanoparticles with nanocarbon materials improves their dispersion, catalytic efficiency, and stability.

**$CO_2$ Reduction:** Pd nanoparticles supported on carbon black and graphene have been shown to efficiently convert $CO_2$ to formate with a high Faradaic efficiency (FE) of over 90% at low overpotentials as shown in Figure 7.6. This performance is attributed to the conductive carbon matrix, which facilitates rapid electron transport (Min & Kanan, 2015).

**Alcohol Oxidation:** Palladium nanoparticles, with an average diameter of less than 5 nm, are deposited onto nickel (Ni)-CNTs, where the Ni metal nanoparticles are primarily present as single atoms or in sizes ranging from 1–2 nm. These mono-dispersed Pd nanoparticles are uniformly anchored on both synthesized Ni- and nitrogen (N)-doped CNTs, as well as on commercially available N-doped CNTs, utilizing

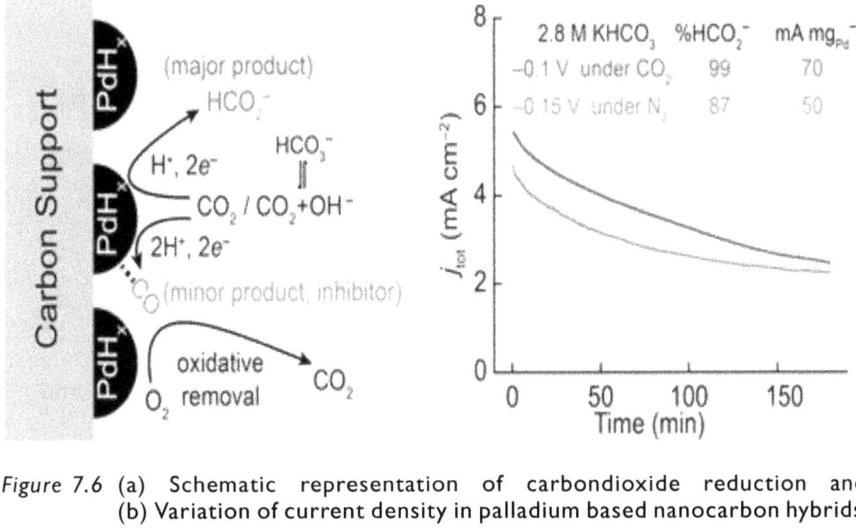


*Figure 7.6* (a) Schematic representation of carbondioxide reduction and (b) Variation of current density in palladium based nanocarbon hybrids.

**(Reprinted with permission from (Min & Kanan, 2015). Copyright 2015 American Chemical Society.)**

poly (diallyldimethylammonium) chloride (PDDA) as a key bonding agent. The resulting materials exhibit enhanced electrocatalytic performance in various reactions, including the HER, ORR, OER, and methanol oxidation reaction (MOR). Some of these samples show superior HER performance under acidic conditions and OER activity under basic conditions compared to commercial Pd/C (40 wt.%) catalysts. These findings suggest a promising strategy for developing highly efficient and cost-effective catalytic materials (X. Zhang et al., 2021).

### 7.7.3 Gold-based nanocarbon hybrids

Gold (Au) exhibits unique catalytic properties, particularly in reactions involving electrochemical sensing and environmental catalysis. The integration of Au nanoparticles with nanocarbon materials enhances their activity by providing a conductive and chemically stable support.

**Electrochemical Sensing:** Au-carbon hybrids are widely used in sensors due to their high sensitivity and selectivity. For instance, Au nanoparticles deposited on CNTs have been employed for glutathione detection in human blood samples. This performance is attributed to the synergistic effect of Au's catalytic sites and the high conductivity of CNTs (X. Chen et al., 2023) (Figure 7.7).

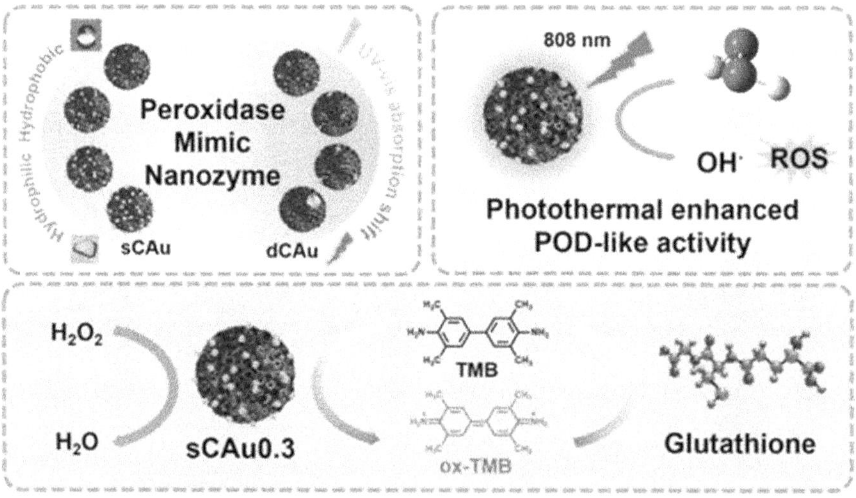


*Figure 7.7* Schematic representation of electrochemical sensing in Au – carbon hybrids.

**(Reprinted with permission from (X. Chen et al., 2023). Copyright 2023 American Chemical Society.)**

**Environmental Catalysis:** Au-graphene hybrids are effective for electrochemical reduction of nitrites and nitrates, pollutants commonly found in water systems. The hybrid structure promotes efficient electron transfer and high catalytic activity (Zheng et al., 2022).

### 7.7.4 Mechanisms of enhanced performance

The improved performance of non-hybrid metal-carbon hybrids arises from several key factors:

1. **Uniform Dispersion:** Nanocarbon materials prevent the agglomeration of metal nanoparticles, ensuring uniform distribution and increased exposure of active sites.
2. **Enhanced Conductivity:** The high electrical conductivity of nanocarbon substrates facilitates rapid electron transfer, which is crucial for electrochemical reactions.
3. **Strong Metal-Support Interaction:** Nitrogen-doped carbon materials, in particular, form strong interactions with metal nanoparticles, improving their stability and catalytic efficiency (Ma et al., 2015).
4. **Stability and Durability:** The structural integrity of nanocarbon materials ensures that the hybrid catalysts remain stable under harsh electrochemical conditions.

### 7.7.5 Emerging applications

#### *7.7.5.1 Fuel cells*

Recent advancements in fuel cell technology have focused on Pt-containing alloys, non-Pt catalysts, and nanocarbon composites for improving electrolyte membranes and corrosion protection. Key developments include the role of heteroatom doping in boosting reactivity, the use of organic-inorganic composite proton exchange membranes to enhance proton conductivity, and the application of amorphous carbon films and conductive polymer-nanocarbon composites for corrosion resistance. These innovations highlight the potential of nanocarbon-based materials to increase the performance, reliability, and lifespan of fuel cell components, facilitating broader commercial adoption (Agrawal et al., 2024).

#### *7.7.5.2 Energy storage*

The electrochemical energy storage capabilities of a porous carbon/gold nanoparticle composite were assessed using a symmetric two-electrode configuration. An aqueous symmetric supercapacitor utilizing a G10/C composite electrode achieved a gravimetric specific capacitance of 80 F $g^{-1}$ at a current density of 1 A $g^{-1}$. Remarkably, the device maintained 93% of its initial capacitance after 5000 charge-discharge cycles, demonstrating excellent cycling stability and promising potential for durable energy storage applications (Anjana et al., 2019).

#### *7.7.5.3 CO oxidation*

Pd-based carbon hybrids are effective catalysts for CO oxidation, a critical reaction for eliminating CO poisoning in fuel cell electrodes. For instance, Pd/rGO catalysts achieve nearly 100% CO conversion at room temperature (Velinova et al., 2023).

Nanocarbon-based electrocatalysts with non-hybrid metals like Pt, Pd, and Au are at the forefront of advancements in sustainable energy technologies. By leveraging the unique properties of nanocarbon supports, these hybrids exhibit enhanced catalytic activity, stability, and durability. Future research focusing on cost-effective synthesis, scaling up production, and expanding their applications in diverse electrochemical systems will further establish their role in next-generation energy solutions.

## 7.8 FUTURE SCOPE

Although substantial progress has been made in developing carbon-based catalysts for the OER, the field continues to offer significant opportunities for innovation and improvement. One major objective moving forward is to enhance both the catalytic activity and long-term durability of these materials under realistic and demanding working conditions, such as those encountered in industrial-scale water electrolysis.

Future efforts should prioritize the design of catalysts that combine the advantages of heteroatom doping with the presence of transition metal centers. The interaction between these components can lead to improved charge transfer properties, better adsorption of intermediates, and optimized electronic structures, all of which contribute to more efficient OER performance. Exploring combinations such as nitrogen or sulfur dopants with metals like cobalt, iron, or nickel could lead to breakthroughs in catalyst functionality.

Equally important is the development of advanced fabrication methods that offer precise control over material characteristics at the nanoscale. Techniques that enable the tuning of pore structures, introduction of active defects, and modulation of surface functional groups will play a vital role in tailoring catalytic behavior. Creating edge-enriched, defect-dense carbon architectures may prove especially effective in increasing active site density and improving reaction kinetics.

The integration of computational tools, particularly those driven by machine learning and data science, is expected to revolutionize catalyst discovery. These technologies can analyze vast datasets, predict promising material combinations, and guide experimental design, thereby reducing development time and cost. By leveraging artificial intelligence, researchers can more efficiently identify optimal compositions and structural features that maximize catalytic performance.

Another critical challenge lies in scaling up the production of these advanced carbon-based materials in a cost-effective and environmentally sustainable manner. While many lab-scale prototypes demonstrate excellent performance, translating these successes into commercially viable solutions requires scalable synthesis techniques, consistent material quality, and compatibility with existing electrochemical systems. Pilot-scale studies and performance assessments under real-world conditions will be essential to bridge this gap.

Ultimately, addressing these challenges will require a multi-disciplinary approach that combines materials science, electrochemistry, engineering, and computational modeling. By advancing knowledge across these domains, carbon-based OER electrocatalysts can become a key component of future renewable energy systems, contributing significantly to clean hydrogen production and sustainable energy conversion technologies.

## 7.9 CONCLUSION

Investigating heterogeneous electrocatalysts based on nanostructured carbon for the OER in alkaline media is a major step toward affordable and sustainable energy solutions. This chapter analyzes the evolution of these materials and highlights their potential to replace expensive and scarce noble metal catalysts such as $IrO_2$ and $RuO_2$. The exceptional electrocatalytic efficacy of graphitized carbon catalysts, which is fueled by their low cost, unique physicochemical features, and tunability, makes them excellent candidates for improving renewable energy technologies, particularly electrochemical water splitting. The debate began with metal-free nanocarbon electrocatalysts, with a focus on single and multi-element doping (e.g., N, S, P) and the design of edge- and defect-rich structures. These changes improve the electrical characteristics and catalytic activity of carbon frameworks by enhancing the adsorption of OER intermediaries. Transition metal and heteroatom-doped nanocarbons, such as Co-N-C, Ni-N-C, and Fe-N-C, improve performance by harnessing the synergy between metal centers and nitrogen-doped carbon matrices, attaining efficiencies comparable to noble metal benchmarks at more affordable prices. Furthermore, hybrid electrocatalysts combining nanocarbons with TMNx, TMSx, TMSex, and TMPx, as well as non-hybrid metals (e.g., Pt, Pd, Au), exhibit improved conductivity, stability, and active site availability, broadening their applicability across OER, HER, ORR, and energy storage systems. The requirement for scalable synthesis techniques, enhanced long-term stability in challenging electrochemical environments, and a better comprehension of the catalytic mechanisms at work are still obstacles despite these developments. Computational technologies like machine learning (ML) and density functional theory (DFT) have been used to speed up catalyst design by providing predictive insights into the best material structures and compositions. By addressing current limitations and harnessing their full potential, these materials can pave the way for affordable, efficient, and sustainable electrochemical energy systems, contributing significantly to the global transition toward a carbon-neutral future.

## REFERENCES

Abghoui, Y., & Skúlason, E. (2017). Hydrogen evolution reaction catalyzed by transition-metal nitrides. *The Journal of Physical Chemistry C*, *121*(43), 24036–24045. https://doi.org/10.1021/acs.jpcc.7b06811

Adalati, R., Sharma, M., Sharma, S., Kumar, A., Malik, G., Boukherroub, R., & Chandra, R. (2022). Metal nitrides as efficient electrode material for supercapacitors: A review. *Journal of Energy Storage*, *56*, 105912. https://doi.org/10.1016/j.est.2022.105912

Agrawal, P., Ebrahim, S., & Ponnamma, D. (2024). Advancements in nanocarbon-based catalysts for enhanced fuel cell performance: A comprehensive review. *International Journal of Energy and Water Resources*. https://doi.org/10.1007/s42108-024-00324-w

Anjana, P. M., Bindhu, M. R., & Rakhi, R. B. (2019). Green synthesized gold nanoparticle dispersed porous carbon composites for electrochemical energy storage. *Materials Science for Energy Technologies*, *2*(3), 389–395. https://doi.org/10.1016/j.mset.2019.03.006

Chen, X., Qi, Y., He, B., Liang, Y., Lei, Y., & Sun, J. (2023). Fabrication of adjustable Au/Carbon hybrid nanozymes with photothermally enhanced peroxidase activity and ultra-sensitivity for glutathione detection. *ACS Applied Materials & Interfaces*, *15*(17), 20788–20799. https://doi.org/10.1021/acsami.3c02420

Chen, X., Xue, Z., Zheng, Y., Liu, X., & Zhang, Y. (2021). Uniformly dispersed platinum nanoparticles over nitrogen-doped reduced graphene oxide as an efficient electrocatalyst for the oxygen reduction reaction. *RSC Advances*, *11*(54), 34125–34131. https://doi.org/10.1039/D1RA04857K

Chen, Y., Jiang, T., Tian, C., Zhan, Y., Adabifiroozjaei, E., Kempf, A., Molina-Luna, L., Hofmann, J. P., Riedel, R., & Yu, Z. (2023). Molybdenum phosphide quantum dots encapsulated by P/N-doped carbon for hydrogen evolution reaction in acid and alkaline electrolytes. *ChemSusChem*, *16*(20), e202300479. https://doi.org/10.1002/cssc.202300479

Ding, R., Chen, Y., Li, X., Rui, Z., Hua, K., Wu, Y., Duan, X., Wang, X., Li, J., & Liu, J. (2022). Atomically dispersed, low-coordinate Co–N sites on carbon nanotubes as inexpensive and efficient electrocatalysts for hydrogen evolution. *Small*, *18*(4), 2105335. https://doi.org/10.1002/smll.202105335

Downes, C. A., Van Allsburg, K. M., Tacey, S. A., Unocic, K. A., Baddour, F. G., Ruddy, D. A., LiBretto, N. J., O'Connor, M. M., Farberow, C. A., Schaidle, J. A., & Habas, S. E. (2022). Controlled synthesis of transition metal phosphide nanoparticles to establish composition-dependent trends in electrocatalytic activity. *Chemistry of Materials*, *34*(14), 6255–6267. https://doi.org/10.1021/acs.chemmater.2c00085

Gan, L., Liu, J., Han, L., Li, J., Zhao, J., Chen, K., Jiang, D., & Ren, Y. (2024). Highly active CoP embedded in hollow nitrogen-doped carbon as an efficient bifunctional electrocatalyst for rechargeable zinc-air batteries. *International Journal of Hydrogen Energy*, *51*, 1391–1399. https://doi.org/10.1016/j.ijhydene.2023.09.197

Gao, C., Yao, H., Wang, P., Zhu, M., Shi, X.-R., & Xu, S. (2024). Carbon-based composites for oxygen evolution reaction electrocatalysts: Design, fabrication, and application. *Materials*, *17*(10), 2265. https://doi.org/10.3390/ma17102265

Gómez, M. J., Loiácono, A., Pérez, L. A., Franceschini, E. A., & Lacconi, G. I. (2019). Highly efficient hybrid Ni/Nitrogenated graphene electrocatalysts for hydrogen evolution reaction. *ACS Omega*, *4*(1), 2206–2216. https://doi.org/10.1021/acsomega.8b02895

Gong, L., Xia, F., Zhu, J., Mu, X., Chen, D., Zhao, H., Chen, L., & Mu, S. (2024). Hydrogen evolution reactivity of pentagonal carbon rings and p-d orbital hybridization effect with Ru. *Angewandte Chemie International Edition*, *63*(50), e202411125. https://doi.org/10.1002/anie.202411125

Gupta, V., Chandra, N., Tari, V., & Kannan, K. (2024). A sustainable approach for the green synthesis of silver nanoparticles using *Brassica oleracea* subsp. *botrytis* (L.) leaves and its antibacterial activity. *Physics and Chemistry of Solids*, *25*(2), 303–310. https://doi.org/10.15330/pcss.25.2.303-310

Gurushankar, K., Karuppaiya, C., Tari, V., & Kannan, K. (2024). Exploring MXenes and MBenes in metal sulfur and selenium battery technologies. In K. Kannan & V. Tari (Eds.), *Novel energy storage and conversion technologies for two-dimensional MXenes and MBenes*. IGI Global. https://doi.org/10.4018/979-8-3693-9546-2.ch005

Hu, X., Gao, Y., Luo, X., Xiong, J., Chen, P., & Wang, B. (2024). Insight into the intrinsic activity of various transition metal sulfides for efficient hydrogen evolution reaction. *Nanoscale*, *16*(9), 4909–4918. https://doi.org/10.1039/D3NR06456E

Ju, Y., & Li, F. (2006). A novel carbon nanotube-supported NiP amorphous alloy catalyst and its hydrogenation activity. *Journal of Natural Gas Chemistry*, 15(4), 313–318. https://doi.org/10.1016/S1003-9953(07)60012-7

Kannan, K., & Tari, V. (Eds.). (2025). *Novel energy storage and conversion technologies for two-dimensional MXenes and MBenes*. https://www.amazon.com/dp/9798369395462

Li, G., Li, L., & Lin, Z. (2021). CoP-anchored high N-doped carbon@graphene sheet as bifunctional electrocatalyst for efficient overall water splitting. *International Journal of Hydrogen Energy*, *46*(35), 18224–18232. https://doi.org/10.1016/j.ijhydene.2021.02.219

Li, H., Pan, Y., Wu, L., He, R., Qin, Z., Luo, S., Yang, L., & Zeng, J. (2023). $IrO_2$ deposited on $RuO_2$ as core-shell structured $RuO_2$@$IrO_2$ for oxygen evolution reaction in electrochemical water electrolyzer. *Molecular Catalysis*, *551*, 113619. https://doi.org/10.1016/j.mcat.2023.113619

Li, X., Bergljót Gunnarsdóttir, A., Pershina, V., Höskuldsson, Á. B., Hidalgo, M. F., Skúlason, E., Dögg Flosadóttir, H., & Giordano, C. (2024). Towards improved stability of transition metal nitrides in aqueous solutions. *Progress in Solid State Chemistry*, *75*, 100474. https://doi.org/10.1016/j.progsolidstchem.2024.100474

Li, Y., Yin, Z., Cui, M., Liu, X., Xiong, J., Chen, S., & Ma, T. (2021). Interface engineering of transitional metal sulfide–$MoS_2$ heterostructure composites as effective electrocatalysts for water-splitting. *Journal of Materials Chemistry A*, *9*(4), 2070–2092. https://doi.org/10.1039/D0TA10815D

Liang, Y., Li, Y., Wang, H., & Dai, H. (2013). Strongly coupled inorganic/nanocarbon hybrid materials for advanced electrocatalysis. *Journal of the American Chemical Society*, *135*(6), 2013–2036. https://doi.org/10.1021/ja3089923

Liu, Q., Liu, K., Huang, J., Hui, C., Li, X., & Feng, L. (2024). A review of modulation strategies for improving the catalytic performance of transition metal sulfide self-supported electrodes for the hydrogen evolution reaction. *Dalton Transactions*, *53*(9), 3959–3969. https://doi.org/10.1039/D3DT04244H

Ma, J., Habrioux, A., Luo, Y., Ramos-Sanchez, G., Calvillo, L., Granozzi, G., Balbuena, P. B., & Alonso-Vante, N. (2015). Electronic interaction between platinum nanoparticles and nitrogen-doped reduced graphene oxide: Effect on the oxygen reduction reaction. *Journal of Materials Chemistry A*, *3*(22), 11891–11904. https://doi.org/10.1039/C5TA01285F

Meng, Z., Zheng, S., Luo, R., Tang, H., Wang, R., Zhang, R., Tian, T., & Tang, H. (2022). Transition metal nitrides for electrocatalytic application: Progress and rational design. *Nanomaterials*, *12*(15), 2660. https://doi.org/10.3390/nano12152660

Min, X., & Kanan, M. W. (2015). Pd-catalyzed electrohydrogenation of carbon dioxide to formate: High mass activity at low overpotential and identification of the deactivation pathway. *Journal of the American Chemical Society*, *137*(14), 4701–4708. https://doi.org/10.1021/ja511890h

Muhammad, P., Zada, A., Rashid, J., Hanif, S., Gao, Y., Li, C., Li, Y., Fan, K., & Wang, Y. (2024). Defect engineering in nanocatalysts: From design and synthesis to applications. *Advanced Functional Materials*, *34*(29), 2314686. https://doi.org/10.1002/adfm.202314686

Namsheer K, Pramoda, K., Kumar, K. S. S., Radhakrishnan, S., & Rout, C. S. (2023). Molybdenum sulfo-selenide nanocomposites with carbon nanotubes and reduced graphene oxide for photocatalytic hydrogen evolution reaction. *Energy Advances*, *2*(10), 1724–1734. https://doi.org/10.1039/D3YA00219E

Ni, L., Gallenkamp, C., Wagner, S., Bill, E., Krewald, V., & Kramm, U. I. (2022). Identification of the catalytically dominant iron environment in iron- and nitrogen-doped carbon catalysts for the oxygen reduction reaction. *Journal of the American Chemical Society*, *144*(37), 16827–16840. https://doi.org/10.1021/jacs.2c04865

Pi, C., Huang, C., Yang, Y., Song, H., Zhang, X., Zheng, Y., Gao, B., Fu, J., Chu, P. K., & Huo, K. (2020). In situ formation of N-doped carbon-coated porous MoP nanowires: A highly efficient electrocatalyst for hydrogen evolution reaction in a wide pH range. *Applied Catalysis B: Environmental*, *263*, 118358. https://doi.org/10.1016/j.apcatb.2019.118358

Porter, W. N., Turaczy, K. K., Yu, M., Mou, H., & Chen, J. G. (2024). Transition metal nitride catalysts for selective conversion of oxygen-containing molecules. *Chemical Science*, *15*(18), 6622–6642. https://doi.org/10.1039/D4SC01314J

Sajjad, M., Amin, M., Javed, M. S., Imran, M., Hu, W., Mao, Z., & Lu, W. (2021). Recent trends in transition metal diselenides ($XSe_2$: X = Ni, Mn, Co) and their composites for high energy faradic supercapacitors. *Journal of Energy Storage*, *43*, 103176. https://doi.org/10.1016/j.est.2021.103176

Singh, H., Marley-Hines, M., Chakravarty, S., & Nath, M. (2022). Multi-walled carbon nanotube supported manganese selenide as a highly active bifunctional OER and ORR electrocatalyst. *Journal of Materials Chemistry A*, *10*(12), 6772–6784. https://doi.org/10.1039/D1TA09864K

Song, J., Wei, C., Huang, Z.-F., Liu, C., Zeng, L., Wang, X., & Xu, Z. J. (2020). A review on fundamentals for designing oxygen evolution electrocatalysts. *Chemical Society Reviews*, *49*(7), 2196–2214. https://doi.org/10.1039/C9CS00607A

Tari, V., Kannan, K., Koliar, J., Gurushankar, K., & Wahyuni, D. K. (2025). Influence of zinc oxide nanomaterials on crop plants: A prospective analysis. In D. Jabbrova, D. Sarkar, Rakesh, S. Singh, & R. Datta (Eds.), *Smart nanomaterials in agroforestry systems*. Springer. https://doi.org/10.1007/978-981-96-1337-3_8

Tari, V., Kannan, K., & Vishwakarma, V. (2022). Emerging antiviral technologies. In P. Nunthavarawong, S. M. Rangappa, S. Siengchin, & M. Thoppil-Mathew (Eds.), *Antimicrobial and antiviral materials: Polymers, metals, ceramics, and applications* (1st ed., pp. 137–150). CRC Press. https://doi.org/10.1201/9781003143093-8

Theerthagiri, J., Lee, S. J., Murthy, A. P., Madhavan, J., & Choi, M. Y. (2020). Fundamental aspects and recent advances in transition metal nitrides as electrocatalysts for hydrogen evolution reaction: A review. *Current Opinion in Solid State and Materials Science*, *24*(1), 100805. https://doi.org/10.1016/j.cossms.2020.100805

Tiwari, S. K., Pandey, S. K., Pandey, R., Wang, N., Bystrzejewski, M., Mishra, Y. K., & Zhu, Y. (2023). Stone–Wales defect in graphene. *Small*, *19*(44), 2303340. https://doi.org/10.1002/smll.202303340

Velinova, R., Naydenov, A., Kichukova, D., Tumbalev, V., Atanasova, G., Kovacheva, D., & Spassova, I. (2023). Highly efficient RGO-supported Pd catalyst for low temperature hydrocarbon oxidation. *Catalysts*, *13*(8), 1224. https://doi.org/10.3390/catal13081224

Wu, L., Guo, T., & Li, T. (2022). Data-driven high-throughput rational design of double-atom catalysts for oxygen evolution and reduction. *Advanced Functional Materials*, *32*(31), 2203439. https://doi.org/10.1002/adfm.202203439

Xing, H., He, W., Liu, Y., Long, G., Sun, Y., Feng, J., Feng, W., Zhou, Y., Zong, Y., Li, X., Zhu, X., & Zheng, X. (2021). Ultrathin and highly crumpled/porous CoP nanosheet arrays anchored on graphene boosts the capacitance and their synergistic effect toward high-performance battery-type hybrid supercapacitors. *ACS Applied Materials & Interfaces*, *13*(22), 26373–26383. https://doi.org/10.1021/acsami.1c04921

Yang, T., Du, M., Zhu, H., Zhang, M., & Zou, M. (2015). Immobilization of Pt nanoparticles in carbon nanofibers: Bifunctional catalyst for hydrogen evolution and electrochemical sensor. *Electrochimica Acta*, *167*, 48–54. https://doi.org/10.1016/j.electacta.2015.03.077

Yang, X., Yu, G., & Chen, W. (2024). Realizing a high OER activity in new single-atom catalysts formed by introducing $TMN_x$ ( x = 3 and 4) units into carbon nanotubes using high-throughput calculations. *Nanoscale*, *16*(1), 273–283. https://doi.org/10.1039/D3NR04396G

Yi, Y., Hai, F., Guo, J., Tian, X., Zheng, S., Wu, Z., Wang, T., & Li, M. (2022). Progress and prospect of practical lithium-sulfur batteries based on solid-phase conversion. *Batteries*, *9*(1), 27. https://doi.org/10.3390/batteries9010027

Youn, D. H., Bae, G., Han, S., Kim, J. Y., Jang, J.-W., Park, H., Choi, S. H., & Lee, J. S. (2013). A highly efficient transition metal nitride-based electrocatalyst for oxygen reduction reaction: TiN on a CNT–graphene hybrid support. *Journal of Materials Chemistry A*, *1*(27), 8007. https://doi.org/10.1039/c3ta11135k

Yu, M., Guo, X., Chang, X., Ma, X., & Zhang, M. (2022). Assembled cobalt phosphide nanoparticles on carbon nanofibers as a bifunctional catalyst for hydrogen evolution reaction and oxygen evolution reaction. *Sustainable Energy & Fuels*, *6*(21), 5000–5007. https://doi.org/10.1039/D2SE01128J

Zhang, H., Maijenburg, A. W., Li, X., Schweizer, S. L., & Wehrspohn, R. B. (2020). Bifunctional heterostructured transition metal phosphides for efficient electrochemical water splitting. *Advanced Functional Materials*, *30*(34), 2003261. https://doi.org/10.1002/adfm.202003261

Zhang, L., Xiao, J., Wang, H., & Shao, M. (2017). Carbon-based electrocatalysts for hydrogen and oxygen evolution reactions. *ACS Catalysis*, 7(11), 7855–7865. https://doi.org/10.1021/acscatal.7b02718

Zhang, X., Xu, G., & Zhang, L. (2023). Solvent-induced structural regulation over $Ni_2$ P/CNT hybrids towards boosting the performance of supercapacitors. *Dalton Transactions*, *52*(20), 6763–6772. https://doi.org/10.1039/D3DT00517H

Zhang, X., Yang, P., & Jiang, S. P. (2021). Pd nanoparticles assembled on Ni- and N-doped carbon nanotubes towards superior electrochemical activity. *International Journal of Hydrogen Energy*, *46*(2), 2065–2074. https://doi.org/10.1016/j.ijhydene.2020.10.096

Zhao, H., Gao, G., Wang, Y., Chen, R., Du, Y., Wang, M., Li, Z., Liu, Y., & Wang, L. (2022). Growth of carbon nanotubes coated CoP as electrocatalyst for hydrogen evolution reaction under acidic and alkaline solutions. *Journal of Alloys and Compounds*, *927*, 167057. https://doi.org/10.1016/j.jallcom.2022.167057

Zheng, L., Xu, P., Zhao, Y., Peng, J., Yang, P., Shi, X., & Zheng, H. (2022). Unique core-shell $Co_2(OH)_2CO_3$@MOF nanoarrays with remarkably improved cycling life for high performance pseudocapacitors. *Electrochimica Acta*, *412*, 140142. https://doi.org/10.1016/j.electacta.2022.140142

Zhong, Y., Xia, X., Shi, F., Zhan, J., Tu, J., & Fan, H. J. (2016). Transition metal carbides and nitrides in energy storage and conversion. *Advanced Science*, *3*(5), 1500286. https://doi.org/10.1002/advs.201500286

Zhu, W., Cheng, Y., Wang, C., Pinna, N., & Lu, X. (2021). Transition metal sulfides meet electrospinning: Versatile synthesis, distinct properties and prospective applications. *Nanoscale*, *13*(20), 9112–9146. https://doi.org/10.1039/D1NR01070K

Zhuang, Z., Giles, S. A., Zheng, J., Jenness, G. R., Caratzoulas, S., Vlachos, D. G., & Yan, Y. (2016). Nickel supported on nitrogen-doped carbon nanotubes as hydrogen oxidation reaction catalyst in alkaline electrolyte. *Nature Communications*, 7(1), 10141. https://doi.org/10.1038/ncomms10141

Chapter 8

# rGO/MXene hybrid exhibit fast oxygen reduction reaction catalytic activity

*Vipin Kumar*

## 8.1 INTRODUCTION

Petroleum and its derivatives are non-renewable, and their exploitation has a harmful impact on the environment (Tari and Patil, 2017a; Tari and Patil, 2017b). In recent years, fossil fuel-based energy has also imposed a significant economic burden on society. Therefore, the development of sustainable and eco-friendly energy technologies has become one of the most pressing concerns of modern civilization (Tari et al., 2025; Gupta et al., 2024, Tari et al., 2022). With the increasing demand for economically viable and renewable energy sources, scientists have shown keen interest in energy storage and conversion technologies (Tong et al., 2021; Gurushankar et al., 2024; Kannan and Tari, 2025).

Among these, the oxygen reduction reaction (ORR) and oxygen evolution reaction (OER) are critical in designing high-performance bifunctional electrocatalysts, which are essential for developing renewable energy systems. However, the high overpotentials associated with these reactions significantly hinder the efficiency of energy conversion and storage devices. ORR and OER are fundamental to the operation of many energy systems, playing pivotal roles in electrochemical processes (Zhao et al., 2023).

In energy conversion systems, the overpotentials of ORR and OER govern the efficiency of electrochemical transformations for generating renewable energy (Wu et al., 2024; Mao et al., 2014). Improving the kinetics of these reactions is crucial for reducing overpotentials. Modern systems employ advanced catalysts for ORR and OER with lower overpotentials than conventional noble metal-based catalysts such as $IrO_2$ and Pt (Shan et al., 2023). Alternatives include MXenes (Cano et al., 2018), transition metal carbides/nitrides (Wang et al., 2018), and graphene/reduced graphene oxide (rGO)(Luo et al., 2014).

Although the structural design of ORR/OER/HER (hydrogen evolution reaction) systems can be complex, their basic functional principles are straightforward. These systems contain two electrodes: the cathode supports the HER or ORR, while the anode facilitates the OER or the

DOI: 10.1201/9781003584742-8

oxidation of chemical fuels. One major challenge to real-world implementation is the slow kinetics of the OER, which requires a high overpotential ($\eta$). This is because OER involves complex electron-proton coupling, while ORR primarily involves electron transfer (Suen et al., 2017).

In order to analyze its behavior in cathodic activities and in the closing of the electron transfer circuit, ORR is thoroughly investigated at the cathode of the fuel cell and the oxygen-metal battery. Several adsorption/desorption procedures and reactive stages, including oxygen-containing molecules such as O, $O^{2-}$, OH, $HO^{2-}$, and $H_2O_2$ are typically included in ORR. However, because of the high bond energy of the $O_2$ molecule, ORR is frequently the slowest process in oxygen batteries and fuel cells, necessitating the employment of catalysts (Yoo et al., 2022).

Developing efficient energy storage and conversion devices necessitates novel "cutting-edge" materials. Hierarchical composites that combine the distinct advantages of each material layer have emerged as promising candidates for next-generation "energy materials".

MXenes, known for their excellent electrical conductivity and unique structure, when combined with rGO—which offers superior mechanical strength and flexibility—form composites with significantly enhanced performance. These rGO/MXene composites hold great promise across various scientific applications. Firstly, they offer a comprehensive overview of the current state-of-the-art, demonstrating their versatile roles and potential advantages in energy systems.

Since the successful isolation of graphene from graphite in 2004, interest in two-dimensional (2D) nanomaterials has grown rapidly due to their unique chemical and structural properties (Li et al., 2020). Owing to its remarkable electronic, mechanical, and surface chemical characteristics (Lu et al., 2020), graphene's 2D structure is highly promising for applications in energy storage and electrocatalysis.

A new class of 2D transition metal carbides, nitrides, and carbonitrides—collectively termed MXenes—was discovered in 2011. The "ene" suffix links MXenes with other 2D materials like graphene, silicene, and phosphorene (Wei et al., 2020). MXenes are characterized by a layered structure of transition metal atoms bonded to carbon and/or nitrogen. They offer tunable properties, such as surface area and electron transport rate, and their composition is highly versatile. Terminal functional groups such as –O and –P can further enhance their conductivity and stability (Intikhab et al., 2017).

Unlike the predecessors of graphene or other 2D substances that have been layered and keep their strength from weak van der Waals forces, the connections that exist between polynary layered in the MXene phase are too robust to be destroyed by simple mechanical procedures.

Kan et al. (2021) thoroughly investigated the ORR and OER features of the 26 regenerated catalysts with just one atom on MXene based on first-principles research; their terminal F and O groups of MXene enhanced the stability of their structures and activity as catalysts. The transition metal carbons (TMCs), like g-$C_3N_4$, have been used widely in addition to MXenes due to their low cost, excellent chemical stability, and good conductivity to electricity (Mansor et al., 2016).

The synergistic characteristics of rGO and MXene are examined in this chapter, with a focus on how they could cooperate to develop energy conversion innovations. New developments in hybrid electrocatalyst design using MXene have been reported recently. On the OER and ORR performance of rGO/MXene hybrid electrocatalysts, the influence of molecular species, terminal functional groups, defects, and dimensions ought to be discussed.

The final step ought to show how to modify Gr, Gra, graphene oxide (GO), and rGO (graphene family)/MXene-based electrocatalysts. The two main strategies are increasing the total number of those with active sites to improve kinetics and boosting the intrinsic catalyst activity of the sites that are active to improve thermodynamics. Specific strategies for boosting MXene catalytic activity include functionality development, atom substitutes, defect engineering, morphological control, and heterojunction formation employing graphene family and LSHs/layered double hydroxides (LDHs). Finally, we have presented the challenges and future prospects of these materials and then offered our conclusions and perspectives.

### 8.1.1 (Gr, GO, rGO, Gra, GDY)/MXene hybrids

Due to the growing global demand for renewable energy, researchers are focusing on developing adaptive hybrid 2D materials and their composites. Among 2D materials, transition metal carbides/nitrides (MXenes), LDHs, and GO have demonstrated enormous potential in energy storage because of their unique morphological characteristics and excellent thermomechanical, electrical, electronic, magnetic, and electromagnetic properties.

Many organizations have developed heterogeneous structures combining graphene and MXenes to improve catalytic activity. Geng et al. (2017) fabricated $Mo_2C$/MXene on graphene sheets, which enhanced the OER and ORR performance, owing to the distinct bilayer structure of MXene and a variety of transition-metal connections. Zhou et al. (2018) discovered the electronic connection between graphitic carbon and MXene through a detailed analysis of the catalytic mechanisms of MXenes and N-doped carbon-based materials.

Moreover, the heterogeneous structural features had a synergistic effect on catalytic performance. It is relevant to examine various graphitic (Gra)/MXene heterostructures as catalytic elements with enhanced properties. Chen *et al.* integrated density functional theory (DFT) simulations with machine learning (ML) to predict the catalytic activity of 78 Gra/MXene ($V_2C$) heterostructures. To accelerate catalyst design and improve prediction accuracy, they identified that the $CoN_4$-$V_2C$ heterostructure functioned as an efficient ORR/OER bifunctional catalyst (Chen et al., 2022). Since nitrogen doping reduces OER activity, MXenes' electrochemical stability can be improved, suggesting that N-atom-doped MXenes may offer better charge transfer than C-atom-doped forms.

The goal of that study was to complex design and analyze 153 $MN_4$-graphene (Gra)/MXene ($M_2NO$) electrocatalysts for OER/ORR catalytic activity using ML along with DFT. The DFT results showed that $CoN_4$-Gra/$Ti_2NO$ demonstrated excellent OER (0.30 V) and ORR (0.37 V) overpotentials, while $MN_4$-Gra/$Cr_2NO$ and $TiN_4$-Gra/$M_2NO$ had significant overpotentials. Orbital spin polarization and d-band centers far from the Fermi energy level affect the adsorption energy of oxygen intermediates, lowering catalytic performance.

The ML outcomes demonstrated that the gradient boosting regression (GBR) model accurately predicted the overpotentials of monofunctional catalysts like $RuN_4$-Gra/$W_2NO$ (OER, 0.45 V) and $RhN_4$-Gra/$Ti_2NO$ (ORR, 0.39 V), as well as those of the bifunctional catalyst $RuN_4$-Gra/$W_2NO$ (ORR, 0.39 V; OER, 0.45 V). Combining these GO-based 2D materials with GO/MXene can enhance thermomechanical stability, increase interlayer spacing, and promote the formation of a network-like 2D/2D structure while minimizing the risk of layer agglomeration. These combined characteristics significantly improve GO/MXene's potential for energy storage and conversion (Bai et al., 2024).

2D material composites like graphene and MXene are drawing increased attention due to their exceptional electrical, mechanical, and electronic properties, making them ideal for advanced electromagnetic applications (Cao et al., 2024). Their high conductivity, flexibility, tunable band gaps, and strong light–matter interactions make them suitable for flexible devices, high-frequency electronics, and optoelectronics (Kumbhakar et al., 2023).

For instance, new 2D heterostructure electrodes, particularly those that mix graphdiyne (GDY) and MXenes, are gaining a lot of attention in the area of novel materials. These new combinations demonstrate great potential for enhancing ionic and electron mobility. The unique properties of GDY, such as its high conductivity to electricity and mechanical stability, enhance MXenes, which are renowned for their remarkable conductivity and hydrophilicity. By merging these substances, researchers have created a synergy that could lead to better electrode performance. The effectiveness of devices such as supercapacitors and batteries could be enhanced by these materials' potential for improving their charge transport mobility values (Zhou et al., 2024).

For instance, new 2D heterostructure electrodes-particularly those that combine GDY and MXenes are gaining significant attention in the field of advanced materials. These novel combinations show great potential for enhancing both ionic and electronic mobility. The unique properties of GDY, such as its high electrical conductivity and mechanical flexibility, complement the exceptional conductivity and hydrophilicity of MXenes. By merging these materials, researchers have created a synergistic effect that could lead to improved electrode performance. The effectiveness of devices such as supercapacitors and batteries may be significantly enhanced by the improved charge transport mobility offered by these materials (Zhou et al., 2024).

### 8.1.2 3D MXene/graphene hybrids

Assembling MXene sheets into a three-dimensional (3D) porous material remains challenging due to their low gelation ability (Zhang et al., 2017). Numerous significant efforts have been made to address this issue, such as modifying the hydrophilic-to-hydrophobic ratio of $Ti_3C_2T_x$ at water–oil interfaces. (Bian et al., 2018) assembled $Ti_3C_2T_X$-based architectures using an ice-template freezing technique, followed by annealing or chemical reduction with hydroiodic acid.

As a supercapacitor material, Yue et al. (2018) designed a $Ti_3C_2T_x$/rGO aerogel. Using GO, $Ti_3C_2T_x$ sheets were hydrothermally assembled to form highly porous, conductive, 3D structures. This process was followed by directional storage and freeze-drying. When the intrinsic structure of $Ti_3C_2T_x$ was well-retained, the resulting $Ti_3C_2T_x$ MXene/rGO hybrid aerogel (MGA) with an aligned core–shell structure provided its epoxy-based nanocomposite with remarkable thermal stability (695.9 S $m^{-1}$) and electromagnetic interference (EMI) shielding effectiveness greater than 50 dB at a modest $Ti_3C_2T_x$ ratio of 0.74 vol%. The microscopic structure of the hybrid aerogel formed by $Ti_3C_2T_x$ and rGO sheets was analyzed, and the proposed assembly mechanism was illustrated. As shown in Figure 8.1 [30], an MGA was fabricated using a GO-assisted bidirectional freezing process followed by freeze-drying.

$Ti_3C_2T_x$ sheets were assembled into a 3D porous structure using GO sheets, which possess strong gelation properties (Tong et al., 2021). Stable aqueous $Ti_3C_2T_x$/GO suspensions were easily prepared due to their similar hydrophilic nature and negative surface charges. Although MXene sheets

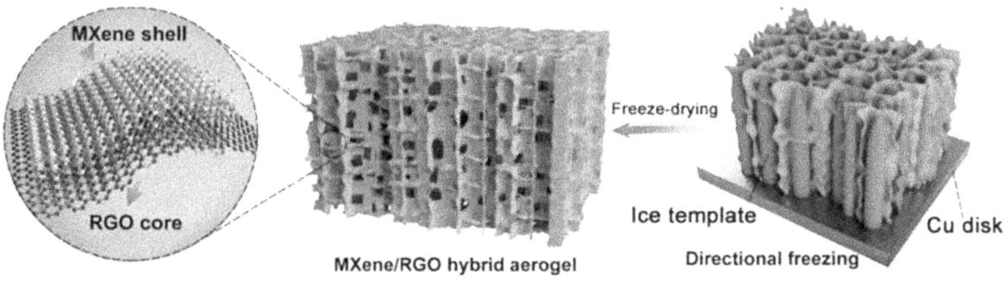


*Figure 8.1* Shows shematic of synthesis 3D $Ti_3C_2T_x$ MXene/rGO hybrid composite structure by frozen process [30].

**(Reproduced with get permission from ACS Society, Copyright 2018.)**

tend to oxidize easily in water at high temperatures, mild hydrothermal treatment—especially at low temperatures was chosen to minimize oxidation and preserve the structure of $Ti_3C_2T_x$. During this hydrothermal process, GO is reduced using ascorbic acid, while $Ti_3C_2T_x$ sheets are anchored onto the surfaces of the rGO framework to form an MXene/rGO hybrid. A 3D structure is formed as the rGO sheets interconnect.

### 8.1.3 Properties and applications of rGO/MXene composites

GO/MXene and rGO/MXene composite materials have emerged as a highly promising research area in recent decades. These composites improve performance and offer a wide range of applications, including in medical technology, energy storage, water purification, catalysis, electromagnetic shielding, and sensing technologies (Li et al., 2020).

By combining MXene with GO, the advantages of each material can be utilized to overcome their individual limitations. MXene's strong permeability enhances charge transfer and electrostatic properties, while GO's excellent dispersibility facilitates uniform composite formation (Figure 8.2). Additionally, the large surface area of GO provides more active sites for

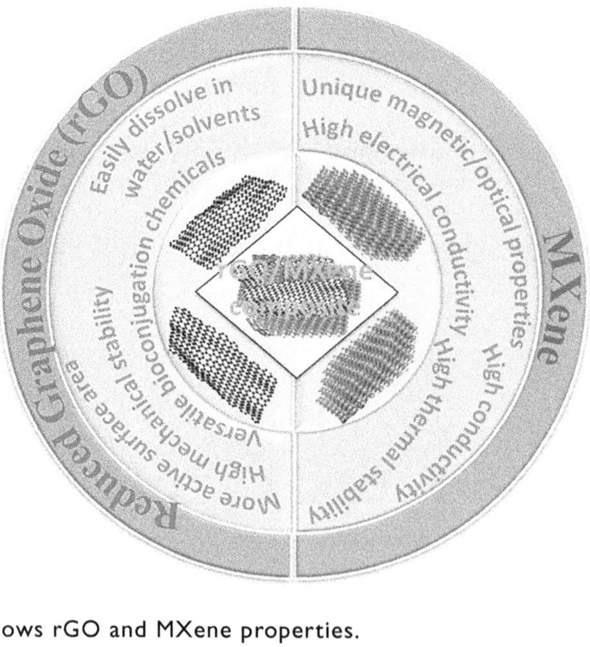


*Figure 8.2* Shows rGO and MXene properties.

reactions, and the mechanical strength of MXene contributes to the nanocomposite's overall structural stability (Sandhu et al., 2025).

The innovative integration of MXene and GO enables the development of nanocomposites with unique properties, making them suitable for advanced applications in energy storage, catalysis, sensors, and beyond. The detailed exploration and design of their individual and combined properties underscore the significance of their synergistic interaction and the potential for improved performance and multifunctionality across various scientific domains (Liu et al., 2020).

One of the most promising applications for rGO/MXene hybrids is in energy conversion and storage (Xiao et al., 2023). Their excellent electrical conductivity, high surface area, and outstanding (Iravani & Varma, 2022) electrochemical stability make them ideal candidates for supercapacitors and lithium-ion batteries. Furthermore, GO/MXene hybrids show great promise as catalysts for diverse chemical reactions due to their unique interface chemistry and high catalytic activity. They could be employed in environmental remediation and industrial chemical synthesis (Iravani & Varma, 2022).

Additionally, rGO/MXene hybrids have significant potential in biosensing technologies. These features make them excellent composite materials for sensor development, capable of enabling highly sensitive and selective detection of various analytes (Jia et al., 2021; Ma et al., 2016). Given their electrochemical properties, rGO/MXene composites are well-suited for electrochemical sensing applications (Bilibana, 2023). Moreover, their excellent adsorption capacity and selectivity make them effective adsorbents for water treatment. These materials can help remove harmful substances, such as toxic chemicals, dyes, and heavy metals from contaminated water, contributing to cleaner and safer water systems (Li et al., 2022).

### 8.1.4 Applications of rGO/MXene composites in membranes and electronics

In membrane science, rGO/MXene composites have shown great potential to enhance the filtration efficiency of membranes (Yin et al., 2021). Incorporating such nanomaterials into membrane matrices can improve water purification technologies by increasing permeability, fouling resistance, and separation efficiency (Liu et al., 2020) . Beyond water treatment and sensing applications, these composites are promising for a broad range of domains. Their exceptional electrical conductivity and mechanical flexibility make them suitable for smart electronics, including wearable technologies and flexible display screens.

Moreover, due to their strength and lightweight nature, rGO/MXene hybrids are being considered for aerospace applications, particularly in structural elements and components for aircraft. These materials also exhibit excellent potential for microwave absorption and electromagnetic interference (EMI) shielding, offering robust protection from radiation emitted by electronic devices and communication networks.

Their mechanical stability makes GO/MXene composites suitable for strain-sensing applications, enabling health monitoring, soft robotics, actuators, and wearable electronics. These materials are capable of detecting and quantifying mechanical deformation and dynamic pressure. For instance, Jia et al. (2021) developed an electrically conductive, flexible multilayer GO/ $Ti_3C_2T_x$ MXene film for humidity-driven sensor applications. The mechanical performance could be finely tuned by adjusting the GO-to-MXene ratio and by controlling oxygen functional groups and interlayer spacing.

GO also plays a critical role in stabilizing MXene, preventing its reduction and enhancing cycle durability. These composite actuators, designed to simulate natural movements—such as flower blooming—demonstrated capabilities for wireless motion control and switch mechanisms. Due to their biocompatibility and reliable response to moisture changes, they also serve well in respiratory monitoring applications (Achakulwisut et al., 2023).

However, many efficient electrocatalysts rely on expensive noble metals, making scalability difficult. Thus, developing cost-effective, high-performance catalysts remains a top priority (Guo et al., 2017). 2D materials like GO and MXenes have attracted considerable attention for their electrocatalytic potential in both photo- and electrochemical applications (Kalia et al., 2022).

A significant breakthrough was achieved by Qiao et al., who developed a $C_3N_4/Ti_3C_2$ hybrid film that demonstrated OER activity with an overpotential of 0.42 V at 10 mA $cm^{-2}$, marking the first application of an MXene-based hybrid in electrocatalysis (Birdja et al., 2019).

### 8.1.5 Electrochemical reactions

Electrochemical processes are necessary for cleaning energy expansion. Batteries powered by lithium-ion batteries have begun to utilize these factors permeable graphene elements (Jaramillo et al., 2007).

A common technique for producing exceptionally well composites, which accede to the inherent functionality of the MXene while also having innovative areas that catalyze the OER and ORR, is the growth of active species in situ on rGO/MXenes.

### 8.1.6 Mechanism of OER reactions

The OER is a critical anodic half-reaction in total water splitting. Due to its complex four-electron transfer mechanism, it demands higher overpotential compared to the simpler two-electron hydrogen evolution reaction (HER) (Liu & Dong, 2021).

**In acidic media**

$$2H_2O \rightarrow O_2 + 4H^+ + 4e^- \quad (8.1)$$

**In alkaline media**

$$4OH^- \rightarrow O_2 + 2H_2O + 4e^- \quad (8.2)$$

A widely accepted mechanism involves electrochemical oxide formation, where an oxide layer grows on the catalyst's surface during OER, increasing the number of active oxidation sites (Liu & Dong, 2021). This is represented by:

$$H_2O + {}^* \rightarrow {}^*OH + H^+ + e^- \quad (8.3)$$

$$^*OH \rightarrow {}^*O + H^+ + e^- \quad (8.4)$$

$$2^*O \rightarrow O_2 + 2^* \quad (8.5)$$

Alternate mechanisms, such as associative peroxide pathways, have also been proposed:

$$H_2O + {}^* \rightarrow {}^*OH + H^- + e^- \quad (8.6)$$

$$^*OH \rightarrow {}^*O + H^+ + e^- \quad (8.7)$$

$$^*O + H_2O \rightarrow {}^*OOH + H^+ + e^- \quad (8.8)$$

$$^*OOH + O_2 \rightarrow {}^* + H^+ + e^- \quad (8.9)$$

$IrO_2$ and $RuO_2$ have demonstrated the highest OER catalytic activity, but their cost limits widespread adoption (Liu & Dong, 2021). Therefore, identifying non-noble metal catalysts is crucial.

To address these challenges, (Gadipelli et al., 2016) introduced a simple synthesis method to produce highly active Co–N–C catalytic centers for OER. Their strategy involved precise graphitization and control over cobalt distribution within nitrogen-doped graphitic networks and cobalt nanoparticles. These structural features significantly improved OER kinetics, producing performance comparable or superior to $IrO_2$/C.

Mao et al. (2014) used cobalt oxide nanocrystals with crumpled graphene (CG) to enhance catalytic surface exposure by minimizing graphene sheet agglomeration. They further improved performance by nitrogen doping. The resulting composite of N-doped reduced graphene oxide (N-rGO) and $Co_3O_4$ exhibited outstanding OER activity due to strong synergistic interactions between the two components. Interestingly, while individual N-rGO and $Co_3O_4$ materials showed poor OER activity, their combination produced exceptional catalytic performance (Liang et al., 2011).

The majority of the extensive study on OER catalysts currently available has been on alkaline electrolytes because non-noble metal OER catalysts often have poor endurance in alkaline environments. Among the non-noble metal OER catalysts developed are MOFs, metal-free compounds, LDH, TM oxides, TM chalcogenides, TM borides, TM hydroxides, TM nitrides, and TM carbide compounds (Hao et al., 2019). However, the modest active surface area and intrinsic weak conductivity of electricity. Since it is well known that increasing the conductivity and surface area in substances enhances the OER performance of catalysts, mixing with other conductive materials—such as rGO, nanofibers, porous carbon, and MXenes—becomes an essential improvement strategy.

The remarkable ability of metal-free catalysts, such as carbon nitride ($C_3N_4$) and modified carbon materials, to stop catalytic corrosion in the OER process has led to a great deal of research into these compounds. Combining MXene with metal-free substances is another essential technique for producing OER catalysts with inexpensive prices, excellent durability, and exceptional efficiency.

Qiao's group reported using $Ti_3C_2T_x$ and $C_3N_4$ nanosheets together as an OER electrode component (Ma et al., 2016). With a $\eta_{10}$ of 0.42 V, the porous $Ti_3C_2/C_3N_4$ film, which was produced by sonication-assisted exfoliation and filtering, performed better in terms of OER capability in the alkaline electrolyte than both $C_3N_4$ and $Ti_3C_2$.

The remarkable OER performance might be attributed to the connection between Ti and $C_3N_4$, which reduced the electron density of Ti atoms and hence promoted the adsorption of $OH^-$ in the alkaline electrolyte. These investigations provided a simple and effective way to produce a stable electrocatalyst for OER having remarkable electrochemical activity. The hybrid material showed better catalytic activity, fast kinetics, and high durability toward OER in alkaline environments compared to $RuO_2$ and its graphene-based counterpart (FeNi-LDH/rGO).

Additionally, there were a lot of OER uses for MXene-based materials. When 2D $Ti_3C_2$ MXene nanosheets were joined with 2D g-$C_3N_4$ via Ti-$N_x$ contact, for example, a pore-independent surface with a high hydrophilic and conductive frame was formed, demonstrating the remarkable catalytic activity of OER (Hu et al., 2021). The surface of MXene could be used to create 2D materials in addition to mixing them mechanically.

### 8.1.7 Electrocatalysts for ORR

The ORR plays a pivotal role in renewable energy technologies, such as fuel cells and metal-air batteries (Ahmad Junaidi et al., 2021). To meet real-world demands, electrocatalysts must exhibit high catalytic activity and long-term stability. MXenes, when coupled with catalytically active materials, can redistribute electron density at the active sites, increase the number of active sites, and improve performance.

Although pristine MXenes exhibit limited catalytic activity for ORR, their performance improves significantly after delamination. However, research on MXenes as ORR catalysts in alkaline media is still limited (Ahmad Junaidi et al., 2021). Tailoring the electronic structure or bonding state of the active center in MXenes can significantly enhance ORR efficiency.

Composites of rGO and MXenes are particularly effective ORR catalysts, owing to rGO's high surface area, excellent carrier mobility, and chemical stability (Ding et al., 2023). Introducing rGO as an interlayer spacer mitigates restacking of MXene nanosheets, increasing the accessible surface area for electrolyte interaction.

Due to its sluggish kinetics and multi-electron transfer steps, ORR suffers from low mass transfer efficiency (Ding et al., 2023). Therefore, developing catalysts that combine high ORR activity with a large surface area is critical. Using machine learning models, (Bai et al., 2024) designed 19 $MN_4$-Gra/MXene ($M_2NO$) heterostructures to predict ORR/OER activity and stability. These catalysts were analyzed through the four-electron ORR pathway:

$$^{*}+O_2(g)+H^{+}+e^{-}\rightarrow \Delta G_1 \rightarrow ^{*}OOH \tag{8.10}$$

$$^{*}OOH+H^{+}+e^{-}\rightarrow \Delta G_2 \rightarrow ^{*}O+H_2O(l) \tag{8.11}$$

$$^{*}O+H^{+}+e^{-}\rightarrow \Delta G_3 \rightarrow ^{*}OH \tag{8.12}$$

$$^{*}OH+H^{+}+e^{-}\rightarrow \Delta G_4 \rightarrow H_2O(l)+^{*} \tag{8.13}$$

Here, * indicates the active site on the $MN_4$-Gra/MXene catalyst (Figure 8.3).

Ding et al. (2023) enhanced MXene's conductivity and reduced its charge transfer impedance from 14.37 Ω to 5.37 Ω by compounding it with FePc and rGO. This hybrid showed improved electron transfer rates and lower electrochemical resistance.

Figure 8.4b and c highlight the ORR behavior of FePc-MXene-rGO with varying compositions. Among them, the FePc-M-$rGO_{211}$ sample exhibited the highest ORR activity, evidenced by a peak potential of 0.94 V.

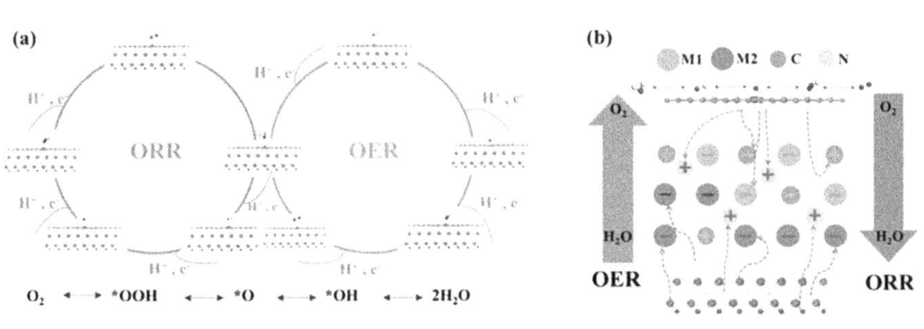


*Figure 8.3* (a) The ORR/OER reactions, (b) Schematic design of the interlayer electron transmission kinetics [33].

**(Reproduced with get permission from Elsevier, Copyright 2024.)**

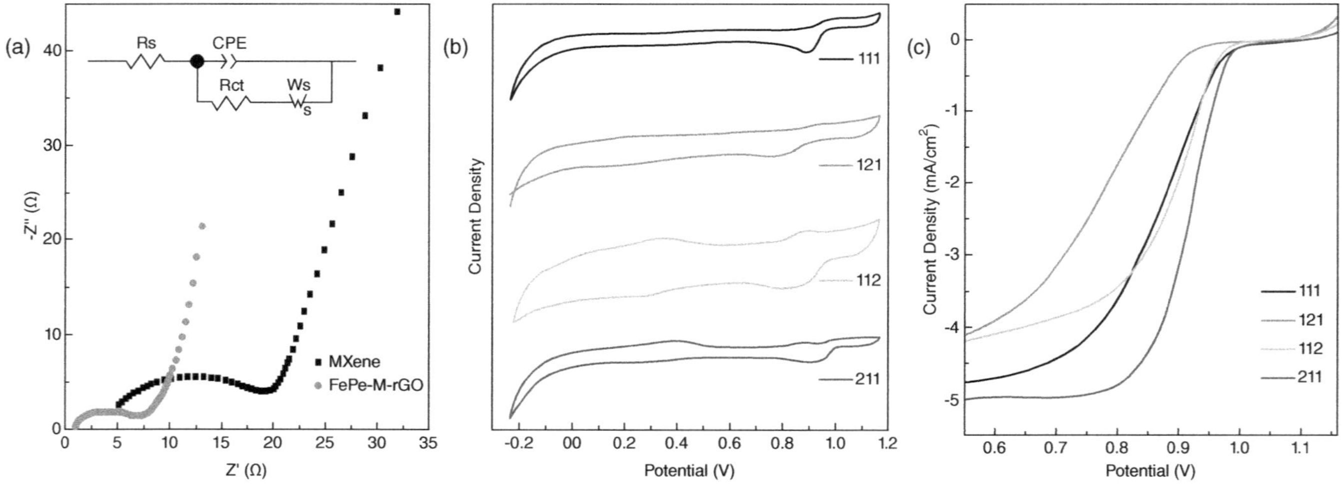


*Figure 8.4* (a) EIS, (b) CV and (c) LSV curves of four catalysts compounded according to different mass ratio (scanning rate of CV: 50 mV/s, scanning rate of LSV: 5 mV/s, same as below) [78].

**(Reproduced with get permission from Elsevier, Copyright 2023.)**

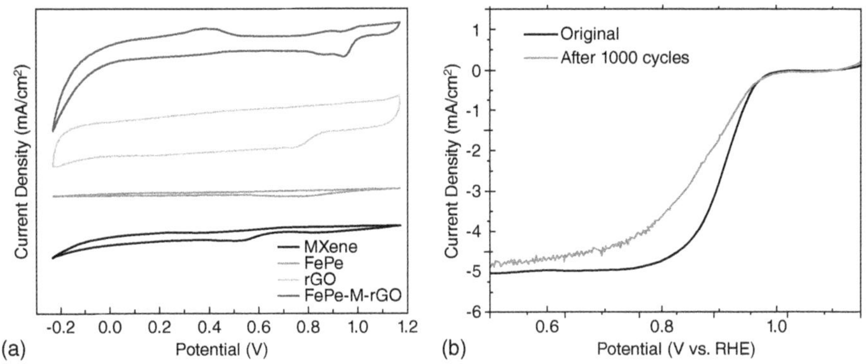


*Figure 8.5* (a) CV and (b) LSV curve of FePc-MXene-rGO after 1000 cycles [78]. **(Reproduced with get permission from Elsevier. Copyright 2023.)**

The improved dispersion of $FeN_4$ active sites, supported by rGO and MXene preventing aggregation, enhanced the catalyst's performance.

When compared to commercial Pt/C catalysts (Figure 8.5), FePc-M-rGO displayed high electron transfer numbers (~3.94), confirming its operation through a 4-electron reduction pathway. Long-term stability tests showed only a minor drop in half-wave potential ($E_{1/2}$ = 0.86 V after 1000 cycles), indicating excellent durability (Ostadhossein et al., 2019).

To understand MXene's effect in Li–$O_2$ batteries, the ORR rate on MXene was compared to graphene. Simulations revealed that void-rich MXene surfaces provided the highest ORR activity. In contrast, excessive functionalization (e.g., with $F^-$ or $O^-$) diminished catalytic efficiency by reducing the number of available active sites. Rapid $Li_2O_2$ formation was observed within 40 ps, reaching equilibrium around 150 ps. The ORR pathway included a disproportionation reaction that transformed $LiO_2$ into $Li_2O$ and $O_2$, influenced by the surface chemistry of MXene (Gallant et al., 2013).

Sambyal et al. (2019) [80] fabricated porous $Ti_3C_2T_x$/CNT thin film through electrostatic self-assembly. The 2D/1D hybrid structure minimized restacking and improved electrolyte penetration. Compared to $Ti_3C_2$/graphene systems, these porous MXene/CNT films exhibited enhanced electrocatalytic properties. These results confirm MXene's potential to modulate active site electron configurations, generate new active centers, and serve as a high-conductivity, hydrophilic support for catalyst frameworks. Furthermore, the 2D nature of MXene enables the construction of 3D architectures, promoting fast charge/mass transport and reducing interlayer restacking.

Wu et al. (2024) utilized ultrasonic-assisted aerosol spray drying to transform 2D $Ti_3C_2$ colloids into 3D MXene networks with large surface areas (Figure 8.6). The resulting structure improved material handling and enhanced the number of exposed active sites. Such synergetic integration of 3D MXenes with catalytic materials opens pathways for next-generation catalysts with high site density and mass transfer capabilities.

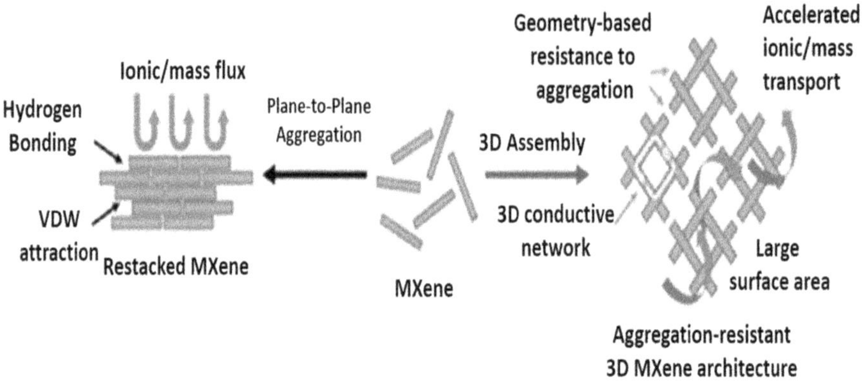


*Figure 8.6* 2D MXene to construct into a 3D structure to prevent restacking between neighboring layers.

**(Reproduced with get permission from Elsevier, Copyright 2024.)**

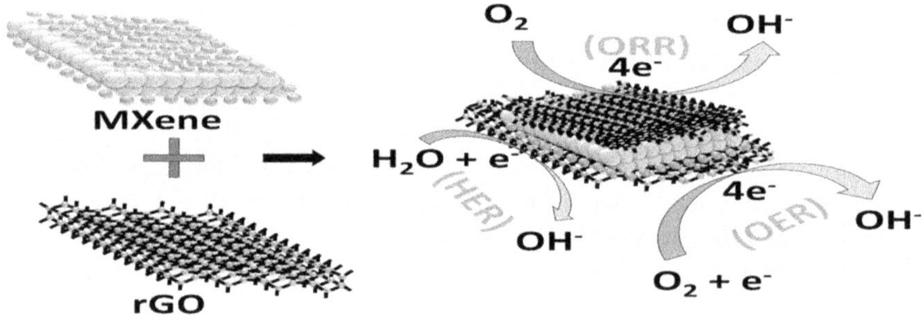


*Figure 8.7* Shows schematic of $4e^-$ ORR/OER electrochemical reaction on MXene/rGO composite.

Figure 8.7 illustrates the OER/ORR mechanism on rGO/MXene composites. The enhanced activity is likely due to abundant –OH and –O functional groups on the surface, contributing to improved interaction with reaction intermediates and higher electrocatalytic efficiency. Superior electrochemical OER/ORR reactions based on GO/MXene have been summarized in Table 8.1.

The above-discussed characterization suggests that MXene could connect with active molecules to alter the electrical structure associated with the active center, create new active regions, and ultimately boost the catalytic efficiency of the resultant catalysts. Meanwhile, MXenes have the potential to be excellent electrocatalyst framework components due to their high conductivity and hydrophilicity. Crucially, 2D MXene had the capability to create a 3D structure that inhibited restacking between layers that were

*Table 8.1* Comparison of OER/ORR reaction of graphene/MXene based electrocatalysts materials

| *Electrocatalysts* | *Electrolyte* | *Overpotential* $\eta_{10}$ *(V)* | *Tafel slope (mV dec$^{-1}$)* | *Onset potential* | *Current density* | *Ref.* |
|---|---|---|---|---|---|---|
| **OER kinetics** | | | | | | |
| $RuN_4$-Gra/$W_2$NO | 1M KOH | 0.45 | - | - | - | Bai et al. (2024) |
| $CoN_4$-Gra/$Ti_2$NO | 1M KOH | 0.30 | - | - | - | Bai et al. (2024) |
| $Ti_3C_2T_x$/g-$C_3N_4$ | 1 M KOH | 0.42 | 74.6 | - | - | Ma et al. (2016) |
| $Ti_3C_2$/g-$C_3N_4$ | 1 M KOH | 0.42 | 74.6 | - | - | Ma et al. (2016) |
| Iron/nickel doped graphene oxide @ MXene (GMX) | 0.1 $KMnO_4$ | 0.37 | 70 | | 10 mA $cm^{-2}$ | Savan et al. (2022) |
| **ORR kinetics** | | | | | | |
| $CoN_4$-Gra/$Ti_2$NO | 0.1 M KOH | 0.37 | - | - | - | Bai et al. (2024) |
| $RhN_4$–Gra/$Ti_2$NO | 0.1 M KOH | 0.39 | - | - | - | Bai et al. (2024) |
| $PtN_4$–Gra/$Hf_2$NO | 0.1 M KOH | 0.42 | - | - | - | Bai et al. (2024) |
| $CoN_4$–Gra/$Ti_2$NO | 0.1 M KOH | 0.37 | - | - | - | Bai et al. (2024) |
| FePc-M-rGO | 0.1 M KOH | 0.94 | - | - | - | Ding et al. (2023) |
| MXene/GO for Li-$CO_2$ batteries | - | 2.4 | - | - | 100 $mAg^{-1}$ | Bharti et al. (2022) |

adjacent and improved electrocatalytic transport channels and intrinsic catalytic function by a quick charge–mass transport channel and improving catalytic active sites.

### 8.1.8 Issues and challenges

Despite the tremendous potential of MXene-based materials for electrocatalytic applications, several limitations must be addressed to enable their practical deployment. Pure MXenes exhibit inherently low catalytic activity, a strong tendency to restack, and limited antioxidant properties, all of which hinder their long-term performance and stability. Additional concerns include a scarcity of surface functional groups, uneven dispersion, and susceptibility to oxidation.

On the other hand, rGO, while beneficial as a composite material, also suffers from drawbacks such as reduced electrical conductivity and weaker mechanical strength compared to pristine graphene. Although rGO/MXene hybrids have demonstrated promise in energy storage and catalysis, critical challenges remain, including:

- **Optimizing the composition and ratio** of the composite materials.
- **Understanding the interfacial interactions** and bonding mechanisms between MXene and rGO.
- **Improving the stability and scalability** of composite fabrication for real-world applications.

Recent approaches, such as electrolyte-assisted film formation, have enabled the creation of solid rGO/MXene composite films with enhanced compactness and flexibility, suitable for supercapacitor applications. These films offer increased areal capacitance and mechanical adaptability, showcasing a potential pathway for overcoming material limitations in energy devices.

## 8.2 CONCLUSION

While precious metal-based electrocatalysts (e.g., Pd, Ir, Ru, Ag) have exhibited outstanding OER/ORR performance, their high cost and scarcity necessitate alternative materials. In this context, MXene-based hybrids, especially those combined with GO or rGO, have emerged as promising low-cost, high-performance alternatives.

This chapter reviewed the synergistic integration of MXenes and graphene derivatives, emphasizing their complementary characteristics:

- MXenes offer excellent electrical conductivity, a tunable 2D structure, and hydrophilic surfaces.
- GO and rGO provide superior mechanical flexibility, high surface area, and chemical tunability.

These hybrids exhibit enhanced electrochemical performance, particularly in oxygen evolution and reduction reactions (OER/ORR), making them ideal for next-generation energy conversion and storage systems.

Given the increasing body of research, this chapter aims to support ongoing investigations into MXene/graphene composites and inspire further innovation in the field of sustainable electrocatalysis.

## 8.3 FUTURE PERSPECTIVES

To support the sustainable development of energy systems, ongoing research must focus on advancing MXene-based catalysts through both material innovation and device engineering. Key future directions include:

- **Designing new MXene architectures** (e.g., hierarchical 3D structures, surface-functionalized arrays) for optimized electron and mass transport.
- **Creating metalphilic MXene surfaces** to enhance catalytic center stability.
- **Integrating MXene composites with green nanomaterials**, polymers, and biocompatible electrolytes to develop eco-friendly energy devices.
- **Combining MXenes with metal-free catalysts** like rGO to suppress catalytic degradation during OER, thereby improving durability.
- **Exploring MOF/MXene composites**, leveraging the high porosity and tunable chemistry of MOFs with the conductivity of MXenes to yield advanced multifunctional electrocatalysts.

By addressing these challenges and expanding the design space of MXene-based materials, the field can progress toward scalable, high-efficiency, and cost-effective solutions for future renewable energy conversion and storage technologies.

## REFERENCES

Achakulwisut, P., Erickson, P., Guivarch, C., Schaeffer, R., Brutschin, E., & Pye, S. (2023). Global fossil fuel reduction pathways under different climate mitigation strategies and ambitions. *Nature Communications*, *14*(1), 1–15. https://doi.org/10.1038/s41467-023-41105-z

Ahmad Junaidi, N. H., Wong, W. Y., Loh, K. S., Rahman, S., & Daud, W. R. W. (2021). A comprehensive review of MXenes as catalyst supports for the oxygen reduction reaction in fuel cells. *International Journal of Energy Research*, *45*(11), 15760–15782. https://doi.org/10.1002/ER.6899

Bai, X., Lu, S., Song, P., Jia, Z., Gao, Z., Peng, T., Wang, Z., Jiang, Q., Cui, H., Tian, W., Feng, R., Liang, Z., Kang, Q., & Yuan, H. (2024). Heterojunction of MXenes and $MN_4$–graphene: Machine learning to accelerate the design of bifunctional oxygen electrocatalysts. *Journal of Colloid and Interface Science*, *664*, 716–725. https://doi.org/10.1016/J.JCIS.2024.03.073

Bharti, A., Manna, G., Saha, P., Achutarao, G., & Bhattacharyya, A. J. (2022). Probing the function of a Li-$CO_2$ battery with a MXene/graphene oxide composite cathode electrocatalyst. *Journal of Physical Chemistry Letters*, *13*(32), 7380–7385. https://doi.org/10.1021/ACS.JPCLETT.2C01499/SUPPL_FILE/JZ2C01499_SI_002.PDF

Bian, R., Lin, R., Wang, G., Lu, G., Zhi, W., Xiang, S., Wang, T., Clegg, P. S., Cai, D., & Huang, W. (2018). 3D assembly of $Ti_3C_2$-MXene directed by water/oil interfaces. *Nanoscale*, *10*(8), 3621–3625. https://doi.org/10.1039/C7NR07346A

Bilibana, M. P. (2023). Electrochemical properties of MXenes and applications. *Advanced Sensor and Energy Materials*, 2(4), 100080. https://doi.org/10.1016/J.ASEMS.2023.100080

Birdja, Y. Y., Pérez-Gallent, E., Figueiredo, M. C., Göttle, A. J., Calle-Vallejo, F., & Koper, M. T. M. (2019). Advances and challenges in understanding the electrocatalytic conversion of carbon dioxide to fuels. *Nature Energy*, *4*(9), 732–745. https://doi.org/10.1038/S41560-019-0450-Y

Cano, Z. P., Banham, D., Ye, S., Hintennach, A., Lu, J., Fowler, M., & Chen, Z. (2018). Batteries and fuel cells for emerging electric vehicle markets. *Nature Energy* , *3*(4), 279–289. https://doi.org/10.1038/s41560-018-0108-1

Cao, W. Q., Zhang, M., & Cao, M. S. (2024). A perspective of tailoring dielectric genes for 2D materials toward advanced electromagnetic functions. *Advanced Functional Materials*, *34*(52), 2410928. https://doi.org/10.1002/adfm.202410928

Chen, Y. Y., Ross Kunz, M., He, X., & Fushimi, R. (2022). Recent progress toward catalyst properties, performance, and prediction with data-driven methods. *Current Opinion in Chemical Engineering*, *37*, 100843. https://doi.org/10.1016/J.COCHE.2022.100843

Ding, Q., Ji, C., Shan, Q., Yang, C., & Wu, W. (2023). Composite of iron phthalocyanine and reduced graphene oxide with low temperature hydrothermally grown $Ti_3C_2Tx$ Mxene for ORR catalytic application. *Materials Today Communications*, *37*, 107012. https://doi.org/10.1016/J.MTCOMM.2023.107012

Gadipelli, S., Zhao, T., Shevlin, S. A., & Guo, Z. (2016). Switching effective oxygen reduction and evolution performance by controlled graphitization of a cobalt–nitrogen–carbon framework system. *Energy & Environmental Science*, *9*(5), 1661–1667. https://doi.org/10.1039/C6EE00551A

Gallant, B. M., Kwabi, D. G., Mitchell, R. R., Zhou, J., Thompson, C. V., & Shao-Horn, Y. (2013). Influence of $Li_2O_2$ morphology on oxygen reduction and evolution kinetics in Li–$O_2$ batteries. *Energy & Environmental Science*, *6*(8), 2518–2528. https://doi.org/10.1039/C3EE40998H

Geng, D. C., Zhao, X. X., Chen, Z. X., Sun, W. W., Fu, W., Chen, J. Y., Liu, W., Zhou, W., & Loh, K. P. (2017). Direct synthesis of large-area 2D $Mo_2C$ on in situ grown graphene. *Advanced Materials*, *29*(35), 1700072. https://doi.org/10.1002/ADMA.201700072

Guo, X., Zheng, S., Zhang, G., Xiao, X., Li, X., Xu, Y., Xue, H., & Pang, H. (2017). Nanostructured graphene-based materials for flexible energy storage. *Energy Storage Materials*, *9*, 150–169. https://doi.org/10.1016/J.ENSM.2017.07.006

Gupta, V., Chandra, N., Tari, V., & Kannan, K. (2024). A sustainable approach for the green synthesis of silver nanoparticles using *Brassica oleracea* subsp. *botrytis* (L.) leaves and its antibacterial activity. *Physics and Chemistry of Solids*, *25*(2), 303–310. https://doi.org/10.15330/pcss.25.2.303-310

Gurushankar, K., Karuppaiya, C., Tari, V., & Kannan, K. (2024). Exploring MXenes and MBenes in metal sulfur and selenium battery technologies. In K. Kannan & V. Tari (Eds.), *Novel energy storage and conversion technologies for two-dimensional MXenes and MBenes*. IGI Global. https://doi.org/10.4018/979-8-3693-9546-2.ch005

Hao, C., Wu, Y., An, Y., Cui, B., Lin, J., Li, X., Wang, D., Jiang, M., Cheng, Z., & Hu, S. (2019). Interface-coupling of CoFe-LDH on MXene as high-performance oxygen evolution catalyst. *Materials Today Energy*, *12*, 453–462. https://doi.org/10.1016/J.MTENER.2019.04.009

Hu, J., Ding, J., & Zhong, Q. (2021). Ultrathin 2D $Ti_3C_2$ MXene co-catalyst anchored on porous g-$C_3N_4$ for enhanced photocatalytic $CO_2$ reduction under visible-light irradiation. *Journal of Colloid and Interface Science*, *582*, 647–657. https://doi.org/10.1016/J.JCIS.2020.08.047

Intikhab, S., Snyder, J. D., & Tang, M. H. (2017). Adsorbed hydroxide does not participate in the volmer step of alkaline hydrogen electrocatalysis. *ACS Catalysis*, *7*(12), 8314–8319. https://doi.org/10.1021/ACSCATAL.7B02787/SUPPL_FILE/CS7B02787_SI_001.PDF

Iravani, S., & Varma, R. S. (2022). MXene-based photocatalysts in degradation of organic and pharmaceutical pollutants. *Molecules*, *27*(20), 6939. https://doi.org/10.3390/MOLECULES27206939

Jaramillo, T. F., Jørgensen, K. P., Bonde, J., Nielsen, J. H., Horch, S., & Chorkendorff, I. (2007). Identification of active edge sites for electrochemical $H_2$ evolution from $MoS_2$ nanocatalysts. *Science*, *317*(5834), 100–102. https://doi.org/10.1126/SCIENCE.1141483/SUPPL_FILE/JARAMILLO_SOM.PDF

Jia, G., Zheng, A., Wang, X., Zhang, L., Li, L., Li, C., Zhang, Y., Cao, L. (2021). Flexible, biocompatible and highly conductive MXene-graphene oxide film for smart actuator and humidity sensor. *Sensors and Actuators B: Chemical*, *346*, 130507. https://doi.org/10.1016/j.snb.2021.130507

Kalia, S., Rana, D. S., Thakur, N., Singh, D., Kumar, R., & Singh, R. K. (2022). Two-dimensional layered molybdenum disulfide ($MoS_2$)-reduced graphene oxide (rGO) heterostructures modified with $Fe_3O_4$ for electrochemical sensing of epinephrine. *Materials Chemistry and Physics*, *287*, 126274. https://doi.org/10.1016/J.MATCHEMPHYS.2022.126274

Kan, D., Wang, D., Cheng, Y., Lian, R., Sun, B., Chen, K., Huo, W., Wang, Y., Chen, G., & Wei, Y. (2021). Designing of efficient bifunctional ORR/OER Pt single-atom catalysts based on O-terminated MXenes by first-principles calculations. *ACS Applied Materials and Interfaces*, *13*(44), 52508–52518. https://doi.org/10.1021/ACSAMI.1C12893/SUPPL_FILE/AM1C12893_SI_001.PDF

Kannan, K., & Tari, V. (Eds.). (2025). *Novel energy storage and conversion technologies for two-dimensional MXenes and MBenes*. https://www.amazon.com/dp/9798369395462

Kumbhakar, P., Jayan, J. S., Sreedevi Madhavikutty, A., Sreeram, P. R., Saritha, A., Ito, T., & Tiwary, C. S. (2023). Prospective applications of two-dimensional materials beyond laboratory frontiers: A review. *IScience*, *26*(5). https://doi.org/10.1016/J.ISCI.2023.106671

Li, K., Liang, M., Wang, H., Wang, X., Huang, Y., Coelho, J., Pinilla, S., Zhang, Y., Qi, F., Nicolosi, V., & Xu, Y. (2020). 3D MXene architectures for efficient energy storage and conversion. *Advanced Functional Materials*, *30*(47), 2000842. https://doi.org/10.1002/ADFM.202000842

Li, K., Xiong, T., Liao, J., Lei, Y., Zhang, Y., & Zhu, W. (2022). Design of MXene/graphene oxide nanocomposites with micro-wrinkle structure for efficient separating of uranium(VI) from wastewater. *Chemical Engineering Journal*, *433*, 134449. https://doi.org/10.1016/J.CEJ.2021.134449

Liang, Y., Li, Y., Wang, H., Zhou, J., Wang, J., Regier, T., & Dai, H. (2011). $Co_3O_4$ nanocrystals on graphene as a synergistic catalyst for oxygen reduction reaction. *Nature Materials*, *10*(10), 780–786. https://doi.org/10.1038/NMAT3087

Liu, H. J., & Dong, B. (2021). Recent advances and prospects of MXene-based materials for electrocatalysis and energy storage. *Materials Today Physics*, *20*, 100469. https://doi.org/10.1016/J.MTPHYS.2021.100469

Liu, T., Liu, X., Graham, N., Yu, W., & Sun, K. (2020). Two-dimensional MXene incorporated graphene oxide composite membrane with enhanced water purification performance. *Journal of Membrane Science*, *593*, 117431. https://doi.org/10.1016/J.MEMSCI.2019.117431

Lu, Z., Neupane, G. P., Jia, G., Zhao, H., Qi, D., Du, Y., Lu, Y., & Yin, Z. (2020). 2D materials based on main group element compounds: Phases, synthesis, characterization, and applications. *Advanced Functional Materials*, *30*(40), 2001127. https://doi.org/10.1002/ADFM.202001127

Luo, J., Im, J. H., Mayer, M. T., Schreier, M., Nazeeruddin, M. K., Park, N. G., Tilley, S. D., Fan, H. J., & Grätzel, M. (2014). Water photolysis at 12.3% efficiency via perovskite photovoltaics and Earth-abundant catalysts. *Science*, *345*(6204), 1593–1596. https://doi.org/10.1126/SCIENCE.1258307/SUPPL_FILE/LUO-SM.PDF

Ma, T. Y., Cao, J. L., Jaroniec, M., & Qiao, S. Z. (2016). Interacting carbon nitride and titanium carbide nanosheets for high-performance oxygen evolution. *Angewandte Chemie International Edition*, *55*(3), 1138–1142. https://doi.org/10.1002/ANIE.201509758

Mansor, N., Miller, T. S., Dedigama, I., Jorge, A. B., Jia, J., Brázdová, V., Mattevi, C., Gibbs, C., Hodgson, D., Shearing, P. R., Howard, C. A., Corà, F., Shaffer, M., Brett, D. J. L., & McMillan, P. F. (2016). Graphitic carbon nitride as a catalyst support in fuel cells and electrolyzers. *Electrochimica Acta*, *222*, 44–57. https://doi.org/10.1016/J.ELECTACTA.2016.11.008

Mao, S., Wen, Z., Huang, T., Hou, Y., & Chen, J. (2014). High-performance bi-functional electrocatalysts of 3D crumpled graphene–cobalt oxide nanohybrids for oxygen reduction and evolution reactions. *Energy & Environmental Science*, *7*(2), 609–616. https://doi.org/10.1039/C3EE42696C

Ostadhossein, A., Guo, J., Simeski, F., & Ihme, M. (2019). Functionalization of 2D materials for enhancing OER/ORR catalytic activity in Li–oxygen batteries. *Communications Chemistry*, *2*(1), 1–11. https://doi.org/10.1038/s42004-019-0196-2

Sambyal, P., Iqbal, A., Hong, J., Kim, H., Kim, M. K., Hong, S. M., Han, M., Gogotsi, Y., & Koo, C. M. (2019). Ultralight and mechanically robust $Ti_3C_2T_x$ hybrid aerogel reinforced by carbon nanotubes for electromagnetic interference shielding. *ACS Applied Materials and Interfaces*, *11*(41), 38046–38054. https://doi.org/10.1021/ACSAMI.9B12550/SUPPL_FILE/AM9B12550_SI_001.PDF

Sandhu, Z. A., Raza, M. A., Kainat, A. F., Arshid, M. S., Kashf Jaffary, S. D., Arshad, S., Raza, M., Al-Sehemi, A. G., & Dege, N. (2025). Graphene-metal oxide nanocomposites: Empowering next-generation energy storage devices. *Materials Science and Engineering: B*, *313*, 117947. https://doi.org/10.1016/J.MSEB.2024.117947

Shan, P., Bai, X., Jiang, Q., Chen, Y., Lu, S., Song, P., Jia, Z., Xiao, T., Han, Y., Wang, Y., Liu, T., Cui, H., Feng, R., Kang, Q., Liang, Z., & Yuan, H. (2023). Bilayer $MN_4$-O-$MN_4$ by bridge-bonded oxygen ligands: Machine learning to accelerate the design of bifunctional electrocatalysts. *Renewable Energy*, *203*, 445–454. https://doi.org/10.1016/J.RENENE.2022.12.059

Suen, N. T., Hung, S. F., Quan, Q., Zhang, N., Xu, Y. J., & Chen, H. M. (2017). Electrocatalysis for the oxygen evolution reaction: Recent development and future perspectives. *Chemical Society Reviews*, *46*(2), 337–365. https://doi.org/10.1039/C6CS00328A

Tari, V., Kannan, K., Koliar, J., Gurushankar, K., & Wahyuni, D. K. (2025). Influence of zinc oxide nanomaterials on crop plants: A prospective analysis. In D. Jabbrova, D. Sarkar, Rakesh, S. S., & R. Datta (Eds.), *Smart nanomaterials in agroforestry systems*. Springer. https://doi.org/10.1007/978-981-96-1337-3_8

Tari, V., Kannan, K., & Vishwakarma, V. (2022). Chapter 8: Emerging antiviral technologies. In P. Nunthavarawong, S. M. Rangappa, S. Siengchin, & M. Thoppil-Mathew (Eds.), *Antimicrobial and antiviral materials: Polymers, metals, ceramics, and applications* (1st ed., pp. 137–150). CRC Press. https://doi.org/10.1201/9781003143093-8

Tari, V. S., & Patil, P. Y. (2017a). Challenge of contamination of pesticide for Alphonso in Ratnagiri district, Maharashtra, India. *Research Journal of Chemistry and Environment*, *21*(12), 57–67.

Tari, V. S., & Patil, P. Y. (2017b). Transfer factor of heavy metal in soil to plant from pesticide-contaminated area (mango orchards), Ratnagiri district, Maharashtra. *Research Journal of Chemistry and Environment*, *21*(1), 26–32.

Tong, D., Farnham, D. J., Duan, L., Zhang, Q., Lewis, N. S., Caldeira, K., & Davis, S. J. (2021). Geophysical constraints on the reliability of solar and wind power worldwide. *Nature Communications*, *12*(1), 1–12. https://doi.org/10.1038/s41467-021-26355-z

Wang, H. F., Tang, C., & Zhang, Q. (2018). A review of precious-metal-free bifunctional oxygen electrocatalysts: Rational design and applications in Zn–Air batteries. *Advanced Functional Materials*, *28*(46), 1803329. https://doi.org/10.1002/ADFM.201803329

Wei, C., Tao, Y., An, Y., Tian, Y., Zhang, Y., Feng, J., & Qian, Y. (2020). Recent advances of emerging 2D MXene for stable and dendrite-free metal anodes. *Advanced Functional Materials*, *30*(45), 2004613. https://doi.org/10.1002/ADFM.202004613

Wu, X., Wang, Y., & Wu, Z. S. (2024). Recent advancement and key opportunities of MXenes for electrocatalysis. *IScience*, *27*(2). https://doi.org/10.1016/J.ISCI.2024.108906/ASSET/74945573-0942-4622-B7E0-A307E6B514FD/MAIN.ASSETS/SC6.JPG

Xiao, Z., Xiao, X., Kong, L. B., Dong, H., Li, X., Sun, X., He, B., Ruan, S., & Zhai, J. (2023). MXenes and MXene-based composites for energy conversion and storage applications. *Journal of Materiomics*, *9*(6), 1067–1112. https://doi.org/10.1016/J.JMAT.2023.04.013

Yin, Z., Lu, Z., Xu, Y., Zhang, Y., He, L., Li, P., Xiong, L., Ding, L., Wei, Y., & Wang, H. (2021). Supported MXene/GO composite membranes with suppressed swelling for metal ion sieving. *Membranes*, *11*(8), 621. https://doi.org/10.3390/MEMBRANES11080621

Yoo, R., Pranada, E., Johnson, D., Qiao, Z., & Djire, A. (2022). Review—The oxygen reduction reaction on MXene-based catalysts: Progress and prospects. *Journal of The Electrochemical Society*, *169*(6), 063513. https://doi.org/10.1149/1945-7111/AC766E

Yue, Y., Liu, N., Ma, Y., Wang, S., Liu, W., Luo, C., Zhang, H., Cheng, F., Rao, J., Hu, X., Su, J., & Gao, Y. (2018). Highly self-healable 3D microsupercapacitor with MXene-graphene composite aerogel. *ACS Nano*, *12*(5), 4224–4232. https://doi.org/10.1021/ACSNANO.7B07528/SUPPL_FILE/NN7B07528_SI_001.PDF

Zhang, C. J., Pinilla, S., McEvoy, N., Cullen, C. P., Anasori, B., Long, E., Park, S. H., Seral-Ascaso, A., Shmeliov, A., Krishnan, D., Morant, C., Liu, X., Duesberg, G. S., Gogotsi, Y., & Nicolosi, V. (2017). Oxidation stability of colloidal two-dimensional titanium carbides (MXenes). *Chemistry of Materials*, *29*(11), 4848–4856. https://doi.org/10.1021/ACS.CHEMMATER.7B00745/SUPPL_FILE/CM7B00745_SI_002.AVI

Zhao, Y., Adiyeri Saseendran, D. P., Huang, C., Triana, C. A., Marks, W. R., Chen, H., Zhao, H., & Patzke, G. R. (2023). Oxygen evolution/reduction reaction catalysts: From in situ monitoring and reaction mechanisms to rational design. *Chemical Reviews*, *123*(9), 6257–6358. https://doi.org/10.1021/ACS.CHEMREV.2C00515/ASSET/IMAGES/MEDIUM/CR2C00515_0067.GIF

Zhou, Q., Dong, H., Liu, L., Wei, C., Liang, X., Zhang, H., Wang, L., Lu, H., Nie, S., Xu, L., Yang, W., Yang, W., & Yuen, A. C. Y. (2024). In-situ surface growth strategy to synthesize MXene@graphdiyne heterostructure for achieving high capacity and desirable stability in lithium-ion batteries. *Journal of Power Sources*, *603*, 234404. https://doi.org/10.1016/J.JPOWSOUR.2024.234404

Zhou, S., Yang, X., Pei, W., Liu, N., & Zhao, J. (2018). Heterostructures of MXenes and N-doped graphene as highly active bifunctional electrocatalysts. *Nanoscale*, *10*(23), 10876–10883. https://doi.org/10.1039/C8NR01090K

Chapter 9

# MXene supported composite materials for electrocatalytic nitrogen reduction reaction

*V. Navakoteswara Rao, Jyoti Prakash Das, and Kim Sang Jae*

## 9.1 INTRODUCTION

Ammonia, the second most manufactured chemical globally, is essential in agriculture and food sectors for its role in fertilizers (Burdett et al., 1987; Manjunatha et al., 2020; J. Sun et al., 2020; M. Zhao et al., 2024; Tari and Patil, 2017). Furthermore, ammonia is being viewed as a potential energy carrier in the hydrogen economy due to its lack of carbon content. The high hydrogen content of 17.8 wt% by weight and volumetric density of 107 $kgH_2\ m^{-3}$ (at 8 bar and 298 K), along with its cost-effective liquid form transportation, make it an attractive choice for storing and transporting hydrogen (Johnson et al., 2020). Furthermore, the method of extracting hydrogen from ammonia proves to be effective and eco-friendly, with no carbon emissions being generated. This reinforces its viability as a sustainable energy carrier for storing and distributing hydrogen (Amrillah et al., 2021; Q. J. Fang et al., 2023).

Ammonia plays an important role in the global chemical industry as a key ingredient in the production of various materials such as polyamide, explosives, refrigerants, dyes, and cleaning products (Akter et al., 2024; Chowdhury et al., 2023; Ding et al., 2021a; Xiong et al., 2024; Zhang et al., 2024). To meet the increasing global demand, 177 million metric tons of ammonia are produced annually, and more than 96% of this production is based on the Haber-Bosch process. This industrial process combines nitrogen ($N_2$) in the air with hydrogen ($H_2$), which is usually formed from carbon dioxide ($CH_4$) (Kannan and Tari, 2025; Nazari et al., 2024; Wang et al., 2024; Wiskich & Rapson, 2023). The reaction proceeds at a high temperature (400–650°C) and pressure exceeding 150 bar using an iron-based catalyst.

Despite its ability to produce large amounts of ammonia, the Haber-Bosch process has serious environmental drawbacks. It requires a significant amount of energy, accounting for about 2% of global energy consumption, and 1.2% of global emissions, mainly due to the carbon dioxide emitted in

DOI: 10.1201/9781003584742-9

the process. Switching to more sustainable ammonia production methods, particularly those that do not rely on methane, is critical to reducing the environmental impact of this important chemical (J. Li et al., 2024; Liu, Liang, Gao, et al., 2022; H. Liu et al., 2022; X. Wei et al., 2020).

In nature, biological nitrogen fixation provides a model for sustainable ammonia production (Tari and Kannan, 2022). This process, facilitated by the nitrogenase enzyme, converts atmospheric nitrogen into ammonia using hydrogen derived from water, releasing hydrogen gas without relying on methane and thus producing no $CO_2$ emissions (Tari et al., 2024). Mimicking this natural mechanism offers a pathway to synthesizing "green ammonia," positioning it as a pivotal energy and chemical carrier for future renewable energy systems (Figure 9.1). To achieve this, developing efficient green ammonia synthesis methods across different stages and applications is essential (Liu, Liang, Gao, et al., 2022; H. Liu et al., 2022; X. Wei et al., 2020).

### 9.1.1 Importance of ammonia production and its applications

Ammonia ($NH_3$) is one of the most important compounds used extensively as an essential intermediate in agriculture, industry, and environmental management. It is primarily used to manufacture nitrogen-based fertilizers such as urea, ammonium sulfate, and ammonium nitrate, which are essential for improving soil fertility and increasing crop productivity without adverse effects on global food security. Apart from its agricultural role, ammonia is used in several industrial applications. It is used in the manufacture of explosives for use in mining, as a coolant in cooling systems due to its high thermodynamic efficiency, and it can be found in textiles for treatment and fixation of dyestuffs. Ammonia is also an important precursor for chemical processes, including the production of hydrazine for use as a rocket propellant and ammonium salts that are essential components in detergents, pharmaceuticals, or food additives. Ammonia is emerging as a versatile and ideal chemical for sustainable and environmental technologies. Due to its high energy density along with the carbon-free burning, hydrogen is becoming a popular research topic as a possible hydrogen carrier and clean source of energy. In addition, it possesses an alkaline nature, which is efficient for neutralizing acidic waste flows and decreasing nitrogen oxide emissions and therefore contributes to pollution control (L. Li & Xiao, 2023).

### 9.1.2 Challenges and limitations for ammonia production

The production, distribution, and storage of hydrogen as a major energy carrier for the future presents considerable challenges. These challenges are particularly significant in the transportation sector, where specific

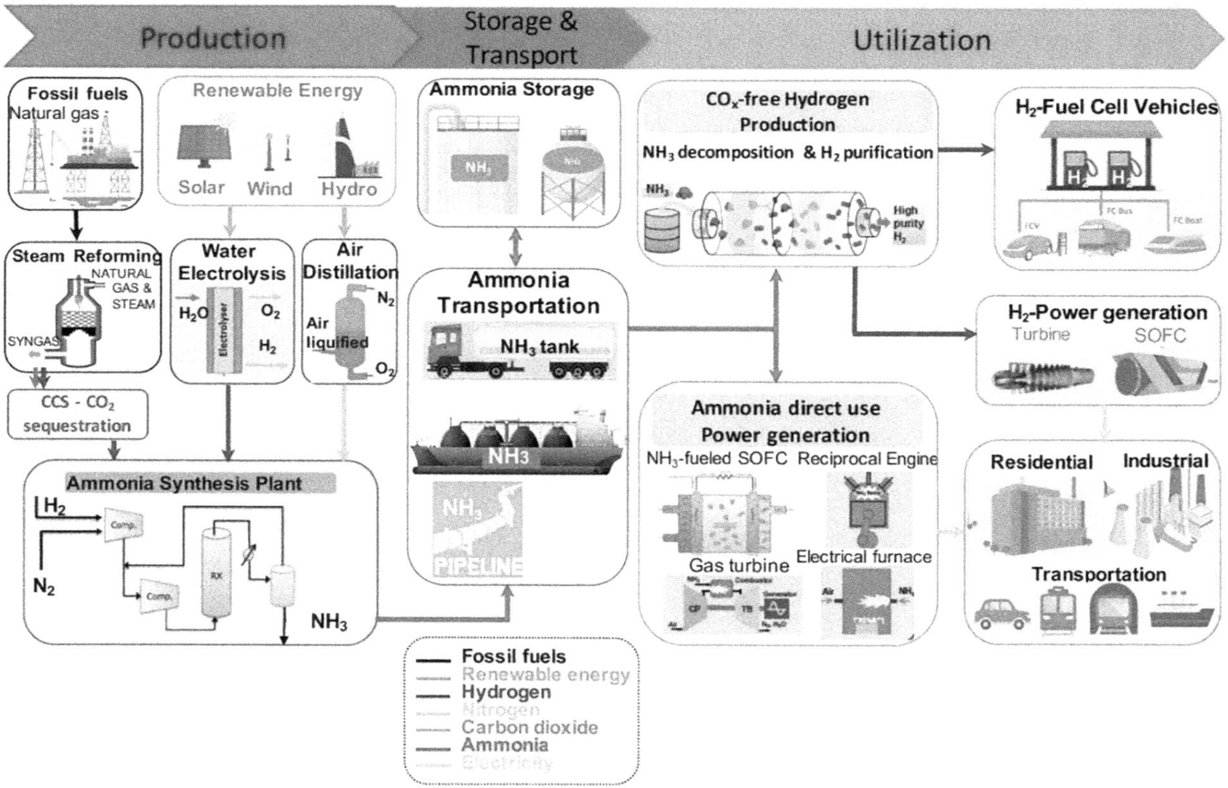


*Figure 9.1* Schematics representation of the importance of the MXene for enhancing ammonia evolution. **(Reproduction with permission from Elsevier 2021 (Morlanés et al., 2021).)**

requirements have been highlighted, such as those outlined in the Freedom CAR 2050 system targets (Akter et al., 2024; Wiskich & Rapson, 2023). These targets include achieving high reversible hydrogen storage capacity (3 kWh per kg or 9 wt.% hydrogen and 2.7 kWh per liter or 0.081 kg $H_2$ per liter system), rapid kinetics (0.02 g $H_2$ per second per kW), excellent reversibility over 1,500 cycles, and operation within ambient temperature ranges (−40 to 60°C). Additionally, these systems must meet criteria for low cost ($2 per kWh or $67 per kg $H_2$), high hydrogen purity (less than 1 ppm CO), and safety standards. Among these requirements, research has largely focused on identifying materials capable of achieving the highest hydrogen storage capacity (Tari et al., 2025; X. Wei et al., 2020; Q. Zhang et al., 2020). However, meeting this goal is a formidable challenge, further complicated by the fact that system integration often results in the loss of up to half of the weight-based storage capacity. Addressing these technical hurdles remains crucial for the successful adoption of hydrogen as a sustainable energy solution (Ahmad Junaidi et al., 2021).

### 9.1.3 Introduction to two-dimensional (2D) heterostructures

Two-dimensional (2D) heterostructures, especially those derived from MXene, have become notable prospects for facilitating $NH_3$ production from $N_2$ has been displayed in Figure 9.2a and b. MXene, a group of 2D transition metal carbides, nitrides, or carbonitrides, exhibits remarkable conductivity, large surface area, and adjustable surface chemistry (T. Guo et al., 2022; Liang et al., 2021; Liu, Liang, Gao, et al., 2022; Liu, Liang, Zhu, et al., 2022; Yu et al., 2020). These characteristics render them flexible platforms for supporting catalytically active sites. By combining MXene with various materials in heterostructure forms, scientists can synergistically merge distinct electronic, chemical, and structural properties to address the fundamental challenges of the nitrogen reduction reaction (NRR), including the high dissociation energy of the N≡N bond and the competing hydrogen evolution reaction (HER). The promise of MXene-based heterostructures in enhancing $NH_3$ synthesis from $N_2$. Special attention is given to their structural flexibility, surface modifications, and electronic characteristics that facilitate targeted and effective $N_2$ activation. Moreover, approaches like co-doping, defect engineering, and hybrid nanocomposite synthesis are emphasized for adjusting the activity and stability of these systems in catalytic environments. The discussion seeks to offer perspectives on the recent progress, hurdles, and future trends in utilizing MXene-based 2D heterostructures for sustainable $NH_3$ synthesis.

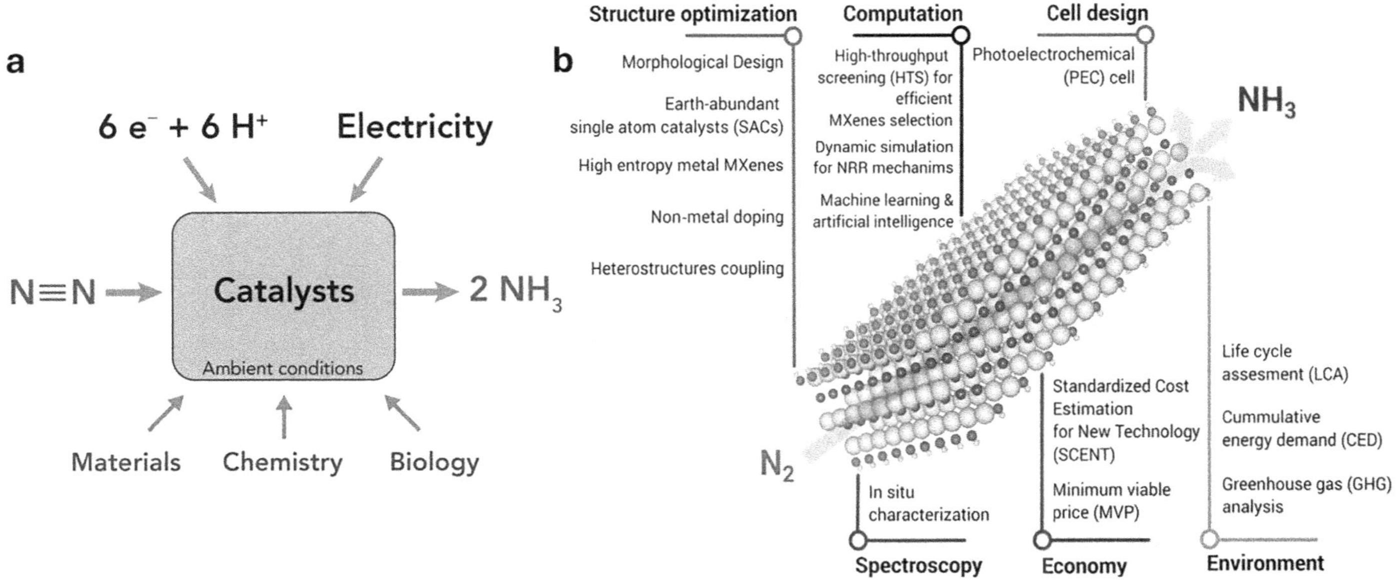


*Figure 9.2* (a) Schematic representation of the basic requirement of the electrochemical ammonia production Copyrights Permission from Cell press 2018 (Deng et al., 2018), (b) Future investigations into $N_2$ reduction to $NH_3$ over MXene and MXene derivatives utilizing electrocatalysts.

**(Reproduction with permission from Elsevier 2021 (Amrillah et al., 2021).)**

#### 9.1.3.1 *Mechanisms and principles of nitrogen reduction reaction (NRR)*

The electrochemical NRR offers a sustainable pathway to convert nitrogen ($N_2$) into ammonia ($NH_3$) under ambient conditions, presenting an alternative to the traditional, energy-intensive Haber-Bosch process (Luo et al., 2019a). This reaction requires breaking the strong triple bond in nitrogen, which involves a series of proton and electron transfer steps (Figure 9.3a). The NRR generally follows two main mechanisms, the associative pathway, where nitrogen is reduced stepwise with both atoms still bonded, and the dissociative pathway, where nitrogen first dissociates before hydrogenation. The associative pathway is preferred under ambient conditions due to the high activation energy required for nitrogen dissociation. The overall NRR reaction can be represented, with each step involving proton/electron transfer and the formation of intermediate species, such as $*N_2H$ and $*N_2H_2$ (Huang et al., 2021; H. Liu et al., 2022). Electrocatalytic conversion of nitrogen to ammonia using heterogeneous catalysts generally follows two primary mechanisms: the dissociative and associative pathways, each involving distinct intermediates. In the dissociative mechanism (Figure 9.3b), the adsorbed $N_2$ molecules undergo the cleavage of the $N{\equiv}N$ triple bond before any hydrogenation occurs. This process requires significant energy due to the inert nature of the $N{\equiv}N$ bond, which explains the harsh conditions necessary for the Haber–Bosch process that operates via this pathway.

In contrast, the associative mechanism maintains the bond between the two nitrogen atoms in the adsorbed $N_2$ molecule until the first ammonia ($NH_3$) molecule is formed. Depending on the sequence of hydrogen addition to the nitrogen atoms, the associative mechanism can proceed via two distinct routes: the distal pathway and the alternating pathway.

In the distal pathway (Figure 9.3c), the nitrogen atom farthest from the adsorption site is preferentially hydrogenated, leading to the release of the first $NH_3$ molecule. The remaining nitrogen atom subsequently undergoes the same hydrogenation process to form the second $NH_3$ molecule. Alternatively, the alternating pathway (Figure 9.3d) involves the sequential hydrogenation of both nitrogen atoms in a stepwise manner, coupled with proton-electron transfer, ultimately releasing two $NH_3$ molecules at the end of the process.

Additionally, the enzymatic pathway exhibits similarities to the alternating pathway but involves both nitrogen atoms binding to the catalyst surface in a side-on coordination mode. This pathway, however, is not typically observed on heterogeneous catalysts.

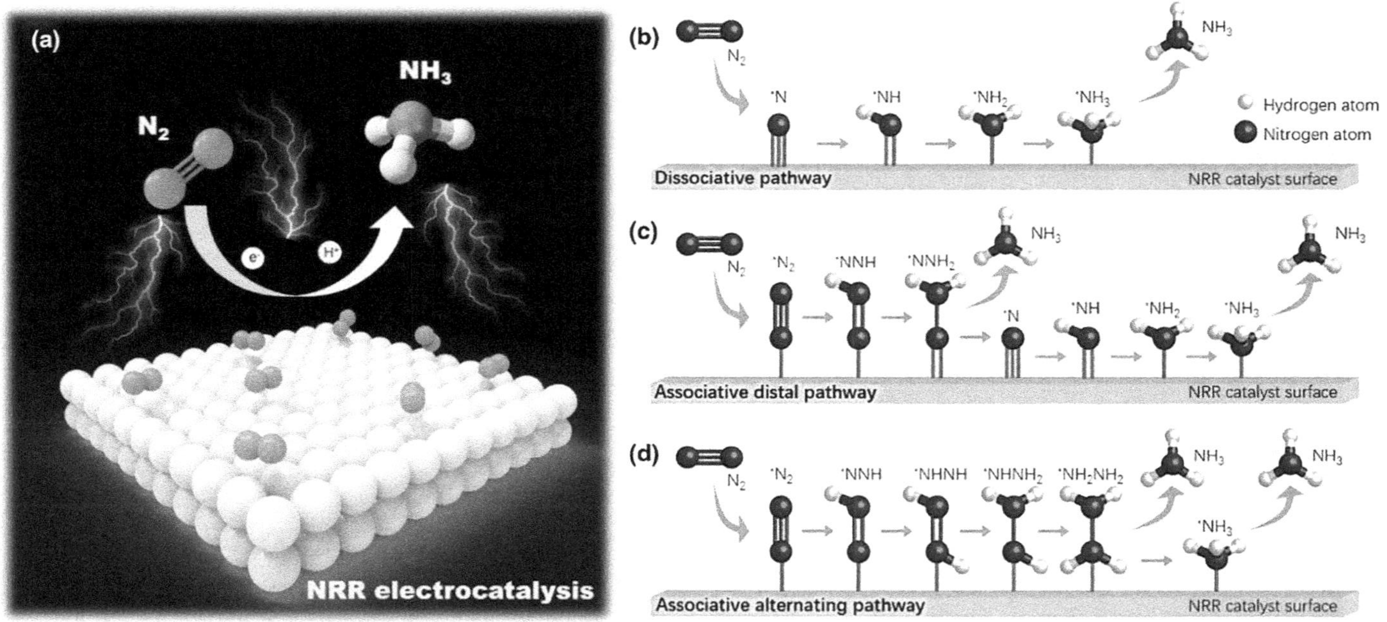


*Figure 9.3* Schematic illustrations of (a) the nitrogen reduction reaction (NRR) electrocatalysis process and its possible mechanisms, including (b) the dissociative pathway, (c) the associative distal pathway, and (d) the associative alternating pathway.

**(Reproduction with permission from Elsevier 2019 (Wan et al., 2019).)**

## 9.2 THEORETICAL UNDERSTANDING OF NITROGEN ADSORPTION, ACTIVATION, AND REDUCTION ON CATALYSTS

Appreciations to advancements in modern experimental and computational methodologies, investigations into MXene and their modifications for electrocatalytic nitrogen ($N_2$) fixation have seen significant growth (Xiong et al., 2024; K. Zhang et al., 2024). Given the inherently stable chemical bonds of $N_2$, overcoming the kinetic barriers associated with ammonia ($NH_3$) production requires considerable energy input. To address this challenge, the development of advanced modifications and the design of highly efficient MXene-based catalysts offer promising pathways to reduce the energy demands of the reaction, enabling the NRR to occur under milder conditions (Gurushankar et al., 2024; Nazari et al., 2024). Density functional theory (DFT) calculations have emerged as a powerful tool to evaluate the feasibility of MXenes as $N_2$ reduction catalysts and to optimize their structural and electronic properties. Through DFT analyses, researchers have been able to identify suitable combinations of elements for designing MXene with enhanced catalytic performance. A notable study demonstrated that middle titanium (Ti) atoms at the edge sites of MXene structures serve as highly active centers for NRR. On the (0001) surface of MXene, nitrogen dissociation was found to proceed efficiently, with the top sites above the metal ions on the MXene nanosheet surface acting as favorable sites for end-on $N_2$ adsorption (Luo et al., 2019a). For instance, $Ti_3C_2$ in its 2D nanosheet form showed a remarkable potential for nitrogen fixation, with catalytic activity strongly associated with its edge planes, as depicted in Figure 9.4a and b. These findings underscore the importance of edge-specific activity and the critical role of computational approaches in guiding the rational design of MXene-based catalysts for sustainable $NH_3$ synthesis (Jiao & Xu, 2019).

### 9.2.1 Two-dimensional heterostructures for ammonia production

2D building blocks exhibit unique physicochemical properties that make them highly advantageous for conversion of the ammonia ($NH_3$) production. These advantages include: (i) a large surface area combined with numerous reactive sites, which promote efficient adsorption, activation, and reduction of nitrogen ($N_2$) molecules into $NH_3$; (ii) their ultrathin structure, which accelerates the dissociation and transport of charge carriers from the bulk material to the surface; (iii) the capability of their extended surface area to establish electronic coupling with other materials, enhancing charge carrier separation and interfacial charge transfer efficiency; (iv) the tunability of their thickness, which enables control over band gap energy through quantum confinement effects, along with adjustments to light absorption properties and the positioning of conduction and valence band edges; and (v) their highly exposed surface atoms, which allow for the application of various engineering approaches such

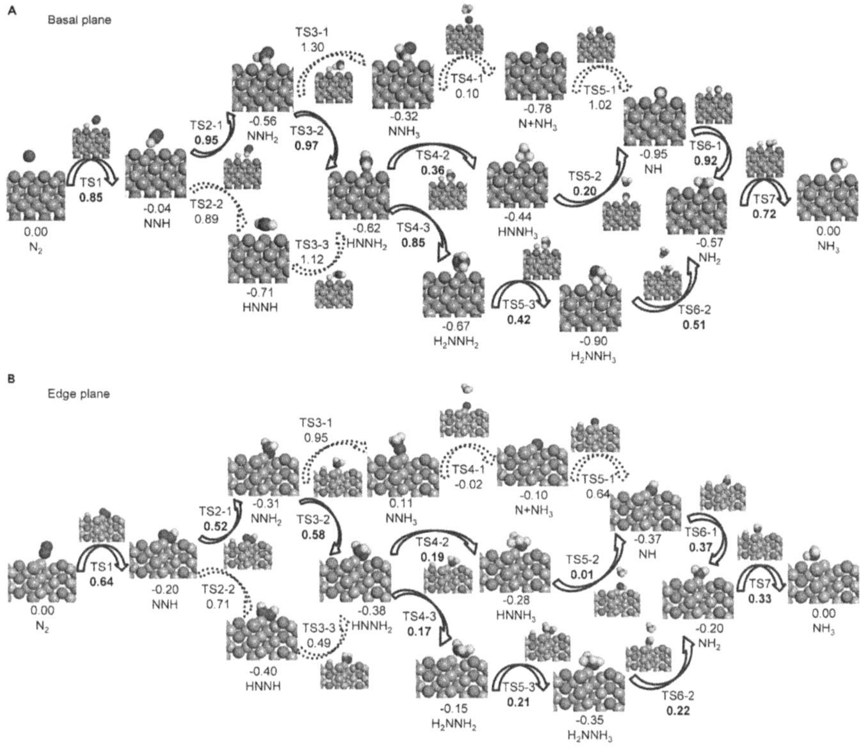


*Figure 9.4* Comparison of the reaction pathways on the basal plane and the edge plane (A and B) Structures and energies of the intermediates and TSs for various NRR paths catalyzed on the (a) basal plane and (b) edge plane (solid arrows, low-energy route; dotted arrow, unfavorable route).

**(Reproduction with permission from Elsevier 2019 (Luo et al., 2019a).)**

as doping, vacancy creation, and single-atom anchoring to optimize desired functionalities and performance. It is important to highlight that diverse active sites, such as cation or anion vacancies, heteroatom incorporation, and single-atom anchoring, have been successfully introduced in 2D materials to improve $N_2$ adsorption, activation, and reduction to $NH_3$.

## 9.3 PREPARATION METHODS FOR MXENE NANOSHEETS

Traditionally, MXene is synthesized through a top-down wet chemical process, where "A" layers (e.g., Al, Si, or Ga) are selectively etched from MAX phases. Modern synthesis techniques now include top-down, bottom-up, and hybrid up-bottom approaches (Figure 9.5). Top-down methods, such as

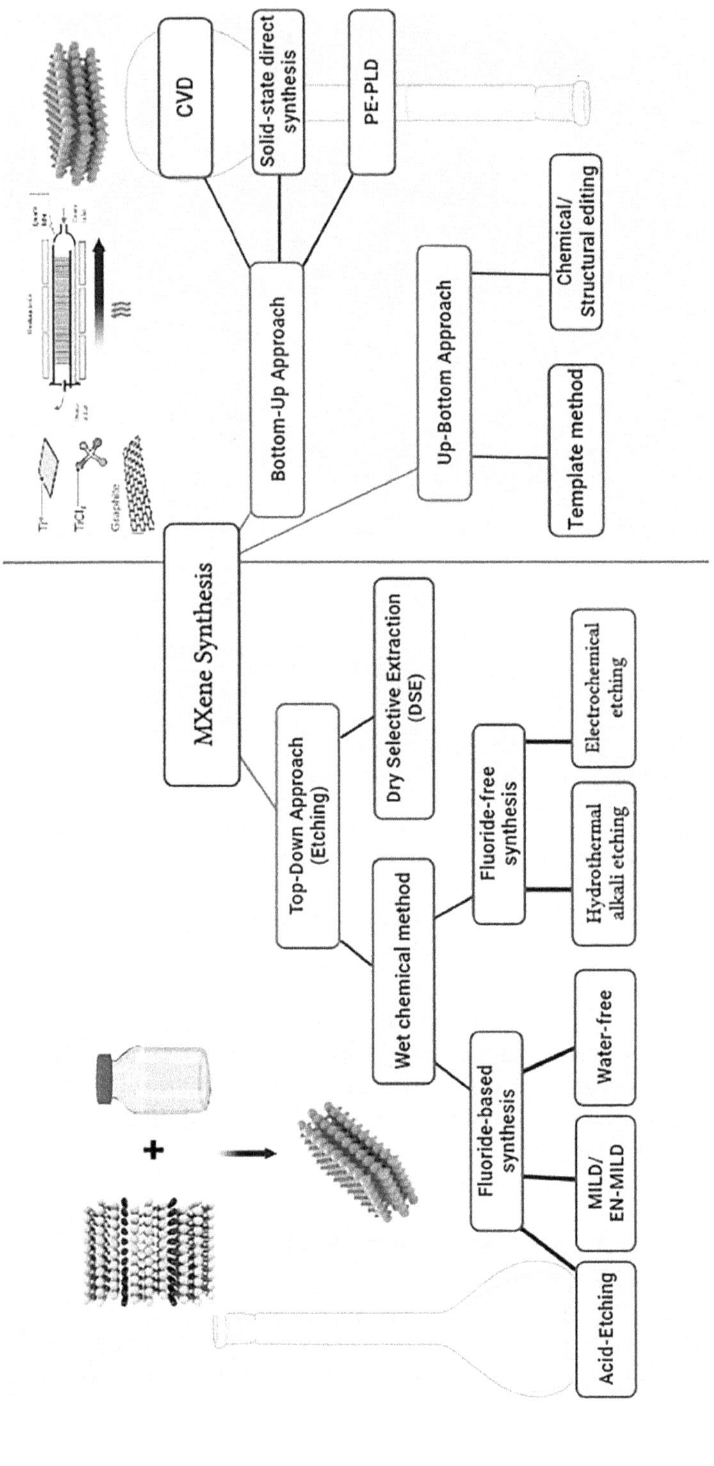


*Figure 9.5* Schematics representation of the MXene preparation top-down approach methods. **(Reproduction with permission from Elsevier 2023.)**

dry selective extraction (DSE) and wet chemical etching, dominate research due to their efficiency. DSE, a fluoride-free approach, uses iodine vapor under optimized conditions (e.g., 350°C) to avoid over-etching. Wet chemical methods remain popular, with fluoride-based techniques relying on HF or in-situ HF generation using fluoride salts (LiF, NaF) combined with milder acids like HCl (MILD method). Modifications such as the "Evaporated-Nitrogen MILD" method have further improved MXene quality, achieving higher conductivity and fewer defects. Despite its effectiveness, concerns regarding HF toxicity have fueled the development of fluoride-free alternatives. Fluoride-free etching methods, including hydrothermal alkali treatment, acid treatment, and electrochemical techniques, offer safer and environmentally friendly pathways. Hydrothermal alkali treatment uses concentrated NaOH to etch MAX phases, though delamination requires additional intercalation and sonication. Acid treatments, such as low-concentration HCl, were initially explored to avoid fluoride terminations, while acid-free electrochemical methods utilize $NH_4Cl$ and TMAOH electrolytes to selectively remove "A" layers. However, fluoride-free approaches still face challenges, including prolonged synthesis times, lower yields (e.g., ~40%), and scalability issues. Further research is needed to optimize these processes for industrial-scale production while maintaining high MXene quality and performance.

## 9.4 SPECIFIC 2D HETEROSTRUCTURES IN NITROGEN REDUCTION REACTION (NRR)

2D heterostructures have been identified as a viable option for the development of new catalysts for the NRR, which is a vital process in the production of ammonia in an environmentally friendly manner. These heterostructures are the stacking or planar integration of two or more 2D materials, and due to their interfaces, they can facilitate efficient charge transfer, nitrogen molecule adsorption, and activation. As opposed to the conventional bulk catalysts, 2D heterostructures possess many active sites, tetravalent electronic structure, and a large surface-area-to-volume ratio and are therefore suitable for NRR. Additionally, their customizable composition allows for the precise engineering of electronic structures and chemical environments to optimize catalytic performance. Among the widely studied 2D materials, heterostructures combining transition metal dichalcogenides (TMDs) like $MoS_2$ or $WS_2$ with other materials such as graphene, MXenes, or black phosphorus have shown significant promise. The synergy between the materials in these heterostructures enhances their catalytic activity by facilitating the adsorption and activation of $N_2$ molecules. For instance, the introduction of an interface between $MoS_2$ and graphene can improve electron conductivity while retaining strong binding with nitrogen intermediates. Similarly, MXene-based heterostructures, when coupled with metal nitrides or carbides, can leverage their high conductivity and active edge sites to promote nitrogen reduction under ambient conditions. Another advantage of 2D heterostructures in the NRR is their ability to suppress competing HER, a common challenge in aqueous environments. By carefully designing the heterostructure interfaces, researchers can create an environment that selectively favors $N_2$ activation over HER. For example, g-$C_3N_4$/TMD hybrids exhibit modulated electronic structures that lower the energy barrier for nitrogen adsorption while minimizing proton reduction. This selectivity is further enhanced through strategies such as doping with transition metals, which create localized active sites with high affinity for nitrogen. Despite their potential, challenges remain in the practical implementation of 2D heterostructures for the NRR. Issues such as long-term stability, scalability of synthesis, and achieving high ammonia yields under mild conditions need to be addressed. Nevertheless, ongoing advancements in material design, characterization techniques, and computational modeling are expected to accelerate the development of more efficient 2D heterostructures. With their unique properties and versatility, these materials are poised to play a pivotal role in advancing green ammonia production technologies.

## 9.5 CATALYTIC MECHANISMS IN NRR USING 2D HETEROSTRUCTURES ROLE OF DEFECT SITES

The growing research and expanding applications of defective catalysts across multiple fields have demonstrated that defects serve more than a single function, such as acting as sites for molecular activation. Despite these findings, the structural roles and mechanisms of defects in various catalytic processes remain underexplored. Additionally, the study of catalyst defects still lacks a robust theoretical framework and has yet to establish a systematic, precise, and comprehensive foundation that can effectively guide the design and synthesis of defective catalysts. These challenges present significant obstacles in advancing this emerging area of research.

Nevertheless, the rise of defect engineering in environmental catalysis highlights its increasing prominence as a critical area of frontier scientific exploration. This chapter focuses on the classification of defects in catalytic materials and discusses their construction strategies and advanced characterization techniques (Figure 9.6a). By emphasizing the roles of defects in electrocatalytic, photocatalytic, and thermal catalytic processes, we aim to analyze the characteristics and underlying reaction mechanisms. Furthermore, the correlation between the structural properties of catalyst defects and their catalytic activity is explored in detail. Electrocatalysis is a promising catalytic subfield with significant potential for sustainable applications, particularly in fostering clean energy development and achieving green environmental goals. Its importance lies in addressing critical challenges related to environmental pollution and energy sustainability. Electrocatalytic technology has been extensively applied in diverse areas,

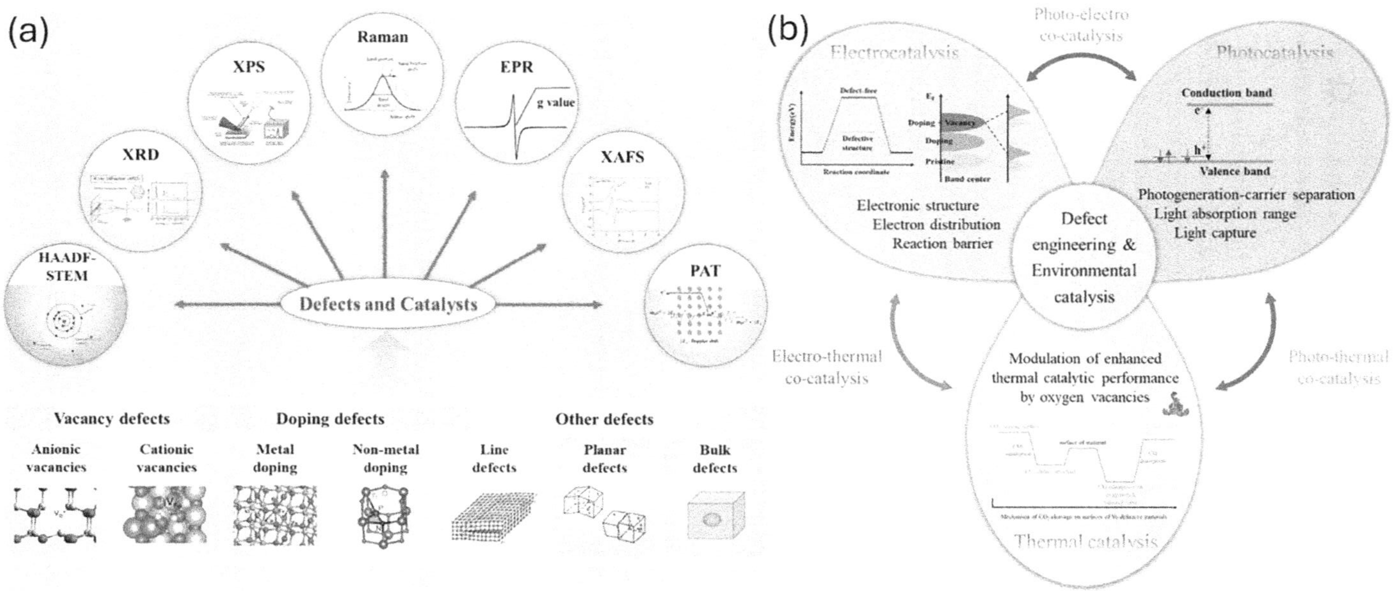


*Figure 9.6* (a) Mechanisms of the impact of defect engineering on environmental catalysis and (b) classification and characterization techniques for defects in catalytic materials.

**(Reproduction with permission from Elsevier 2024 (S. Gao et al., 2024).)**

including the treatment of organic wastewater, degradation of chromium-containing wastewater, flue gas desulfurization, raw coal purification, simultaneous removal of NOx and $SO_2$, and the electrochemical reduction of carbon dioxide ($CO_2$) and nitrogen ($N_2$). As an efficient energy storage and conversion technology, electrocatalysis offers a viable pathway to mitigate energy depletion while combatting environmental pollution. Catalytic combustion, on the other hand, remains the most widely adopted method for managing and controlling volatile organic compounds (VOCs). However, the reliance on fossil fuels for catalytic combustion not only leads to significant energy consumption but also exacerbates the energy crisis and contributes to environmental pollution through fuel emissions. Photocatalysis presents an alternative approach, utilizing the redox potential of photocatalysts activated under light exposure. This technology enables pollutant purification, as well as the synthesis and transformation of various substances. Photocatalytic oxidation typically employs semiconductor materials as catalysts and uses light energy to degrade organic pollutants into harmless byproducts, such as water and carbon dioxide. Due to its environmentally friendly nature and low cost, electrocatalysis holds immense potential for addressing environmental pollution and global warming caused by the continuous combustion of fossil fuels. Similarly, thermal catalysis leverages abundant solar energy by converting light energy into thermal energy through catalyst absorption. This process drives catalytic reactions while minimizing energy consumption and efficiently degrading pollutants. Focusing on the domain of environmental catalysis, this chapter provides a detailed overview of the role of defect engineering in enhancing electrocatalysis and other catalytic processes. The mechanisms through which defects influence catalytic performance are discussed comprehensively, as illustrated in Figure 9.6b.

## 9.6 PARAMETRIC STUDIES AND ANALYSIS

### 9.6.1 Electrolyte salts, temperatures, and pH values

Naturally occurring nitrogenase catalysis exemplifies NRR under ambient conditions, highlighting the potential for electrochemical NRR to function at room temperature, atmospheric pressure, and low salinity. While these mild conditions simplify device design, they may lead to reduced energy efficiency due to the limited electrical conductivity of the electrolyte and the low kinetic efficiency of the reaction. Electrolyte salts primarily serve as conductors, influence hydrogen transport and reactions, and interact with active sites on the electrocatalyst surface, ultimately affecting the selectivity of NRR. Key parameters such as temperature and pH are often optimized to enhance NRR performance. Based on the Arrhenius equation, higher temperatures are expected to increase the reaction rate of NRR. However, this can simultaneously intensify the HER within the same potential range, especially on materials sensitive to HER, thereby adversely impacting NRR efficiency. At low pH, while higher proton concentrations enhance HER, they also facilitate the release of ammonia as ammonium ions ($NH_4^+$) due to protonation. Interestingly, the combined effect of temperature and pH is critical, as HER is typically more pronounced in acidic environments compared to alkaline ones. Consequently, increased temperatures can suppress NRR activity in acidic conditions. Recent studies on MXene-based catalysts are summarized in Table 9.1.

### 9.6.2 Evaluation of stability, selectivity, and Faradaic efficiency in ammonia synthesis

The electrochemical NRR for $NH_3$ production is a promising alternative to conventional ammonia synthesis, and MXene-based nanocomposites have demonstrated immense potential as advanced electrocatalysts. These materials possess high electrical conductivity, tunable surface terminations, and abundant active sites for nitrogen adsorption and activation. Modifications, such as doping with elements like nitrogen or phosphorus or incorporating dual-metal sites (e.g., alloy-based composites), significantly enhance catalytic performance by promoting selective NRR pathways while suppressing the HER. Surface functional groups (–OH, –F, –O) also play a crucial role in modulating electron density, enabling efficient nitrogen fixation under electrochemical conditions. Electrolyte composition and pH are critical factors in optimizing NRR efficiency. Alkaline electrolytes often reduce HER by limiting proton availability, while acidic conditions facilitate $NH_4^+$ formation but increase HER activity. The ionic strength and type of electrolyte salts, such as KOH or $Na_2SO_4$, influence ionic conductivity and interaction with the catalyst surface. Moreover, conducting NRR under low-salinity conditions mimics environmentally friendly scenarios but requires improved conductivity to maintain high energy efficiency.

Careful selection and tuning of these parameters ensure enhanced nitrogen fixation with minimal side reactions. Temperature and applied potential significantly impact the NRR process. Elevated temperatures improve reaction kinetics as per the Arrhenius equation but may also amplify HER on certain electrocatalysts, particularly in acidic environments. The potential range must be carefully optimized to minimize HER competition while maximizing nitrogen reduction activity. High current densities indicate enhanced catalytic performance, but excessively high values may compromise selectivity due to competing reactions. Thus, a balance between these parameters is crucial for achieving efficient $NH_3$ production. The long-term stability and durability of MXene-based nanocomposites are essential for ammonia production and have been summarized in Table 9.2. These materials can suffer from degradation under prolonged electrochemical operation, especially in corrosive or harsh conditions. Protective strategies, such as

*Table 9.1* Recent studies of the MXene-based catalyst for ammonia production with experimental conditions

| *Sr No* | *Catalyst* | *Electrolyte* | *Membrane* | *Cell type* | *Conditions* | *$NH_3$ yield ($\mu gh^{-1}$ $mg^{-1}{}_{cat}$)* | *FE (%), potential vs. RHE* | *$N_2$ reduction mechanism* | *By product* | *Ref* |
|---|---|---|---|---|---|---|---|---|---|---|
| **1** | MXene/$TiFeO_x$ | 0.05 M $H_2SO_4$ | Nafion 117 | H | RT, 1atm | 21.9 | 25.44% at −0.2V | NA | $N_2H_4$ | Y. Guo et al. (2020) |
| **2** | $Ti_3C_2$ nanoribbons | 0.5M KOH | Nafion | H | 20°C, 1atm | 14.76 | 3.1%at −0.5V | NA | $N_2H_4$ | H. Wei et al. (2020) |
| **3** | $Ti_3C_2T_x$ QDs | 0.1M HCl | Nafion | H | RT, 1atm | 62.94 | 13.30% at −0.5V | Distal pathway | $N_2H_4$ | Jin et al. (2020) |
| **4** | 1T-$MoS_2$@$Ti_3C_2$ | 0.1M HCl | Nafion | H | RT, 1atm | 30.33 | 10.94% at −0.3V | NA | $N_2H_4$ | Xu et al. (2020) |
| **5** | $V_2CT_x$ | 0.1 M $Na_2SO_4$ | Nafion | H | RT, 1atm | 12.6 | 4% at −0.7V | Distal pathway | $N_2H_4$ | Xia et al. (2021) |
| **6** | Cu/$Ti_3C_2T_x$ | 0.1 M KOH | Na | RDE | RT, 1atm | 3.04 | 7.31% at −0.5V | Distal pathway | $N_2H_4$ | A. Liu et al. (2021) |
| **7** | ZIF-67@$Ti_3C_2$ | 0.1 M KOH | Na | 3E Cell | RT, 1atm | 6.52 | 20.2% ar −0.4V | NA | NA | Yao et al. (2018) |
| **8** | N-S doped $Ti_3C_2T_X$ | 0.05 M $H_2SO_4$ | Nafion 117 | H | RT, 1atm | 34.23 | 6.6% at −0.55V | NA | $N_2H_4$ | Zeng et al. (2021) |
| **9** | Ni/$V_4C_3T_x$ | 0.1 M KOH | Nafion 117 | H | RT, 1atm | 21.29 | 14.86% at −0.55V | Altering pathway | NA | Du et al. (2021) |
| **10** | $Ti_3C_2T_x$ | 0.01 M $Na_2SO_4$ | Nafion 117 | H | RT, 1atm | 0.1 | 7.4% at −0.55V | NA | NA | Ding et al. (2021b) |
| **11** | $Cr_3C_2$@C nanofiber | 0.1M HCl | Nafion | H | RT, 1atm | 23.9 | 8.6% at −0.3V | Distal pathway | $N_2H_4$ | G. Yu et al. (2019) |
| **12** | $MnO_2$@$Ti_3C_2$TX | 0.1M HCl | Nafion | H | RT, 1atm | 34.12 | 11.39% at −0.55V | Distal pathway | $N_2H_4$ | Kong et al. (2019) |
| **13** | $Ti_3C_2T_x$ | 0.01M HCl | Nafion 211 | H | RT, 1atm | 4.72 | 4.62% at −0.1V | Distal pathway | $N_2H_4$ | Luo et al. (2019b) |
| **14** | $Ti_3C_2T_x$ | 0.1M HCl | Nafion 117 | H | RT, 1atm | 36.9 | 9.1% at −0.3V | NA | $N_2H_4$ | T. Li et al. (2019) |

| | | | | | | | | | | |
|---|---|---|---|---|---|---|---|---|---|---|
| **15** | Au/$Ti_3C_2$ | 0.1M HCl | Nafion 117 | H | RT,1atm | 30.06 | 18.34% at −0.2V | Altering pathway | $N_2H_4$ | D. Liu et al. (2019) |
| **16** | $TiO_2$/$Ti_3C_2T_x$ | 0.1M HCl | Nafion | H | RT,1atm | 26.32 | 8.42% at −0.6V | NA | $N_2H_4$ | Ma et al. (2020) |
| **17** | Oxygen vacancy rich $TiO_2$/ $Ti_3C_2T_X$ | 0.1M HCl | Nafion 211 | H | RT,1atm | 32.17 | 16.07% at −0.45V | Distal pathway | $N_2H_4$ | Y. Fang et al. (2019) |
| **18** | Ru@$Mo_2CT_x$ | 0.5 M $K_2SO_4$ | Nafion 211 | H | RT,1atm | 40.57 | 25.77% at −0.3V | Altering pathway | $N_2H_4$ | Peng et al. (2020) |
| **19** | Ru@$Ti_3C_2T_x$ | 0.1M KOH | Nafion 211 | RDE | RT,1atm | 2.3 | 13.13% at −0.4V | Distal pathway | $N_2H_4$ | A. Liu et al. (2020) |
| **20** | $Ti_3C_2$ | 0.1M KOH | Nafion | H | 20°C ,1atm<br>60°C,1atm | 12.46 | 7.01% at −0.2V<br>9.03% at −0.2V | Distal pathway | $N_2H_4$ | Xia et al. (2020) |
| **21** | $Ti_3C_2T_X$ | 0.1M HCl | Nafion 211 | H | RT,1atm | 20.4 | 9.3% at −0.4V | Distal pathway | $N_2H_4$ | J. Zhao et al. (2018) |

*Table 9.2* Stability of the MXene-based catalyst for ammonia production

| Sr No | Catalyst | Gas | Temperature | Scavenger | Light source Xe Lamp (W) | Test method | $NH_3$ yield µmol $h^{-1}g^{-1}_{cat.}$ | Stability (h) | Yield (%) | Ref |
|---|---|---|---|---|---|---|---|---|---|---|
| 1 | $Ti_3C_2T_x/TiO_2$ | $N_2$ | 25°C | No | 250 | Nessler's reagent Indophenol blue | 3.88 | 10 | 90 | Hou et al. (2020) |
| 2 | $CdS@Ti_3C_2$ | $N_2$ | RT | $CH_3OH$ | 300 | Nessler's reagent | 45 | 6 | 97.94 | B. Sun et al. (2021) |
| 3 | $Ti_3C_2$/N-defect g-$C_3N_4$ | $N_2$ | 35°C | $CH_3OH$ | 300 | Nessler's reagent | 88.79 | 18 | 83 | C. Sun et al. (2021) |
| 4 | Co/MXene/ $TiO_2$ | $N_2$ | NA | No | 300 | Nessler's reagent | 422 | 18 | 99 | W. Gao et al. (2021); Gao et al. (2021) |
| 5 | P25/$Ti_3C_2$ | $N_2$ | RT | $CH_3OH$ | 300 | Ion chromatography | 293.06 | NA | NA | Liao et al. (2020) |
| 6 | $AgInS_2$/MXene | $N_2$ | RT | $CH_3OH$ | 300 | Indophenol blue | 5.792 | 15 | 73 | Qin et al. (2019) |
| 7 | $RuO_2$@ $TiO_2$-MXene | $N_2$ | RT | No | 100 | Salicylic acid-hypochlorite | 110 | 6 | 84.4 | Hao et al. (2020) |
| 8 | $Ti_3C_2$-QD/ Ni-MOF | $N_2$ | 25°C | $Na_2SO_3$ | 300 | Nessler's reagent | 43.44 | 9 | 98 | Qin et al. (2020) |

fluorine passivation and the formation of anti-corrosion layers, help maintain active sites and improve longevity. Faradaic efficiency (FE) and $NH_3$ yield are critical evaluation metrics, with ongoing research focusing on maximizing these outputs while minimizing byproducts. Advanced studies integrating computational modeling and experimental validation are paving the way for scalable, sustainable NRR technologies. The electrochemical behavior of phase-pure few-layered (FL) MXene material was further explored by evaluating its performance in the NRR. Since protons are essential for converting nitrogen ($N_2$) into ammonia ($NH_3$), acidic media are generally preferred for NRR (D. Johnson & Djire, 2023). However, the HER often competes under these conditions. Consequently, adjusting the pH value—and thereby the proton concentration—can significantly influence the FE of the process.

Chronoamperometry (CA) experiments conducted over 4 hours were analyzed to measure $NH_3$ and hydrazine production using the indophenol blue and Watt-Chrisp methods, respectively. Calibration curves were used for quantification. Due to the catalyst's instability in highly acidic conditions and the limitations of the Berthelot method for ammonia detection, data for pH 0 were not collected. Yield and FE results for all tested pH levels are summarized in Figure 9.7a–d. Across all systems, FE declined sharply as more cathodic potentials were applied, consistent with HER dominance at these potentials. This was further supported by the higher currents observed in the CA curves. Notably, no significant hydrazine production was detected using the Watt-Chrisp method.

For the pH 1 electrolyte, the maximum $NH_3$ yield ($10.05 \pm 1.49$ μg $cm^{-2}$ $h^{-1}$) occurred at an applied potential of −0.25 V versus reversible hydrogen electrode (RHE), while the highest FE ($11.39 \pm 1.00\%$) was achieved at −0.15 V versus RHE. This discrepancy can be attributed to HER becoming more dominant as the applied potential becomes more cathodic (Figure 9.7a). Similar trends were observed in the pH 2 electrolyte, where the highest yield ($5.70 \pm 1.07$ μg $cm^{-2}$ $h^{-1}$) and FE ($12.37 \pm 1.75\%$) were obtained at −0.25 V and −0.15 V versus RHE, respectively (Figure 9.7b). However, for pH 3, both the maximum yield ($9.91 \pm 1.05$ μg $cm^{-2}$ $h^{-1}$) and FE ($12.22 \pm 2.74\%$) were achieved at the same applied potential of −0.15 V versus RHE (Figure 9.7c). This shift is likely due to the reduced $H^+$ concentration at higher pH levels, which diminishes HER activity, as evidenced by linear sweep voltammetry (LSV) data. To gain a deeper understanding of how pH influences NRR performance, the maximum yield and FE values at varying pH levels were plotted and are presented in Figure 9.7d. The results reveal that yield does not exhibit a linear relationship with pH but instead shows a distinct minimum at a pH of 2. This observation is notable, as it deviates from expectations based on the NRR onset potential corresponding to these pH values. Conversely, FE demonstrates a slight upward trend as pH increases. This modest increase aligns with the hypothesis that a higher pH, indicative

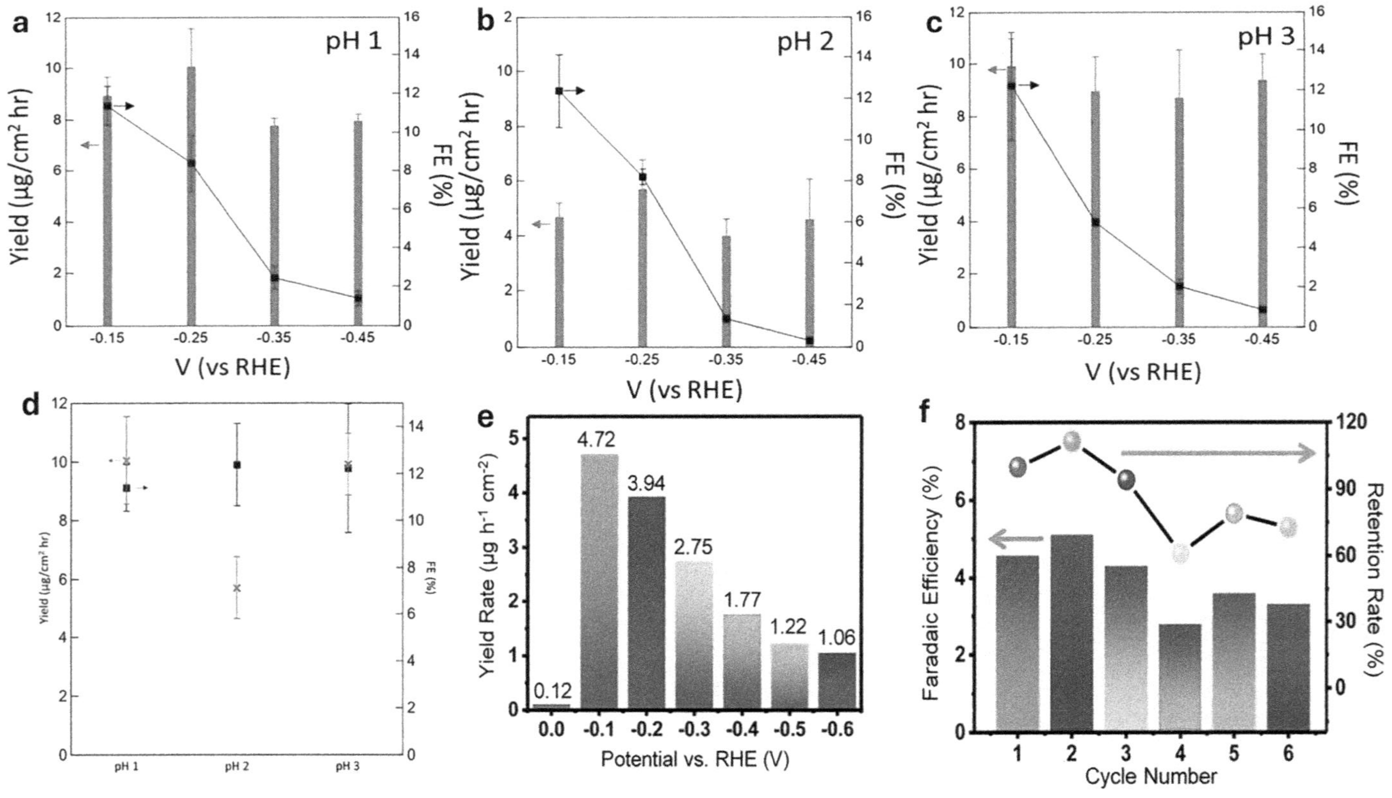


*Figure 9.7* (a–d) Impact of pH on the electrochemical properties and nitrogen reduction reaction performance of MXene (Adoption from the willey 2023) (d. Johnson & Djire, 2023). (e–f) Applied nitrogen reduction and cycling stability results (0.1 V versus RHE).

**(Reproduction with permission from the cell press 2019 (Luo et al., 2019a).)**

of a reduced $H^+$ concentration in the solution, should enhance FE. However, the improvement is less pronounced than initially anticipated.

This phenomenon can be attributed to the requirement for higher currents to draw the lower-concentration $H^+$ ions to the catalyst surface, thereby reducing FE. While the FE values reported are still significantly below the Department of Energy's targets, they provide valuable insights into the relationship between the bulk pH of the electrochemical system and NRR performance, an area that has often been overlooked. Previous studies have partially explored this relationship, and the MXene material shows comparable or superior FE values under acidic conditions compared to pH-independent catalysts like high-entropy alloys, CoVP@NiFeV-layered double hydroxide nanotubes, and single-atom copper-doped nitrogen-doped carbon. Notably, these pH-universal catalysts demonstrate significantly higher FE values as the environment shifts toward neutral or basic pH. This trend highlights the potential of MXene in diverse electrochemical settings, suggesting promising avenues for future research.

Luo et al. (2019a) investigated the NRR performance of MXene deposited on stainless steel mesh (SSM) through detailed electrochemical experiments. Using constant potential electrolysis under nitrogen, the tests were carried out across a potential range from 0 V to −0.6 V versus the RHE. Ammonia concentrations were quantified via the Nessler assay. The results showed that the FE of ammonia synthesis varied significantly applied potential dropped below −0.1 V (versus RHE), and the FE (4.62 %) declined sharply as HER began to dominate. A similar trend was observed for the ammonia production rates, with a peak yield of 4.72 $\mu g\ h^{-1}\ cm^{-2}$ achieved at −0.1 V versus RHE, as illustrated in Figure 9.7e. To confirm that the ammonia was produced via nitrogen reduction catalyzed by MXene, control experiments were performed using argon as the feed gas, $^{15}N_2$ as the isotopically labeled feed gas, and bare SSM as the working electrode. These controls confirmed that ammonia production originated from nitrogen reduction. The catalyst's stability was evaluated through six cycles of chronoamperometric testing, which showed no significant changes in ammonia yield or FE. Performance retention was measured at 72.7%, indicating that MXene exhibits excellent durability during NRR electrocatalysis (Figure 9.7f) (Luo et al., 2019a).

## 9.7 CONCLUSION

In this chapter, we summarize that MXenes and their hybrid structures have a promising prospect to enable sustainable ammonia production through the NRR by electrochemical tools. The unique physicochemical properties of MXenes, such as high conductivity, hydrophilicity, a high number of active sites, and tunable surface chemistry, make MXenes suitable for various applications including the replacement of the energy-demanding

Haber-Bosch process. While advancements in defect engineering, heterostructures, and doping strategies have enhanced catalytic performance, challenges such as low ammonia yield, competing HERs, and scalability persist. Future efforts must focus on optimizing the synthesis of MXenes with tailored properties, improving long-term stability, and achieving higher FEs. The integration of experimental studies with computational modeling will accelerate the rational design of advanced MXene-based catalysts. Addressing these challenges will enable the practical application of MXenes in green ammonia synthesis, contributing significantly to global efforts for sustainable energy and environmental technologies.

## ACKNOWLEDGMENT

This work was supported by the Basic Science Research Program (2023R1A2C3004336) and the Regional Leading Research Center (RS-2024-00405278) through the National Research Foundation of Korea (NRF) grant funded by the Korea Government (MSIT).

## REFERENCES

Ahmad Junaidi, N. H., Wong, W. Y., Loh, K. S., Rahman, S., & Daud, W. R. W. (2021). A comprehensive review of MXenes as catalyst supports for the oxygen reduction reaction in fuel cells. In *International Journal of Energy Research* (Vol. 45, Issue 11, pp. 15760–15782). John Wiley & Sons Ltd. https://doi.org/10.1002/er.6899

Akter, R., Shah, S. S., Ehsan, M. A., Shaikh, M. N., Zahir, M. H., Aziz, M. A., & Ahammad, A. J. S. (2024). Transition-metal-based catalysts for electrochemical synthesis of ammonia by nitrogen reduction reaction: Advancing the green ammonia economy. In *Chemistry - An Asian Journal* (Vol. 19, Issue 16). John Wiley & Sons Ltd. https://doi.org/10.1002/asia.202300797

Amrillah, T., Hermawan, A., Alviani, V. N., Seh, Z. W., & Yin, S. (2021). MXenes and their derivatives as nitrogen reduction reaction catalysts: recent progress and perspectives. In *Materials Today Energy* (Vol. 22). Elsevier Ltd. https://doi.org/10.1016/j.mtener.2021.100864

Burdett, J. K., Hughbanks, T., Miller, G. J., Richardson, J. W., & Smith, J. V. (1987). Structural-electronic relationships in inorganic solids: Powder neutron diffraction studies of the rutile and anatase polymorphs of titanium dioxide at 15 and 295 K. *Journal of the American Chemical Society*, *109*, 3639–3646.

Chowdhury, I. H., Gupta, S., & Rao, V. G. (2023). Covalent organic framework: An emerging catalyst for renewable ammonia production. In *ChemCatChem* (Vol. 15, Issue 11). John Wiley & Sons Inc. https://doi.org/10.1002/cctc.202300243

Deng, J., Iñiguez, J. A., & Liu, C. (2018). Electrocatalytic nitrogen reduction at low temperature. In *Joule* (Vol. 2, Issue 5, pp. 846–856). Cell Press. https://doi.org/10.1016/j.joule.2018.04.014

Ding, Y., Zhang, J., Guan, A., Wang, Q., Li, S., Al-Enizi, A. M., Qian, L., Zhang, L., & Zheng, G. (2021a). Promoting $N_2$ electroreduction to ammonia by fluorine-terminating $Ti_3C_2T_x$ MXene. *Nano Convergence*, *8*(1). https://doi.org/10.1186/s40580-021-00264-9

Ding, Y., Zhang, J., Guan, A., Wang, Q., Li, S., Al-Enizi, A. M., Qian, L., Zhang, L., & Zheng, G. (2021b). Promoting $N_2$ electroreduction to ammonia by fluorine-terminating $Ti_3C_2T_x$ MXene. *Nano Convergence*, *8*(1), 14. https://doi.org/10.1186/s40580-021-00264-9

Du, C.-F., Yang, L., Tang, K., Fang, W., Zhao, X., Liang, Q., Liu, X., Yu, H., Qi, W., & Yan, Q. (2021). Ni nanoparticles/$V_4C_3T_x$ MXene heterostructures for electrocatalytic nitrogen fixation. *Materials Chemistry Frontiers*, *5*(5), 2338–2346. https://doi.org/10.1039/D0QM00898B

Fang, Q. J., Zhang, W., Zhang, X.-J., Wang, J.-H., Zhao, S.-T., Zhang, X.-L., Chen, W.-X., & Zhuang, G.-L. (2023). Rational design of bimetallic MXene solid solution with High-Performance electrocatalytic $N_2$ reduction. *Journal of Colloid and Interface Science*, *640*, 67–77. https://doi.org/10.1016/j.jcis.2023.02.094

Fang, Y., Liu, Z., Han, J., Jin, Z., Han, Y., Wang, F., Niu, Y., Wu, Y., & Xu, Y. (2019). High-performance electrocatalytic conversion of $N_2$ to $NH_3$ using Oxygen-vacancy-rich $TiO_2$ in situ grown on $Ti_3C_2T_x$ MXene. *Advanced Energy Materials*, *9*(16). https://doi.org/10.1002/aenm.201803406

Gao, S., Zhao, S., Tang, X., Sun, L., Li, Q., & Yi, H. (2024). Research on the application of defect engineering in the field of environmental catalysis. *Green Energy & Environment*. https://doi.org/10.1016/j.gee.2024.08.008

Gao, W., Li, X., Luo, S., Luo, Z., Zhang, X., Huang, R., & Luo, M. (2021). In situ modification of cobalt on MXene/$TiO_2$ as composite photocatalyst for efficient nitrogen fixation. *Journal of Colloid and Interface Science*, *585*, 20–29. https://doi.org/10.1016/j.jcis.2020.11.064

Guo, T., Zhou, D., & Zhang, C. (John). (2022). Perspectives on electrochemical nitrogen fixation catalyzed by two-dimensional MXenes. In *Materials Reports: Energy* (Vol. 2, Issue 1). KeAi Communications Co. https://doi.org/10.1016/j.matre.2021.100076

Guo, Y., Wang, T., Yang, Q., Li, X., Li, H., Wang, Y., Jiao, T., Huang, Z., Dong, B., Zhang, W., Fan, J., & Zhi, C. (2020). Highly efficient electrochemical reduction of nitrogen to Ammonia on surface termination modified $Ti_3C_2T_x$ MXene nanosheets. *ACS Nano*, *14*(7), 9089–9097. https://doi.org/10.1021/acsnano.0c04284

Gurushankar, K., Karuppaiya, C., Tari, V., & Kannan, K. (2024). Exploring MXenes and MBenes in metal sulfur and selenium battery technologies. In K. Kannan & V. Tari (Eds.), *Novel Energy Storage and Conversion Technologies for Two-Dimensional MXenes and MBenes*. IGI Global. https://doi.org/10.4018/979-8-3693-9546-2.ch005

Hao, C., Liao, Y., Wu, Y., An, Y., Lin, J., Gu, Z., Jiang, M., Hu, S., & Wang, X. (2020). $RuO_2$-loaded $TiO_2$–MXene as a high performance photocatalyst for nitrogen fixation. *Journal of Physics and Chemistry of Solids*, *136*, 109141. https://doi.org/10.1016/j.jpcs.2019.109141

Hou, T., Li, Q., Zhang, Y., Zhu, W., Yu, K., Wang, S., Xu, Q., Liang, S., & Wang, L. (2020). Near-infrared light-driven photofixation of nitrogen over $Ti_3C_2T_x$/$TiO_2$ hybrid structures with superior activity and stability. *Applied Catalysis B: Environmental*, *273*, 119072. https://doi.org/10.1016/j.apcatb.2020.119072

Huang, Y., Zhu, Y., Chen, S., Xie, X., Wu, Z., & Zhang, N. (2021). Schottky junctions with Bi cocatalyst for taming aqueous phase $N_2$ reduction toward enhanced solar Ammonia production. *Advanced Science*, *8*(6). https://doi.org/10.1002/advs.202003626

Jiao, F., & Xu, B. (2019). Electrochemical ammonia synthesis and ammonia fuel cells. *Advanced Materials*, *31*(31). https://doi.org/10.1002/adma.201805173

Jin, Z., Liu, C., Liu, Z., Han, J., Fang, Y., Han, Y., Niu, Y., Wu, Y., Sun, C., & Xu, Y. (2020). Rational design of Hydroxyl-Rich $Ti_3 C_2 T_x$ MXene quantum dots for high-performance electrochemical $N_2$ reduction. *Advanced Energy Materials*, *10*(22). https://doi.org/10.1002/aenm.202000797

Johnson, D., & Djire, A. (2023). Effect of pH on the electrochemical behavior and nitrogen reduction reaction activity of $Ti_2N$ Nitride MXene. *Advanced Materials Interfaces*, *10*(10). https://doi.org/10.1002/admi.202202147

Johnson, L. R., Sridhar, S., Zhang, L., Fredrickson, K. D., Raman, A. S., Jang, J., Leach, C., Padmanabhan, A., Price, C. C., Frey, N. C., Raizada, A., Rajaraman, V., Saiprasad, S. A., Tang, X., & Vojvodic, A. (2020). MXene Materials for the Electrochemical Nitrogen Reduction-Functionalized or Not? *ACS Catalysis*, *10*(1), 253–264. https://doi.org/10.1021/acscatal.9b01925

Kannan, K., & Tari, V. (Eds.). (2025). Novel energy storage and conversion technologies for two-dimensional MXenes and MBenes. https://www.amazon.com/dp/9798369395462

Kong, W., Gong, F. (Frank), Zhou, Q., Yu, G., Ji, L., Sun, X., Asiri, A. M., Wang, T., Luo, Y., & Xu, Y. (2019). An $MnO_2$ –$Ti_3C_2T_x$ MXene nanohybrid: an efficient and durable electrocatalyst toward artificial $N_2$ fixation to $NH_3$ under ambient conditions. *Journal of Materials Chemistry A*, *7*(32), 18823–18827. https://doi.org/10.1039/C9TA04902A

Li, J., Valenza, R., & Haussener, S. (2024). In situ synthesis of CuxO/N doped Graphdiyne with Pyridine N configuration for ammonia production via nitrate reduction. *Small*, *20*(33). https://doi.org/10.1002/smll.202310467

Li, L., & Xiao, J. (2023). Activity trend and selectivity of electrochemical ammonia synthesis in reverse artificial Nitrogen cycle. *ChemSusChem*, *16*(22). https://doi.org/10.1002/cssc.202300593

Li, T., Yan, X., Huang, L., Li, J., Yao, L., Zhu, Q., Wang, W., Abbas, W., Naz, R., Gu, J., Liu, Q., Zhang, W., & Zhang, D. (2019). Fluorine-free $Ti_3C_2T_x$ (T = O, OH) nanosheets (~50–100 nm) for nitrogen fixation under ambient conditions. *Journal of Materials Chemistry A*, *7*(24), 14462–14465. https://doi.org/10.1039/C9TA03254A

Liang, X., Ren, X., Yang, Q., Gao, L., Gao, M., Yang, Y., Zhu, H., Li, G., Ma, T., & Liu, A. (2021). A two-dimensional MXene-supported metal-organic framework for highly selective ambient electrocatalytic nitrogen reduction. *Nanoscale*, *13*(5), 2843–2848. https://doi.org/10.1039/d0nr08744k

Liao, Y., Qian, J., Xie, G., Han, Q., Dang, W., Wang, Y., Lv, L., Zhao, S., Luo, L., Zhang, W., Jiang, H.-Y., & Tang, J. (2020). 2D-layered $Ti_3C_2$ MXenes for promoted synthesis of $NH_3$ on P25 photocatalysts. *Applied Catalysis B: Environmental*, *273*, 119054. https://doi.org/10.1016/j.apcatb.2020.119054

Liu, A., Gao, M., Ren, X., Meng, F., Yang, Y., Yang, Q., Guan, W., Gao, L., Liang, X., & Ma, T. (2020). A two-dimensional Ru@MXene catalyst for highly selective ambient electrocatalytic nitrogen reduction. *Nanoscale*, *12*(20), 10933–10938. https://doi.org/10.1039/D0NR00788A

Liu, A., Liang, X., Gao, M., Ren, X., Gao, L., Yang, Y., Zhu, H., Li, G., & Ma, T. (2022). Ru and Fe alloying on a two-dimensional MXene support for enhanced electrochemical synthesis of ammonia. *ChemCatChem*, *14*(7). https://doi.org/10.1002/cctc.202101775

Liu, A., Liang, X., Yang, Q., Ren, X., Gao, M., Yang, Y., & Ma, T. (2021). Electrocatalytic synthesis of ammonia using a 2D $Ti_3$ $C_2$ MXene loaded with copper nanoparticles. *ChemPlusChem*, *86*(1), 166–170. https://doi.org/10.1002/cplu.202000702

Liu, A., Liang, X., Zhu, H., Ren, X., Gao, L., Gao, M., Yang, Y., Li, G., & Ma, T. (2022). Two-dimensional MXene supported bismuth for efficient electrocatalytic nitrogen reduction. *ChemCatChem*, *14*(7). https://doi.org/10.1002/cctc.202101683

Liu, D., Zhang, G., Ji, Q., Zhang, Y., & Li, J. (2019). Synergistic electrocatalytic nitrogen reduction enabled by confinement of nanosized Au particles onto a two-dimensional $Ti_3C_2$ substrate. *ACS Applied Materials & Interfaces*, *11*(29), 25758–25765. https://doi.org/10.1021/acsami.9b02511

Liu, H., Lang, X., Zhu, C., Timoshenko, J., Rüscher, M., Bai, L., Guijarro, N., Yin, H., Peng, Y., Li, J., Liu, Z., Wang, W., Cuenya, B. R., & Luo, J. (2022). Efficient electrochemical nitrate reduction to ammonia with copper-supported Rhodium cluster and single-atom catalysts. *Angewandte Chemie - International Edition*, *61*(23). https://doi.org/10.1002/anie.202202556

Luo, Y., Chen, G. F., Ding, L., Chen, X., Ding, L. X., & Wang, H. (2019a). Efficient electrocatalytic $N_2$ fixation with MXene under ambient conditions. *Joule*, *3*(1), 279–289. https://doi.org/10.1016/j.joule.2018.09.011

Luo, Y., Chen, G.-F., Ding, L., Chen, X., Ding, L.-X., & Wang, H. (2019b). Efficient electrocatalytic $N_2$ fixation with MXene under ambient conditions. *Joule*, *3*(1), 279–289. https://doi.org/10.1016/j.joule.2018.09.011

Ma, Z., Chen, J., Luo, D., Thersleff, T., Dronskowski, R., & Slabon, A. (2020). Structural evolution of CrN nanocube electrocatalysts during nitrogen reduction reaction. *Nanoscale*, *12*(37), 19276–19283. https://doi.org/10.1039/D0NR04981F

Manjunatha, R., Karajić, A., Liu, M., Zhai, Z., Dong, L., Yan, W., Wilkinson, D. P., & Zhang, J. (2020). A review of composite/hybrid electrocatalysts and photocatalysts for nitrogen reduction reactions: Advanced materials, mechanisms, challenges and perspectives. In *Electrochemical Energy Reviews* (Vol. 3, Issue 3, pp. 506–540). Springer Science and Business Media B.V. https://doi.org/10.1007/s41918-020-00069-0

Morlanés, N., Katikaneni, S. P., Paglieri, S. N., Harale, A., Solami, B., Sarathy, S. M., & Gascon, J. (2021). A technological roadmap to the ammonia energy economy: Current state and missing technologies. *Chemical Engineering Journal*, *408*. https://doi.org/10.1016/j.cej.2020.127310

Nazari, S., Sun, J., Baharfar, M., Poulin, P., Kalantar-Zadeh, K., Jalili, A., & Esrafilzadeh, D. (2024). High-throughput ammonia production from nitrate using liquid metal synthesized bismuth nano-catalyst. *Advanced Energy Materials*, *14*(28). https://doi.org/10.1002/aenm.202304287

Peng, W., Luo, M., Xu, X., Jiang, K., Peng, M., Chen, D., Chan, T., & Tan, Y. (2020). Spontaneous atomic ruthenium doping in $Mo_2$ $CT_X$ MXene defects enhances electrocatalytic activity for the nitrogen reduction reaction. *Advanced Energy Materials*, *10*(25). https://doi.org/10.1002/aenm.202001364

Qin, J., Hu, X., Li, X., Yin, Z., Liu, B., & Lam, K. (2019). 0D/2D $AgInS_2$/MXene Z-scheme heterojunction nanosheets for improved ammonia photosynthesis of $N_2$. *Nano Energy*, *61*, 27–35. https://doi.org/10.1016/j.nanoen.2019.04.028

Qin, J., Liu, B., Lam, K.-H., Song, S., Li, X., & Hu, X. (2020). 0D/2D MXene quantum Dot/Ni-MOF ultrathin nanosheets for enhanced $N_2$ photoreduction. *ACS Sustainable Chemistry & Engineering*, *8*(48), 17791–17799. https://doi.org/10.1021/acssuschemeng.0c06388

Sun, B., Qiu, P., Liang, Z., Xue, Y., Zhang, X., Yang, L., Cui, H., & Tian, J. (2021). The fabrication of 1D/2D CdS nanorod@$Ti_3C_2$ MXene composites for good photocatalytic activity of hydrogen generation and ammonia synthesis. *Chemical Engineering Journal*, *406*, 127177. https://doi.org/10.1016/j.cej.2020.127177

Sun, C., Chen, Z., Cui, J., Li, K., Qu, H., Xie, H., & Zhong, Q. (2021). Site-exposed $Ti_3C_2$ MXene anchored in N-defect g-$C_3N_4$ heterostructure nanosheets for efficient photocatalytic $N_2$ fixation. *Catalysis Science & Technology*, *11*(3), 1027–1038. https://doi.org/10.1039/D0CY01955K

Sun, J., Kong, W., Jin, Z., Han, Y., Ma, L., Ding, X., Niu, Y., & Xu, Y. (2020). Recent advances of MXene as promising catalysts for electrochemical nitrogen reduction reaction. *Chinese Chemical Letters*, *31*(4), 953–960. https://doi.org/10.1016/j.cclet.2020.01.035

Tari, V., Gupta, V., Joshi, A., Kurve, P., & Chandra, N. (2024). Biodiversity loss due to forest diversion. In A. Rathore (Ed.), *Impact of Societal Development and Infrastructure on Biodiversity Decline* (pp. 137–165). IGI Global. https://doi.org/10.4018/979-8-3693-6950-0.ch010

Tari, V., & Kannan, K. (2022). Microbial bioremediation of polythenes and plastics – A green sustainable approach. In *Microbial Biotechnology in Green Remediation* (1st ed.). Elsevier. https://doi.org/10.1016/B978-0-323-90452-0.00003-7

Tari, V., Kannan, K., Koliar, J., Gurushankar, K., & Wahyuni, D. K. (2025). Influence of zinc oxide nanomaterials on crop plants: A prospective analysis. In D. Jabbrova, D. Sarkar, Rakesh, S. Singh, & R. Datta (Eds.), *Smart Nanomaterials in Agroforestry Systems*. Springer. https://doi.org/10.1007/978-981-96-1337-3_8

Tari, V. S., & Patil, P. Y. (2017). Challenge of contamination of pesticide for Alphonso in Ratnagiri district, Maharashtra, India. *Research Journal of Chemistry and Environment*, *21*(12), 57–67

Wan, Y., Xu, J., & Lv, R. (2019). Heterogeneous electrocatalysts design for nitrogen reduction reaction under ambient conditions. In *Materials Today* (Vol. 27, pp. 69–90). Elsevier B.V. https://doi.org/10.1016/j.mattod.2019.03.002

Wang, Z., Song, J., Zhang, H., Deng, K., Yu, H., Xu, Y., Wang, H., & Wang, L. (2024). Electrocatalytic valorization of nitrate and polyester plastic for simultaneous production of ammonia and glycolic acid. *Small*. https://doi.org/10.1002/smll.202404124

Wei, H., Jiang, Q., Ampelli, C., Chen, S., Perathoner, S., Liu, Y., & Centi, G. (2020). Enhancing $N_2$ fixation activity by converting $Ti_3C_2$ MXenes nanosheets to nanoribbons. *ChemSusChem*, *13*(21), 5614–5619. https://doi.org/10.1002/cssc.202001719

Wei, X., Vogel, D., Keller, L., Kriescher, S., & Wessling, M. (2020). Microtubular gas diffusion electrode based on ruthenium-carbon nanotubes for ambient electrochemical nitrogen reduction to ammonia. *ChemElectroChem*, *7*(22), 4679–4684. https://doi.org/10.1002/celc.202001370

Wiskich, A., & Rapson, T. (2023). Economics of emerging ammonia fertilizer production methods – A role for on-farm synthesis? *ChemSusChem*, *16*(22). https://doi.org/10.1002/cssc.202300565

Xia, J., Guo, H., Yu, G., Chen, Q., Liu, Y., Liu, Q., Luo, Y., Li, T., & Traversa, E. (2021). 2D Vanadium Carbide (MXene) for electrochemical synthesis of ammonia under ambient conditions. *Catalysis Letters*, *151*(12), 3516–3522. https://doi.org/10.1007/s10562-021-03589-6

Xia, J., Yang, S.-Z., Wang, B., Wu, P., Popovs, I., Li, H., Irle, S., Dai, S., & Zhu, H. (2020). Boosting electrosynthesis of ammonia on surface-engineered MXene $Ti_3C_2$. *Nano Energy*, *72*, 104681. https://doi.org/10.1016/j.nanoen.2020.104681

Xiong, Y., Wang, Y., Zhou, J., Liu, F., Hao, F., & Fan, Z. (2024). Electrochemical nitrate reduction: Ammonia synthesis and the beyond. In *Advanced Materials* (Vol. 36, Issue 17). John Wiley & Sons Inc. https://doi.org/10.1002/adma.202304021

Xu, X., Sun, B., Liang, Z., Cui, H., & Tian, J. (2020). High-performance electrocatalytic conversion of $N_2$ to $NH_3$ Using 1T-$MoS_2$ anchored on $Ti_3C_2$ MXene under ambient conditions. *ACS Applied Materials & Interfaces*, *12*(23), 26060–26067. https://doi.org/10.1021/acsami.0c06744

Yao, Y., Zhu, S., Wang, H., Li, H., & Shao, M. (2018). A spectroscopic study on the nitrogen electrochemical reduction reaction on gold and platinum surfaces. *Journal of the American Chemical Society*, *140*(4), 1496–1501. https://doi.org/10.1021/jacs.7b12101

Yu, G., Guo, H., Liu, S., Chen, L., Alshehri, A. A., Alzahrani, K. A., Hao, F., & Li, T. (2019). $Cr_3$ $C_2$ nanoparticle-embedded carbon nanofiber for artificial synthesis of $NH_3$ through $N_2$ fixation under ambient conditions. *ACS Applied Materials & Interfaces*, *11*(39), 35764–35769. https://doi.org/10.1021/acsami.9b12675

Yu, L., Qin, J., Zhao, W., Zhang, Z., Ke, J., & Liu, B. (2020). Advances in two-dimensional MXenes for nitrogen electrocatalytic reduction to ammonia. In *International Journal of Photoenergy* (Vol. 2020). Hindawi Limited. https://doi.org/10.1155/2020/5251431

Zeng, Y., Du, X., Li, Y., Guo, Y., Xie, Y., Huang, J., Rao, G., Lei, T., Gong, C., Wang, X., & Sun, B. (2021). Synergistic performance of nitrogen and sulfur co-doped $Ti_3C_2TX$ for electrohydrogenation of $N_2$ to $NH_3$. *Journal of Alloys and Compounds*, *869*, 159335. https://doi.org/10.1016/j.jallcom.2021.159335

Zhang, K., Sun, P., Huang, Y., Tang, M., Zou, X., Pan, Z., Huo, X., Wu, J., Lin, C., Sun, Z., Wan, Y., Zhang, X., & An, L. (2024). Electrochemical nitrate reduction to ammonia on CuCo nanowires at practical level. *Advanced Functional Materials*. https://doi.org/10.1002/adfm.202405179

Zhang, Q., Liu, B., Yu, L., Bei, Y., & Tang, B. (2020). Synergistic promotion of the electrochemical reduction of nitrogen to ammonia by phosphorus and potassium. *ChemCatChem*, *12*(1), 334–341. https://doi.org/10.1002/cctc.201901519

Zhao, J., Zhang, L., Xie, X.-Y., Li, X., Ma, Y., Liu, Q., Fang, W.-H., Shi, X., Cui, G., & Sun, X. (2018). $Ti_3$ $C_2$ $T_x$ (T = F, OH) MXene nanosheets: conductive 2D catalysts for ambient electrohydrogenation of $N_2$ to $NH_3$. *Journal of Materials Chemistry A*, *6*(47), 24031–24035. https://doi.org/10.1039/C8TA09840A

Zhao, M., Zeng, Y., Zhang, X., & Liang, H. (2024). Recent advances in electrochemical cell design for concurrent chemical production and electricity generation. In *ChemElectroChem* (Vol. 11, Issue 11). John Wiley & Sons Inc. https://doi.org/10.1002/celc.202400192

Chapter 10

# Electrocatalytic conversion of carbon dioxide into fuels and value-added products

*Swati Rani and Anil K. Yadav*

## 10.1 INTRODUCTION

Carbon dioxide ($CO_2$) emissions in the atmosphere have emerged as a major global challenge due to their considerable influence on climate change and global warming (Tari, 2015a). $CO_2$ concentrations in the atmosphere have skyrocketed to an alarming stage caused by the rapid increase in human activities such as fossil fuel combustion, deforestation, industrial activities, and other activities (Tari et al., 2025). Total $CO_2$ emissions are recorded to be 41.6 billion tons in 2024, making it the hottest year according to the World Meteorological Organization's (WMO's) Global Greenhouse Gas Bulletin. Considering its harmful effects on the environment and public health globally, addressing this issue is essential to minimizing $CO_2$ emissions (Tari, 2015b). The carbon capture, utilization, and storage (CCUS) strategy is one of the promising approaches that uses waste and harmful $CO_2$ and utilizes it by converting it into fuels. This approach for $CO_2$ mitigation, particularly the electrocatalytic conversion of $CO_2$ into fuels, is explored by the scientific community around the world. This technique harnesses renewable energy sources, such as solar and wind power, to drive the electrochemical reduction of $CO_2$ utilizing a catalyst, also providing a sustainable pathway for carbon recycling (Tari et al., 2024). By integrating renewable energy, electrochemical $CO_2$ conversion not only reduces greenhouse gas emissions but also creates a circular carbon economy, i.e., $CO_2$ from the atmosphere is reduced to valuable products or fuels, and produced fuels can be used as energy sources. A variety of value-added chemicals are obtained through electrochemical process such as carbon monoxide (CO), methane ($CH_4$), ethylene ($C_2H_4$), methanol ($CH_3OH$), ethanol ($CH_3CH_2OH$), formic acid (HCOOH), acetic acid ($CH_3COOH$), formaldehyde ($CH_2O$), acetaldehyde ($CH_3CHO$), and other hydrocarbons (Hori, 2008; Kortlever, Shen, et al., 2015; Kannan and Tari, 2025). In this chapter, the process of direct $CO_2$ reduction reaction ($CO_2RR$) into valuable compounds is discussed, followed by key components of an electrochemical conversion system. Thereafter, a comprehensive overview is provided of factors affecting the $CO_2RR$ such as efficiency, selectivity, stability, etc., and the various types of electrocatalysts.

 DOI: 10.1201/9781003584742-10

Furthermore, the chapter covers recent advancements in electrocatalyst design, reduction reaction, and electrolyte optimization, and insights into the future research of electrocatalytic $CO_2$ reduction and its potential to contribute to sustainable development.

## 10.2 CONVERSION OF $CO_2$ INTO FUELS AND TYPES OF CONVERSIONS

The conversion of $CO_2$ takes place through chemical reactions in various types of conversion. Depending on the energy source, there are four main conversion processes of $CO_2$, which are electrochemical conversion, photochemical conversion, thermochemical conversion, and biochemical conversion as depicted in Figure 10.1. In electrochemical conversion, electric energy is used in a catalytic reaction to reduce $CO_2$; therefore, it is also known as the electrocatalytic conversion process, and the catalyst utilized in this process is commonly called electrocatalyst. Here, $CO_2$ is reduced at the surface of an electrode coated with an electrocatalyst, which is a part of the electrochemical device used to perform the electrocatalysis (Okoye-Chine et al., 2022). Whereas in photochemical conversion, light energy or photons participate in the catalytic reaction to drive $CO_2$ reduction; the catalyst used in the reaction is known as a photocatalyst and the reaction is a photocatalytic conversion process. The photocatalyst is deposited on the electrode materials, which are exposed to the light initiated by the photocatalytic conversion process. Similarly, the thermochemical conversion converts $CO_2$ into various fuels such as methane, formic acid, methanol, or synthetic gasoline utilizing heat and chemical reactions through catalysis (Okoye-Chine et al., 2022). To compare different biomass conversion methods, this process can be energy efficient and offers advantages in certain industrial applications. However, electrochemical conversion is gaining merits in the broader context of sustainability and adaptability to renewable energy integration due to its high efficiency, selectivity, scalability, and compatibility with renewable electricity sources (Saravanan et al., 2021). In biochemical conversion, microorganisms and enzymes are used to convert

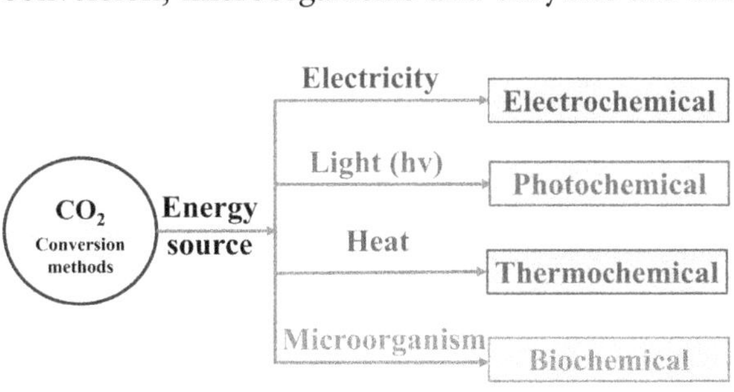


*Figure 10.1* Types of $CO_2$ conversion methods.

$CO_2$ into valuable products like ethanol, in a biochemical system (Nisar et al., 2021; Okoye-Chine et al., 2022). In the extension of this conversion, $CO_2$ is also converted into fuels by two approaches, light-dependent and light-independent (Jha et al., 2024). Light-independent approach functions without direct light, typically using microbiological systems, and can yield larger biomass than the light-dependent approach, which operates on light energy in a bio-electrochemical system. Despite this, light-independent approach suffers from poor life cycle analysis (Vieira et al., 2013). In the end, each conversion method has its advantages and challenges, and among these, electrochemical conversion is becoming prominent due to its potential for high efficiency, scalability, high selectivity, and direct integration with renewable electricity sources (Saravanan et al., 2021).

### 10.2.1 Electrochemical conversion via $CO_2$ reduction reaction ($CO_2$RR)

Electrochemical conversion of $CO_2$ occurs via a chemical reaction driven by electricity commonly called $CO_2$RR. Along with electricity, the $CO_2$RR involves a catalyst, called an electrocatalyst in this method, for the reduction of $CO_2$ at the cathode of an electrochemical cell. This chemical reaction is known as a redox reaction, which is a combination of oxidation and reduction reactions occurring simultaneously. The process begins with water oxidation and then $CO_2$ reduction onto the electrocatalyst surface, which is deposited on the surface of the cathode, followed by multiple electron and proton transfer steps. The overall reaction can yield a variety of products such as CO, $CH_4$, $C_2H_4$, $CH_3OH$, HCOOH, and other hydrocarbons (Hori, 2008; Kortlever, Shen, et al., 2015), depending on the number of electrons involved in the reaction, the electrocatalyst material, and reaction conditions (e.g., applied potential, electrolyte pH, and temperature). The $CO_2$RR includes equations (10.1) and (10.2), which are combinedly written as equation (10.3), given below:

$$\textbf{Water oxidation}: 2H_2O \rightarrow O_2 + 4H^+ + 4e^- \qquad (10.1)$$

This oxidation reaction occurs at the anode, where water is oxidized by losing the electrons ($e^-$) and releasing oxygen gas ($O_2$) along with protons ($H^+$). These $H^+$ and $e^-$ then participate in the $CO_2$ reduction.

$$\mathbf{CO_2}\,\textbf{reduction}: CO_2 + m\left(H^+ + e^-\right) \rightarrow C_xH_yO_z + nH_2O \qquad (10.2)$$

This reduction reaction occurs at the cathode, where $CO_2$ molecules are converted into carbon-rich compounds ($C_xH_yO_z$) by gaining the $e^-$ from the oxidation reaction and releasing water. Here, m denotes the number of $H^+$ and $e^-$ transferred.

$$\textbf{Redox reaction}: CO_2 + H_2O \rightarrow C_xH_yO_z + O_2 \qquad (10.3)$$

Overall, these reactions form the basis of electrochemical $CO_2$ conversion systems by producing various value-added products. Electrocatalysts required in the reaction must have high activity; otherwise, hydrogen evolution reaction (HER), as depicted in equation (10.4), can take place hindering the formation of hydrocarbons.

$$\textbf{Hydrogen evolution reaction}: 2\left(H^+ + e^-\right) \rightarrow H_2 \tag{10.4}$$

$CO_2$RR is a complicated reaction due to overpotential caused by differences in anodic and cathodic potentials. Also, $CO_2$ is a highly stable molecule which makes its reduction difficult. Therefore, for $CO_2$ to be converted into useful products and to avoid HER, highly active catalysts, and specific reaction conditions are required. $CO_2$ molecules are converted into carbon-rich compounds ($C_xH_yO_z$), and the final product depends on the number of $e^-$ transferred. The schematic reaction pathways, including intermediate products, especially for multi-electron products, are shown in Table 10.1, and Figure 10.2 depicts the types of final product formation in redox reactions with the number of $e^-$ involved.

## 10.3 ESSENTIAL COMPONENTS OF $CO_2$ ELECTROCHEMICAL CONVERSION SYSTEM AND OPERATION CONDITIONS

### 10.3.1 Electrochemical cell

For electrocatalytic conversion of $CO_2$ to value-added fuels, the electrochemical cell or fuel cell is a fundamental component. Its design and materials used in the process significantly impact the efficiency, current density, and product distribution of the conversion process. Y. Hori developed the first electrochemical cell in the 1980s. The reduction of $CO_2$ occurs through electrochemical reactions driven by electricity (Hori, 2008). Modern electrochemical cells are generally categorized into four types: H-type cells, polymer electrolyte membrane (PEM) flow cells, solid oxide electrolysis cells (SOECs), and differential electrochemical mass spectrometry (DEMS) cells. In addition, more advanced configurations such as membrane electrode assemblies (MEAs) and gas diffusion electrode (GDE) setups have also been introduced to improve scalability and performance. A performance comparison between different types of electrochemical cells, highlighting their respective advantages and limitations is given in Table 10.2.

Among these, the H-type cell is the most widely used configuration, particularly for laboratory-scale research because of its simplicity, low cost, and convenience of operation. It allows researchers to precisely control reaction conditions and systematically study the effect of catalysts, electrolyte composition, and applied potential on $CO_2$ reduction pathways. Although it is

*Table 10.1* Reaction pathways of value-added products including intermediate products and transferred $e^-$ involved in the reaction

| *Final product* | *Overall reaction* | *Key intermediates* | e− |
|---|---|---|---|
| $(COOH)_2$ | $2CO_2 + 2H^+ + 2e^- \rightarrow C_2H_2O_4$ | $CO_2 \rightarrow^* CO_2 \rightarrow^* C_2O_4^- \rightarrow C_2H_2O_4$ | 2 |
| HCOOH | $CO_2 + 2H^+ + 2e^- \rightarrow HCOOH$ | $CO_2 \rightarrow^* CO_2 \rightarrow^* OCHO \rightarrow HCOOH$ | 2 |
| CO | $CO_2 + 2H^+ + 2e^- \rightarrow CO + H_2O$ | $CO_2 \rightarrow^* CO_2 \rightarrow^* COOH \rightarrow^* CO \rightarrow CO$ | 2 |
| HCHO | $CO_2 + 4H^+ + 4e^- \rightarrow HCHO + H_2O$ | $CO_2 \rightarrow^* CO_2 \rightarrow^* COOH \rightarrow^* CO \rightarrow^* CHO \rightarrow HCHO$ | 4 |
| $CH_3OH$ | $CO_2 + 6H^+ + 6e^- \rightarrow CH_3OH + H_2O$ | $CO_2 \rightarrow^* CO_2 \rightarrow^* COOH \rightarrow^* CO \rightarrow^* CHO \rightarrow^* CH_2O \rightarrow^* CH_3O \rightarrow CH_3OH$ | 6 |
| $CH_4$ | $CO_2 + 8H^+ + 8e^- \rightarrow CH_4 + 2H_2O$ | $CO_2 \rightarrow^* CO_2 \rightarrow^* COOH \rightarrow^* CO \rightarrow^* CHO \rightarrow^* CH_2O \rightarrow^* CH_3 \rightarrow CH_4$ | 8 |
| $C_2H_4$ | $2CO_2 + 12H^+ + 12e^- \rightarrow C_2H_4 + 4H_2O$ | $CO_2 \rightarrow^* CO \rightarrow C-C$ coupling $\rightarrow^* OCCO \rightarrow^* OCCH_3 \rightarrow C_2H_4$ | 12 |
| $C_2H_5OH$ | $2CO_2 + 12H^+ + 12e^- \rightarrow C_2H_5OH + 3H_2O$ | $CO_2 \rightarrow^* CO \rightarrow^* CHO \rightarrow C-C$ coupling $\rightarrow^* CH_3CH_2O$ | 12 |

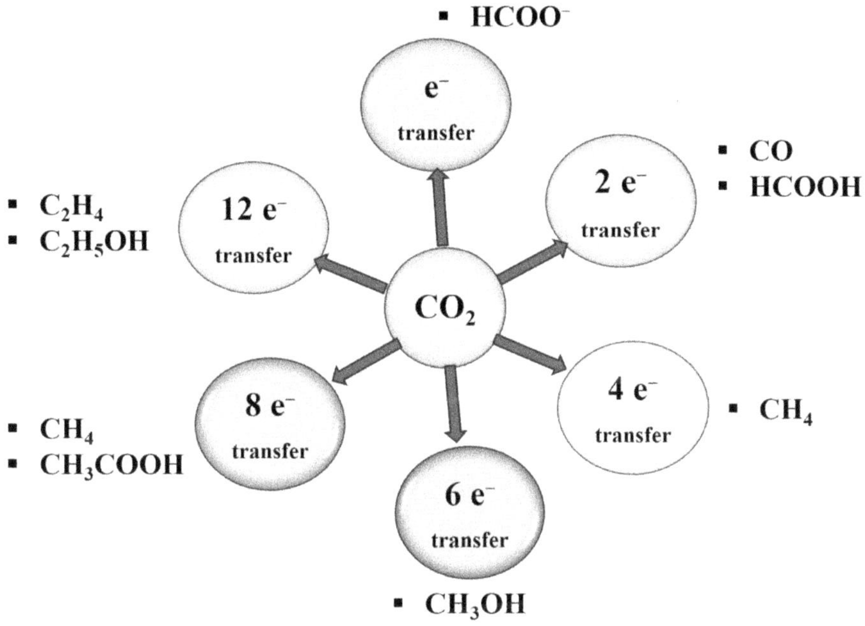


*Figure 10.2* Types of various products in $CO_2$ RR.

*Table 10.2* Types of electrochemical cells, their respective advantages and limitations

| *Sr. No.* | *Electrochemical cells* | *Advantages* | *Limitations* |
|---|---|---|---|
| 1 | **H-type** | Ideal for fundamental research and catalyst screening, Offers easy separation of anodic and cathodic products | Suffers from low current density due to limited $CO_2$ solubility and diffusion in aqueous electrolytes, poor mass transport |
| 2 | **PEM** | Enables continuous flow of $CO_2$ and electrolyte, achieving higher current densities and better scalability | The design is more complex and expensive; membrane durability can also be a challenge |
| 3 | **SOECs** | Operates at high temperatures (600–1000 °C), allowing direct conversion of $CO_2$ to CO or syngas with high efficiency | The high operational temperature leads to material degradation and limits long-term stability |
| 4 | **DEMS** | Designed for real-time product detection, especially gaseous products, using mass spectrometry | Suitable for mechanistic studies but not practical for bulk product generation due to expensive, and complex setup |
| 5 | MEA/GDE Setups | High surface area, industrial relevance, better gas-liquid interface | Complex assembly, flooding or drying issues, difficult product separation |

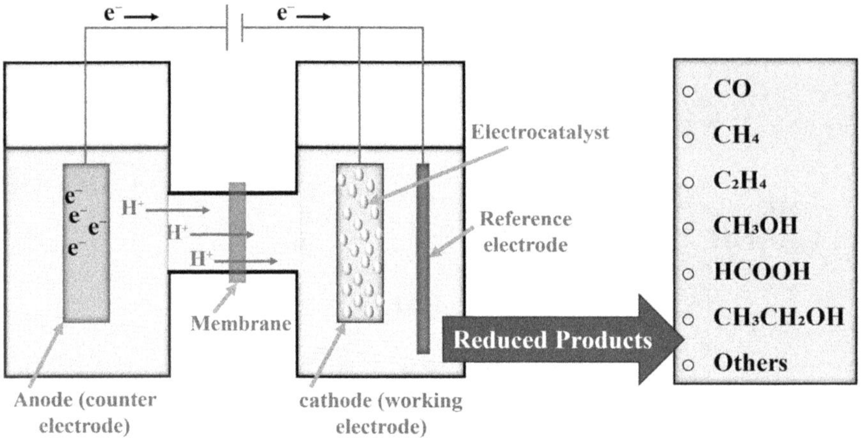


*Figure 10.3* Schematic representation of the H-type electrochemical cell.

not suitable for high-performance or industrial-scale applications, it provides a well-controlled and reliable platform for understanding fundamental electrochemical behavior before scaling up to more advanced systems like PEM or SOEC.

The H-cell is a simple two-compartment cell that consists of three electrodes, the cathode, the anode, and the reference electrode, within these separate compartments divided by an ion-exchange membrane and filled with aqueous electrolyte solution with $CO_2$ dissolved in it. The schematic representation of the H-type electrochemical cell is shown in Figure 10.3. In the electrocatalytic process, a water molecule gets separated into $H^+$, $e^-$, and $O_2$ at the anode. $H^+$ gets transferred to the cathode through an ion-exchange membrane and $e^-$ through an external circuit. At the cathode, they combine with $CO_2$ to give carbon products along with water molecules.

A more comprehensive explanation of the H-cell setup including electrodes, ion-exchange membrane/separator, electrocatalysts, electrolytes, and operation conditions such as temperature and pressure is given in the following sections.

### 10.3.2 Electrodes and ion-exchange membrane/separator

The electrochemical cell consists of two electrodes, the cathode as a working electrode and the anode as a counter electrode. For precise potential measurements, a reference electrode can be added along with the cathode. Usually, silver/silver chloride (Ag/AgCl) or saturated calomel electrode (SCE) are used as reference electrodes. The cathode takes part in the reduction of $CO_2$, whereas the anode takes part in water splitting. The cathodes are often made of conductive materials like copper (Cu), silver (Ag), gold (Au), or carbon-based materials (e.g., graphite, carbon), and the anodes are typically made from inert and stable materials resistant to oxidation, such as platinum (Pt), ruthenium dioxide ($RuO_2$), or graphite. Electrodes play a crucial role in $CO_2$RR; they influence the performance of the reaction depending on their characteristics such as size and morphology. For example, bulky metal electrodes, such as metal foils and meshes, used in early studies have low surface areas and poor catalytic activity (Hori, 2008; Jitaru et al., 1997). Catalyst-deposited electrodes, show improved activity but are negatively influenced by binders in the system. On the other hand, self-supported electrodes avoid these issues but are still problematic since they have low $CO_2$ solubility and mass transport. $CO_2$ solubility is addressed in the GDEs by enabling efficient $CO_2$ transfer through a porous structure that increases the reaction rates and current densities (Martín et al., 2015). Moreover, the ion exchange membrane, which is a crucial part of the cell, is placed between the two compartments of the H-cell. It allows selective ion transport while preventing the transfer of other type of charge and mixing of reaction products. In an electrochemical cell, usually Nafion (proton exchange membrane) is utilized.

### 10.3.3 Electrocatalyst

Electrocatalysts, coated on the cathode surface, are the most vital materials in the electrochemical conversion of $CO_2$. They are responsible for enhancing the reaction rate and efficiency, subsequently producing valuable chemicals (Dong Zhu et al., 2016). Various types of electrocatalysts have been studied for $CO_2$ reduction, including metal-based, nanostructured, metal complexes, transition metals, and carbon-based electrocatalysts (Dong Zhu et al., 2016). Among these, metal-based catalysts, especially Cu and its complexes, have attracted large attention due to their ability to produce a variety of value-added fuels such as CO, HCOOH, $CH_4$, and $C_2H_4$. Cu has demonstrated effective conversion of $CO_2$ because of its higher capacity to generate hydrocarbons (Zhang et al., 2018). However, Cu faces challenges such as high overpotentials, low reaction rates, poor stability, complex preparation methods, material costs, optimization of Faradic efficiency, low product selectivity, optimization of energy efficiency and current density, and catalyst poisoning issues (Vasileff et al., 2017). To advance electrochemical $CO_2$ reduction, future research should focus on the development of cost-effective catalysts, the search for novel materials, and the improvement in Faradaic efficiency for industrial applications (Zhang et al., 2018).

### 10.3.4 Electrolytes

Electrolyte serves a crucial role in the electrochemical cell. It is an electrically conductive medium that enables the flow of ions between the electrodes during the electrocatalytic process. The types of cations and anions present in an electrolyte, along with its pH and the choice of electrolyte, can greatly influence the overall efficiency, reaction selectivity, and energy efficiency of $CO_2$RR. The pH of the electrolyte directly affects the availability of protons ($H^+$) and electrons ($e^-$). For instance, acidic electrolytes favor HER compared to $CO_2$RR, which significantly reduces the efficiency of $CO_2$RR (Ooka et al., 2017); while neutral and alkaline electrolytes often enhance $CO_2$ conversion efficiency by suppressing HER. Moreover, the type of electrolyte also greatly influences $CO_2$RR. Electrolytes such as potassium bicarbonate ($KHCO_3$) and sodium hydroxide (NaOH) are commonly used due to their good conductivity and buffering capacity, whereas electrolytes such as ionic liquids and organic solvents, particularly sulfuric acid ($H_2SO_4$), have great potential in improving product selectivity by stabilizing $CO_2$ intermediates (Sun et al., 2017). In conclusion, both the pH and the choice of electrolyte are pivotal in determining the efficiency and selectivity of $CO_2$RR. Optimizing these factors is essential for maximizing the yield of desired $CO_2$ reduction products, such as CO, HCOOH, $CH_4$, and $C_2H_4$.

### 10.3.5 Temperature

Temperature is one of the pivotal operation conditions in the $CO_2$RR. It affects the solubility of $CO_2$ in the electrolyte, catalyst performance, selectivity, and reaction kinetics (Hori et al., 1985). The solubility of $CO_2$ in the electrolyte increases as the temperature decreases, which can enhance $CO_2$ availability for reduction reactions. However, temperature variations also influence product distribution and the competition between $CO_2$RR and HER. Kaneco et al. studied the impact of temperature on $CO_2$RR using Au and Ag electrodes in a 0.1 M KOH-methanol electrolyte (Kaneco et al., 2002). Their findings revealed that lowering the electrolyte temperature favored $CO_2$RR while suppressing HER, promoting CO formation as the primary product. Similarly, research on $MoO_2$/Pb electrodes demonstrated enhanced catalytic activity at lower temperatures (around 20°C) compared to room temperature (Oh et al., 2014). At room temperature, the product formed was formate and oxalate, while at lower temperatures, oxalate and CO were predominant. Hence, temperature optimization is essential for achieving higher efficiency and selectivity in $CO_2$RR, as it directly affects catalyst behavior, product formation, and reaction pathways.

### 10.3.6 Pressure

Another important operation condition is the pressure of $CO_2$, as it significantly influences its solubility in electrolyte solutions and directly impacts the efficiency and selectivity of $CO_2$RR (Li et al., 2020; Ramdin et al., 2019). Since $CO_2$ has low solubility in water (0.033 mol/liter at 25°C under 1 atm), delivering high current densities for electrolysis can be challenging. By increasing $CO_2$ pressure, the solubility of $CO_2$ in electrolytes can be enhanced. Previous research has shown that $CO_2$ pressure affects product distribution on Group VIII metal electrodes (Nakagawa et al., 1991). For example, a platinum gas-diffused electrode exhibited enhanced methane production with increased $CO_2$ pressure by suppressing HER activity (Hara & Sakata, 1997). Similarly, a silver electrode achieved industrial-grade CO production in a pressurized alkaline system (Gabardo et al., 2018). This emphasizes that even with a specific electrocatalyst, $CO_2$RR performance, including selectivity, efficiency, and product yield, can still be affected highly by pressure conditions. Therefore, optimization of $CO_2$ pressure is necessary for effective $CO_2$ conversion.

## 10.4 KEY PARAMETERS AFFECTING THE ELECTROCATALYST PERFORMANCE IN $CO_2$RR

Several fundamental parameters must be introduced to evaluate and compare the performance of electrocatalysts for $CO_2$ reduction. The overall efficiency is measured by Faradaic efficiency, overpotential, current density, and other factors, which determine how effectively the input electrical energy is converted into chemical products. In this section, key parameters used in the assessment of electrocatalyst performance in $CO_2$RR are discussed.

### 10.4.1 Thermodynamics of $CO_2$ conversion process

$CO_2$, being the most oxidized form of carbon, is highly stable, which makes $CO_2$ conversion challenging. This unfavorable condition requires efficient energy and optimized conditions for any conversion process (Zangeneh et al., 2011). In this regard, Gibbs free energy $(\Delta G)$ and redox potential $(E^0)$ are two important thermodynamic parameters that dictate the feasibility of the reactions involved in $CO_2$ conversion. Understanding these two thermodynamic parameters will assist in developing more effective catalysts. $\Delta G$ is given by the relation (10.5) as,

$$\Delta G = -nFE^0 \tag{10.5}$$

Where n is the number of electrons transferred, F is Faraday's constant, and $E^0$ is the cell potential. A redox reaction is therefore defined by a negative value of $\Delta G$ and a positive value of $E^0$. A negative Gibbs free energy indicates that a reaction can occur spontaneously, whereas a positive value suggests that external energy is required to drive the reaction. Hence, for $CO_2$RR to occur against any energy barriers, the Gibbs free energy must be sufficiently negative. Furthermore, the redox potential is also an important parameter in the redox reaction, which determines the applied voltage required to achieve electron transfer and initiate the reaction. In other words, redox potential is the measure of the ability to gain or lose electrons

*Table 10.3* Equilibrium potential and Gibbs free energy to convert $CO_2$ into different chemicals (Ramírez-valencia et al., 2021).

| *Reduction reactions* | *Gibbs free energy, ΔG (K J/mol)* | *Standard potential, $E^0$ (V vs. RHE)* |
|---|---|---|
| $2CO_2 + 2H^+ + 2e^- \rightarrow C_2H_2O_4$ | +91.8 | −0.475 |
| $CO_2 + 2H^+ + 2e^- \rightarrow HCOOH$ | +38.4 | −0.199 |
| $CO_2 + 2H^+ + 2e^- \rightarrow CO + H_2O$ | +19.9 | −0.103 |
| $CO_2 + 4H^+ + 4e^- \rightarrow HCHO + H_2O$ | +27.5 | −0.071 |
| $CO_2 + 6H^+ + 6e^- \rightarrow CH_3OH + H_2O$ | −17.3 | +0.030 |
| $CO_2 + 8H^+ + 8e^- \rightarrow CH_4 + 2H_2O$ | −130.8 | +0.169 |

in an aqueous solution. A positive redox potential indicates a tendency to gain electrons i.e., reduction, while a negative potential suggests a tendency to lose electrons i.e., oxidation. Redox potentials are defined relative to a reference electrode since absolute values are hard to measure. It is typically measured in volts (V) referenced to the reversible hydrogen electrode (RHE) (i.e., $E^0$ V vs RHE). Despite favorable thermodynamic conditions at high temperatures, $CO_2$ conversions often face slow kinetics, which hinders practical implementation. Various electrochemical reactions occur at equilibrium potential and Gibbs free energy to convert $CO_2$ into different chemicals. Some of the reactions are given in Table 10.3.

### 10.4.2 Faradaic efficiency (FE)

FE is a crucial parameter in the electrocatalytic conversion of $CO_2$, as it determines the quantity of electrons needed to produce the desired product. FE can be expressed mathematically as:

$$\varepsilon_{\text{faradaic efficiency}}(\%) = \frac{\alpha n F}{Q} \tag{10.6}$$

Where $\alpha$ is the number of electrons transferred in the process, n is the number of moles for the desired product, F denotes Faraday's constant (96,485 C $mol^{-1}$), and Q is the total amount of charge passed through the electrocatalytic process. A high FE indicates that most of the applied charge is mainly used in product formation rather than in any side reactions or energy losses. Therefore, achieving high FE is essential for enhancing the efficiency of $CO_2RR$. Researchers are continuously exploring novel electrocatalysts to maximize FE by enhancing their ability to reduce $CO_2$ selectively.

### 10.4.3 Selectivity

The ability of a catalyst to specifically produce certain products over others is known as selectivity in the electrocatalytic reduction of $CO_2$, and it is critical for the economic viability of $CO_2$ utilization technologies. For example, catalysts that promote the generation of methane are more significant for applications involving synthetic natural gas, while those that favor the production of CO may be needed for the generation of compounds such as acetic acid. Improving the overall performance of electrochemical $CO_2$ reduction requires catalysts that can efficiently and selectively convert $CO_2$ to a single product or a mixture. The shape, kind of active sites, and applied potential are some of the variables that affect the reaction pathways.

### 10.4.4 Current density (j)

Current density is a critical measure of the rate of an electrochemical reaction. The total electrochemical $CO_2$ reduction current density (j) is defined as the ratio of applied current (i) to the geometric area (A) of the working electrode:

$$j = \frac{i}{A} \tag{10.7}$$

The partial current density is obtained by multiplying the overall current density with FE. Higher current densities lead to faster $CO_2$RR rates, but the formation of undesirable byproducts is unavoidable, compromising the selectivity of the reaction. Hence, for efficient $CO_2$ reduction, the current density needs to be balanced with overpotential and selectivity. In large-scale industrial applications, high current densities are desired to enhance the overall performance of conversion systems, but this requires stable and well-designed electrocatalysts to maintain their selectivity over long periods of operation without degradation. Therefore, research needs to be focused on the development of electrocatalysts that can sustain high current densities without sacrificing performance or degradation.

### 10.4.5 Overpotential and onset potential

Overpotential and onset potential are important metrics for evaluating the energy efficiency of an electrocatalyst. The onset potential refers to the minimum voltage at which the catalytic reaction begins, while the overpotential is the additional voltage or energy required beyond the theoretical potential to drive the $CO_2$ reduction reaction at a significant rate. Mathematically, the overpotential ($\eta$) is the difference between the onset and equilibrium standard reduction potentials, as shown in equation (10.8).

$$\eta = E_{applied} - E^0 \tag{10.8}$$

where $E^0$ is the standard reduction potential for $CO_2$. Overpotential is a key factor in determining the efficiency of $CO_2$ electrocatalysis, as it directly affects the energy consumption of the process. Lower overpotentials indicate a more energy-efficient electrocatalyst. Conversely, a high-performing electrocatalyst should minimize the overpotential, reducing energy losses while maintaining effective reaction rates. Electrocatalysts with low overpotentials, such as those based on Cu, have demonstrated the ability to reduce $CO_2$ efficiently to hydrocarbons and alcohols. However, minimizing overpotential without compromising selectivity remains a major challenge in the field of $CO_2$ electrocatalysis (Nitopi, 2019).

### 10.4.6 Energy efficiency

Energy efficiency combines the Faradaic efficiency and the overpotential, giving a holistic measure of energy use. It is defined as:

$$\varepsilon_{\text{energy efficiency}} = \frac{E^0}{E^0 + \eta} \times FE \tag{10.9}$$

This parameter highlights the overall energy efficiency of the catalytic process by considering both the thermodynamic efficiency and the effectiveness of electron utilization. Lower overpotentials combined with high Faradaic efficiency lead to improved energetic efficiency.

### 10.4.7 Stability

Stability is a crucial factor when evaluating electrocatalysts for $CO_2$ reduction, particularly for their commercial viability. The stability of an electrocatalyst refers to its ability to maintain performance over extended periods without significant degradation in activity, selectivity, or efficiency.

## 10.5 TYPES OF ELECTROCATALYSTS FOR $CO_2$ REDUCTION

Following the materials, electrocatalysts can be categorized into five types: Nobel metals, non-noble metals, metal oxides, hybrid or composite materials, and carbon-based materials. This section provides a study on some selected, and most promising catalysts for $CO_2RR$ listed in Table 10.4. A highly effective catalyst is needed in $CO_2RR$ for maximum yield while ensuring excellent selectivity. In this context, noble metal-based electrocatalysts, mainly Gold (Au), Silver (Ag), and Palladium (Pd) have exhibited remarkable FE, which is above 90% and great selectivity with CO and HCOOH as the final products (Klinkova et al., 2016; Lu et al., 2014). These metals, due to their optimal binding energies, are particularly effective in producing intermediates like $HCOO^-$ and CO, as suggested by density functional theory (DFT) calculations. For instance, Au surfaces bind CO weakly, promoting CO desorption, whereas Pd stabilizes $HCOO^-$ intermediates favoring

*Table 10.4* List of some most studied electrocatalysts in each category

| *Sr. No.* | *Types of electrocatalysts* | *Electrocatalysts* | *Final product* | *FE (%)* | *Potential (V vs RHE)* | *References* |
|---|---|---|---|---|---|---|
| 1. | Nobel metals | Au Nanoparticles | CO | 90 | −0.67 | Zhu et al. (2013) |
| | | Nanoporous Ag | CO | 92 | −0.60 | Lu et al. (2014) |
| | | Pd nanoparticles ~100 nm) | HCOOH | 97 | −0.20 | Klinkova et al. (2016) |
| | | Pd nanoparticles (3.7 nm) | CO | 91.2 | −0.89 | Martin et al. (2017) |
| | | Au nanowires | CO | 94 | −0.35 | Zhu et al. (2014) |
| | | Au needles | CO | 95 | −0.35 | Liu et al. (2016a) |
| | | Pd nanoparticles | CO | 91.2 | −0.89 | Gao et al. (2015) |
| | | Pd needles | HCOOH | 91 | −0.2 | Liu et al. (2016b) |
| 2. | Non-noble metals | Cu (100) | $C_2H_4$ | 38.6 | −0.97 | Hahn et al. (2017) |
| | | Co nanoparticles (5nm) | $CH_3OH$ | 71.4 | 0.90 | Huang et al. (2018) |
| | | Co nanosheets | Ethanol | 60 | 0.4 | Yin et al. (2021) |
| | | B-doped Cu | $C_2H_4$ | 80 | −1.1 | Zhou et al. (2018) |
| | | Ni/N-C | CO | 97.5 | −1.0 | Tan et al. (2019) |
| | | Fe nanoparticles /OCNT | Hydrocarbons + alcohols | 35.2 | - | Chew et al. (2014) |
| 3. | Metal oxides | ZnO nanowires | CO | 91.6 | 0.62 | Luo et al. (2020) |
| | | Fe-doped $SnO_2$ electrode | HCOOH | 41 | 0.89 | Savino et al. (2022) |
| | | $SnO_2$ NPs | HCOOH | 64 | 1.21 | Liang et al. (2018) |
| | | $Cu_2O$ film on polished Cu (3.6 μm) | $C_2H_4$ | 34.26 | −0.99 V | Ren et al. (2015) |
| | | Ni-MOF | $CH_3OH$ | 100 | −1.345 | Cao et al. (2021) |
| | | $SnO_2$ | Formate | 93.6 | −1.8 | Zhang et al. (2014) |
| | | Nanoporous ZnO | CO | 92 | −1.66 | Jiang et al. (2016) |

*(Continued)*

*Table 10.4* (Continued)

| *Sr. No.* | *Types of electrocatalysts* | *Electrocatalysts* | *Final product* | *FE (%)* | *Potential (V vs RHE)* | *References* |
|---|---|---|---|---|---|---|
| 4. | Hybrid/ Composites | Co-Polypyrrole-C | CO, HCOOH | 39-35 | −0.6 | Mun et al. (2018) |
| | | $Cu_{1.0}/ZnO_{0.20}$ | $C_2H_4$ | 49 | −0.72 | Zhang et al. (2020) |
| | | NiFe | CO | 98.2 | −0.5 | Iqbal et al. (2022) |
| | | CuAg | $C_2H_4$ | 52 | −1.05 | Gao et al. (2019) |
| | | CuAg alloy | $C_2H_4$ | 60 | −0.7 | Andersen et al. (2017) |
| | | CuAg alloy | $C_2H_5OH$ | 25 | −0.7V | Hoang et al. (2018) |
| | | Bi-Sn | HCOOH | 96 | −1.1 | Wen et al. (2018) |
| | | $Pd_{83}Cu_{17}$ | $CH_3OH$ | 80 | −2.1 | L. Lu et al. (2018) |
| | | PdSn/C | $CH_3OH$ | 99 | −0.43 | Bai et al. (2017) |
| | | $Pd_{70}Pt_{30}$ | HCOOH | 91.8 | −0.4 | Kortlever, Peters, et al. (2015) |
| | | $Cu_2Pd$ | $CH_4$ | 50 | −1.8 | Zhang et al. (2015) |
| | | Cu–In alloy | CO | 90 | −0.5 | Rasul et al. (2015) |
| | | Au–Pd core-shell aerogel | CO | 99.96 | −0.5 | Du et al. (2021) |
| | | NiCo-MOF | $CH_3OH$ | 99 | 1.291 | Tian et al. (2022) |
| 5. | Carbon-based | Nafion-bonded carbon black | Formic acid | 73.01 ± 3.42 | - | Wang et al. (2017) |
| | | Carbon coated Ni NPs | CO | 94 | 0.7 | Jia et al. (2018) |
| | | Ni NPs encapsulated in N-doped carbon nanohybrid substrates | CO | 93.1 | 0.9 | Yuan et al. (2019) |
| | | N-doped mesoporous carbon-CNT/Co | $C_2H_5OH$ | 60 | −0.32 | Du et al. (2020) |
| | | Cu/N-doped carbon nanofibers | $C_2H_4$ | 62 | −0.57 | Lee et al. (2020) |
| | | Ni-$N_4$-C | CO | 99 | −0.81 | Li et al. (2017) |
| | | Ni@NCNT/CFM | CO | 100 | −0.67 | Wu et al. (2022) |

HCOOH formation. However, these catalysts face primary challenges due to scarcity and high cost, which limit their suitability for large-scale commercial use in $CO_2$ conversion. Additionally, noble metal catalysts have a high possibility of deactivation over time in an open atmosphere as they are highly reactive.

To counter these limitations, non-noble metals such as cobalt (Co), copper (Cu), nickel (Ni), and iron (Fe) have drawn much attention as cost-effective alternatives. Several studies on non-noble metal-based catalysts including Cu (Zhou et al., 2018), Co (Huang et al., 2018), Ni (Tan et al., 2019), and Fe (Chew et al., 2014) have demonstrated that these materials show good FE and selectivity toward CO and $HCOO^-$ formation. However, their performances were restricted due to low FE and poor selectivity toward hydrocarbons and alcohol formation ($C_2$, $C_3$, and $C_4$ products). Among them, Cu is particularly notable as it facilitates the formation of $C_2$ products, such as ethylene ($C_2H_4$) and ethanol, due to its moderate CO binding energy and ability to stabilize C–C coupling intermediates. DFT studies reveal that engineering surface plans (e.g., Cu(100)) and introducing grain boundaries can significantly alter the electronic structure and promote C–C coupling mechanisms. However, it is still difficult to identify the active sites precisely, requiring sophisticated working conditions in spectroscopy and theoretical modeling. To improve the FE and selectivity of non-noble metals, oxides of some non-noble metals were examined. Reported research including Ni-MOF (Cao et al., 2021), $SnO_2$ (Zhang et al., 2014), and ZnO (Jiang et al., 2016; Luo et al., 2020) exhibited excellent improvement in FE (>90%) and higher selectivity in CO, HCOOH, and $CH_3OH$ production. DFT studies suggest that the presence of oxygen vacancies and surface hydroxyl groups modulate the local electronic structure, facilitating the adsorption and stabilization of key intermediates. In situ spectroscopic analysis and DFT simulations support that $SnO_2$, for example, favors formate production via stabilization of the $HCOO^-$ intermediate.

To enhance the performance and encounter the limitations of noble metals and non-noble metals catalysts, hybrid catalysts have been extensively studied. These catalysts consist of composites of two or more metals, alloys of pure metals, or oxides. Various hybrid materials reported in the literature such as $Pd_{83}Cu_{17}$ (Lu et al., 2018), PdSn/C (Bai et al., 2017), and NiCo-MOF (Tian et al., 2022) have excellent FE (99 %) and higher selectivity for $CH_3OH$. While other composites such as Au–Pd core-shell aerogel (R. Du et al., 2021), Cu–In alloy (Rasul et al., 2015), and NiFe (Iqbal et al., 2022) demonstrated remarkable FE (>90 %) with higher selectivity for CO formation. Bi-Sn (Wen et al., 2018) and $Pd_{70}Pt_{30}$ (Kortlever, Peters, et al., 2015) also show FE > 90 % and are good for HCOOH. However, few of them including $Cu_2Pd$ (S. Zhang et al., 2015), $Cu_{1.0}/ZnO_{0.20}$ (T. Zhang et al., 2020,) CuAg (J. Gao et al., 2019), and CuAg alloy (Andersen et al., 2017) show poor FE (<60 %) but good selectivity for $C_2H_4$. The improved performance is attributed to synergistic effects between metal components that optimize the binding energies of intermediates and lower energy barriers. Cu–In alloys, for example, alter the electronic properties of Cu, which suppresses the HER and enhances $CO_2$RR. In bimetallic systems, DFT results frequently show charge redistribution and orbital hybridization, leading to modified reaction pathways. Despite the benefits of excellent FE and higher selectivity, there are still challenges including complex synthesis procedures, reproducibility, and scalability.

Further, in the attempt to find superior electrocatalysts, carbon-based materials, such as carbon nanotubes, carbon nanofibers, graphene, doped carbon, and porous carbon structures were explored for $CO_2$ reduction. These materials offer high surface areas, tunable electronic properties, and good electrical conductivity. Investigations showed that doping with heteroatoms like nitrogen can improve conversion efficiency as heteroatom doping (N, S, B) introduces charge polarization, creating active sites for $CO_2$ adsorption (Li et al., 2017; Wu et al., 2022). DFT studies indicate that while doped carbons efficiently facilitate single-electron transfer steps, they often lack the required binding energy profiles to promote multi-step hydrocarbon formation. The active site identification remains an active area of research, with theories pointing toward pyridinic or graphitic N-sites as key contributors. However, a major hurdle still exists, which is to obtain high selectivity and activity in carbon-based catalysts compared to metal-based catalysts. Carbon-based materials have demonstrated great potential for $CO_2$ reduction due to their different crystallinity, good conductivity, high surface area, and tunable chemical and physical properties.

Overall, while much progress has been made, a deeper understanding of structure-property relationships, identification of true active sites, and mechanistic pathways via DFT and operational techniques are critical for the logical design of next-generation $CO_2$RR catalysts.

## 10.6 RECENT ADVANCES IN THE ELECTROCATALYSTS FOR $CO_2$RR

As discussed earlier, effective electrocatalysts are essential to achieve high selectivity, efficiency, and stability in $CO_2$RR. This section covers the recent advancements in electrocatalyst design focusing on some critical aspects such as structural engineering, electronic tuning, and emerging materials.

### 10.6.1 Structural engineering of electrocatalysts

The structural configuration of electrocatalytic materials plays a vital role in catalytic performance as it significantly enhances the efficiency and stability of $CO_2$RR. Recent studies have demonstrated that the surface area, porosity, crystallinity, and morphology of electrocatalysts directly influence reaction rates, stability, and durability of $CO_2$RR (Zhang et al., 2024). Tailoring the electronic structure of catalysts directly initiates the formation of reaction intermediates and minimizes energy losses. For example, the optimization of the atomic arrangement on the surface of the catalyst and controlling the nanoparticle size can improve charge transfer, resulting in highly active sites, which are open to catalytic activity (Wang et al., 2022). Therefore, advances in nanostructured catalysts enable the development of highly efficient electrocatalysts with improved properties. There are some nanostructured and single-atom catalysts (SACs), which have shown bright research outcomes. Nanostructured Catalysts: Increased surface area and enhanced active sites of nanostructured catalysts help in the improvement of reaction rates and decrement in overpotentials. These catalysts often include nanoparticles, nanowires, or nanosheets. For instance, Cu nanocubes and nanowires have shown enhanced $C_2$ product formation due to their specific facet exposure (e.g., Cu(100) (Hahn et al., 2017)). Also, defect-rich catalysts such as oxygen vacancies in $Cu_2O$ (Ren et al., 2015) and $SnO_2$ (S. Zhang et al., 2014) create active sites that facilitate $CO_2$ adsorption and intermediate stabilization (Hahn et al., 2017). Single-Atom Catalysts (SACs): SACs consist of metal atoms dispersed at the atomic level on a stable support, ensuring the maximum exposure of active sites. The unique feature of SACs is that each metal atom acts as an isolated catalytic site, providing a highly efficient and tunable catalytic environment (Wang et al., 2022). Recent research has emphasized the development of SACs for various reactions, such as carbon materials, oxides, or nitrides, which enhance both the stability and performance of SACs. Ni and Fe single-atom catalysts anchored on nitrogen-doped carbon matrices such as Ni-MOF (Cao et al., 2021), Ni-$N_4$-C (X. Li et al., 2017), and Ni/N-C (Tan et al., 2019) have shown exceptional selectivity for CO formation with over 90% Faradaic efficiency.

### 10.6.2 Electronic tuning and alloying

Electronic tuning and alloying are key strategies to enhance electrocatalyst performance by modifying their electronic properties. Electronic tuning adjusts the charge distribution around active sites using techniques like doping, optimizing interactions with reaction intermediates, and improving catalytic efficiency. Alloying, which combines metals, alters the electronic structure and surface properties, creating new active sites on catalysts that enhance activity and stability. For example, Cu-Ag and Cu-Au alloys, tune the electronic structure, enhancing $CO_2$ adsorption while preventing excessive hydrogen evolution (Zhang et al., 2017). Core-shell structures, such as Au@Cu core-shell catalysts, improve $CO_2$ conversion at the metal-metal interface. Also, by doping, N- and S-doped carbon-heteroatom can significantly alter the electronic density of the catalyst, favoring $CO_2$ reduction over competing hydrogen evolution.

### 10.6.3 Emerging materials for $CO_2RR$

Some emerging materials that offer significant advancements in $CO_2RR$ are discussed here. These materials show improved catalyst efficiency, selectivity, and stability, which are key factors for developing sustainable $CO_2$ reduction technologies. Metal-Organic Frameworks (MOFs): Co- and Ni-based MOFs, which are Ni-MOF (Cao et al., 2021) and NiCo-MOF (Tian et al., 2022), are gaining attention for $CO_2RR$ due to their excellent FE (100 %), tunable porosity, large surface area, and versatility. These features enhance the adsorption and activation of $CO_2$. 2D Materials: Materials like $MoS_2$, Graphene Oxide, and Cu/N-doped carbon nanofibers (Lee et al., 2020) show promise for $CO_2RR$ because of their unique electronic properties and the potential for defect engineering. The high surface area and adjustable electronic states in 2D materials provide ideal conditions for enhancing $CO_2$ activation and improving catalyst performance. Molecular Catalysts: Co, Mn, and Co-Polypyrrole-C (Mun et al., 2018) are molecular catalysts that offer excellent selectivity for CO formation with minimal overpotential. Their well-defined coordination environments and high catalytic efficiency make them valuable for selective $CO_2$ reduction, enhancing overall reaction efficiency with reduced energy input.

## 10.7 CHALLENGES AND FUTURE DIRECTION IN ELECTROCATALYTIC $CO_2$ REDUCTION

Despite being investigated extensively, there are still major challenges in electrocatalytic $CO_2RR$, which are electrocatalytic degradation, product separation, and scaling the technology for commercialization. The overview of the future directions is introduced in this section.

### 10.7.1 Electrocatalytic degradation

One major challenge in $CO_2RR$ is electrocatalytic degradation. After a while, catalysts undergo surface deactivation due to processes like surface poisoning, restructuring, or agglomeration. These factors decrease the available active sites, resulting in reduced efficiency. Apart from that, material corrosion, in the form of oxidation or leaching of active metal species, can further compromise catalyst performance. Therefore, to address electrocatalytic degradation, emphasis on advanced catalyst design, surface modifications, and material innovations is needed. Protective coatings, non-metal doping, and corrosion-resistant supports such as graphene and MXenes can be further used to improve durability.

### 10.7.2 Product separation

Product separation is another major challenge. $CO_2$ reduction often produces a complex mixture of products, consisting of CO, HCOOH, $CH_4$, and $C_2H_4$, making it difficult to efficiently separate the desired products. The low concentration of these products in the output stream also makes separation even more complicated. Therefore, cost-effective and efficient filtration or distillation techniques are required.

### 10.7.3 Scaling the technology for commercialization

Scaling up from lab-scale reactors to industrial-scale electrochemical cells also poses additional hurdles. This raises the requirement of redesigning larger or modular electrode assembly systems to accommodate higher current densities and larger throughput while maintaining efficiency and stability. The main concern involves the limited electrode area in lab-scale setups limiting the current output and productivity. Providing uniform current distribution and effective mass transport across larger electrodes becomes complex, resulting in reduced efficiency or nonuniform product formation. Furthermore, the cost and energy requirements of capturing, purifying, and compressing $CO_2$, especially from dilute sources such as air or flue gas, are major economic hurdles. Additionally, supplying $CO_2$ from direct air capture can lead to the introduction of impurities that will deactivate catalysts, complicating the process further. These technical and economic hurdles highlight the need for continued research in reactor engineering, catalyst stability, and integrated system design to bridge the gap between laboratory research and commercial viability. Future research must be directed toward making electrocatalytic $CO_2$ reduction a viable and sustainable alternative to improve efficiency and reduce energy consumption through innovations such as selective and nanostructured catalysts and efficient separation methods.

## 10.8 CONCLUSIONS

In conclusion, the electrocatalytic conversion of $CO_2$ into fuels and value-added products represents a promising solution to mitigate the challenges of $CO_2$ emissions and meet the growing demand for sustainable energy. While significant advancements have been made in electrocatalyst design and reaction optimization, there are still challenges such as catalyst degradation and product separation, hindering the large-scale commercialization of this technology. The ongoing research has been discussed in this chapter along with some future scopes. There are numerous possibilities in this research; if one can find a breakthrough in this technology, the successful implementation of $CO_2$RR can drive economic growth, energy security, and environmental sustainability.

## REFERENCES


Andersen, M., Medford, A. J., Nørskov, J. K., & Reuter, K. (2017). Scaling-Relation-Based Analysis of Bifunctional Catalysis: The Case for Homogeneous Bimetallic Alloys. *ACS Catalysis*, *7*(6), 3960–3967. https://doi.org/10.1021/acscatal.7b00482

Bai, X., Chen, W., Zhao, C., Li, S., Song, Y., Ge, R., Wei, W., & Sun, Y. (2017). Exclusive Formation of Formic Acid from $CO_2$ Electroreduction by a Tunable Pd-Sn Alloy. *Angewandte Chemie*, *129*(40), 12387–12391. https://doi.org/10.1002/ange.201707098

Cao, C., Ma, D., Jia, J., Xu, Q., Wu, X., & Zhu, Q. (2021). Divergent Paths, Same Goal: A Pair-Electrosynthesis Tactic for Cost-Efficient and Exclusive Formate Production by Metal–Organic-Framework-Derived 2D Electrocatalysts. *Advanced Materials*, *33*(25). https://doi.org/10.1002/adma.202008631

Chew, L. M., Ruland, H., Schulte, H. J., Xia, W., & Muhler, M. (2014). $CO_2$ Hydrogenation to Hydrocarbons over Iron Nanoparticles Supported on Oxygen-functionalized Carbon Nanotubes. *Journal of Chemical Sciences*, *126*(2), 481–486. https://www.ias.ac.in/article/fulltext/jcsc/126/02/0481-0486

Dong Zhu, D., Long Liu, J., Zhang Qiao, S., Zhu, D. D., Liu, J. L., & Qiao, S. Z. (2016). Recent Advances in Inorganic Heterogeneous Electrocatalysts for Reduction of Carbon Dioxide. *Advanced Materials*, *28*(18), 3423–3452. https://doi.org/10.1002/ADMA.201504766

Du, J., Li, S., Liu, S., Xin, Y., Chen, B., Liu, H., & Han, B. (2020). Selective Electrochemical Reduction of Carbon Dioxide to Ethanol *via* a Relay Catalytic Platform. *Chemical Science*, *11*(19), 5098–5104. https://doi.org/10.1039/D0SC01133A

Du, R., Jin, W., Wu, H., Hübner, R., Zhou, L., Xue, G., Hu, Y., & Eychmüller, A. (2021). Rapid Synthesis of Gold–Palladium Core–Shell Aerogels for Selective and Robust Electrochemical $CO_2$ Reduction. *Journal of Materials Chemistry A*, *9*(32), 17189–17197. https://doi.org/10.1039/D1TA03103A

Gabardo, C. M., Seifitokaldani, A., Edwards, J. P., Dinh, C. T., Burdyny, T., Kibria, M. G., O'Brien, C. P., Sargent, E. H., & Sinton, D. (2018). Combined High Alkalinity and Pressurization Enable Efficient $CO_2$ Electroreduction to CO. *Energy & Environmental Science*, *11*(9), 2531–2539. https://doi.org/10.1039/C8EE01684D

Gao, D., Zhou, H., Wang, J., Miao, S., Yang, F., Wang, G., Wang, J., & Bao, X. (2015). Size-Dependent Electrocatalytic Reduction of $CO_2$ over Pd Nanoparticles. *Journal of the American Chemical Society*, *137*(13), 4288–4291. https://doi.org/10.1021/jacs.5b00046

Gao, J., Zhang, H., Guo, X., Luo, J., Zakeeruddin, S. M., Ren, D., & Grätzel, M. (2019). Selective C–C Coupling in Carbon Dioxide Electroreduction via Efficient Spillover of Intermediates As Supported by Operando Raman Spectroscopy. *Journal of the American Chemical Society*, *141*(47), 18704–18714. https://doi.org/10.1021/jacs.9b07415

Hahn, C., Hatsukade, T., Kim, Y.-G., Vailionis, A., Baricuatro, J. H., Higgins, D. C., Nitopi, S. A., Soriaga, M. P., & Jaramillo, T. F. (2017). Engineering Cu Surfaces for the Electrocatalytic Conversion of $CO_2$: Controlling Selectivity toward Oxygenates and Hydrocarbons. *Proceedings of the National Academy of Sciences*, *114*(23), 5918–5923. https://doi.org/10.1073/pnas.1618935114

Hara, K., & Sakata, T. (1997). Electrocatalytic Formation of $CH_4$ from $CO_2$ on a Pt Gas Diffusion Electrode. *Journal of The Electrochemical Society*, *144*(2), 539–545. https://doi.org/10.1149/1.1837445

Hoang, T. T. H., Verma, S., Ma, S., Fister, T. T., Timoshenko, J., Frenkel, A. I., Kenis, P. J. A., & Gewirth, A. A. (2018). Nanoporous Copper–Silver Alloys by Additive-Controlled Electrodeposition for the Selective Electroreduction of $CO_2$ to Ethylene and Ethanol. *Journal of the American Chemical Society*, *140*(17), 5791–5797. https://doi.org/10.1021/jacs.8b01868

Hori, Y. (2008). Electrochemical $CO_2$ Reduction on Metal Electrodes. In *Modern Aspects of Electrochemistry* (Vol. 42, pp. 89–189). Springer New York. https://doi.org/10.1007/978-0-387-49489-0_3

Hori, Y., Kikuchi, K., & Suzuki, S. (1985). Production of CO and $CH_4$ in Electrochemical Reduction of $CO_2$ at Metal Electrodes in Aqueous Hydrogencarbonate Solution. *Chemistry Letters*, *14*(11), 1695–1698. https://doi.org/10.1246/cl.1985.1695

Huang, J., Guo, X., Yue, G., Hu, Q., & Wang, L. (2018). Boosting $CH_3$ OH Production in Electrocatalytic $CO_2$ Reduction over Partially Oxidized 5 nm Cobalt Nanoparticles Dispersed on Single-Layer Nitrogen-Doped Graphene. *ACS Applied Materials & Interfaces*, *10*(51), 44403–44414. https://doi.org/10.1021/acsami.8b14822

Iqbal, R., Akbar, M. B., Ahmad, A., Hussain, A., Altaf, N., Ibraheem, S., Yasin, G., Khan, M. A., Tabish, M., Kumar, A., Majeed, M. K., Saleem, A., & Ali, S. (2022). Exploring the Synergistic Effect of Novel Ni-Fe in 2D Bimetallic Metal-Organic Frameworks for Enhanced Electrochemical Reduction of $CO_2$. *Advanced Materials Interfaces*, *9*(1). https://doi.org/10.1002/admi.202101505

Jha, H., Sanskrit, U. S. V., & Dubey, B. K. (2024). Conversion of Carbon Dioxide to Fuel, Feed and High Value Chemicals. In *Waste-to-Wealth: Resource Recovery and Value-Added Products for Sustainable Development* (pp. 184–199). CRC Press. https://doi.org/10.1201/9781003327646-13

Jia, M., Choi, C., Wu, T.-S., Ma, C., Kang, P., Tao, H., Fan, Q., Hong, S., Liu, S., Soo, Y.-L., Jung, Y., Qiu, J., & Sun, Z. (2018). Carbon-supported Ni Nanoparticles for Efficient $CO_2$ Electroreduction. *Chemical Science*, *9*(47), 8775–8780. https://doi.org/10.1039/C8SC03732A

Jiang, X., Cai, F., Gao, D., Dong, J., Miao, S., Wang, G., & Bao, X. (2016). Electrocatalytic Reduction of Carbon Dioxide Over Reduced Nanoporous Zinc Oxide. *Electrochemistry Communications*, *68*, 67–70. https://doi.org/10.1016/J.ELECOM.2016.05.003

Jitaru, M., Lowy, D. A., Toma, M., Toma, B. C., & Oniciu, L. (1997). Electrochemical Reduction of Carbon Dioxide on Flat Metallic Cathodes. *Journal of Applied Electrochemistry*, *27*(8), 875–889. https://doi.org/10.1023/A:1018441316386

Kaneco, S., Hiei, N. H., Xing, Y., Katsumata, H., Ohnishi, H., Suzuki, T., & Ohta, K. (2002). Electrochemical Conversion of Carbon Dioxide to Methane in Aqueous $NaHCO_3$ Solution at Less Than 273 K. *Electrochimica Acta*, *48*(1), 51–55. https://doi.org/10.1016/S0013-4686(02)00550-9

Kannan, K., & Tari, V. (Eds.). (2025). *Novel Energy Storage and Conversion Technologies for Two-Dimensional MXenes and MBenes*. https://www.amazon.com/dp/9798369395462

Klinkova, A., De Luna, P., Dinh, C.-T., Voznyy, O., Larin, E. M., Kumacheva, E., & Sargent, E. H. (2016). Rational Design of Efficient Palladium Catalysts for Electroreduction of Carbon Dioxide to Formate. *ACS Catalysis*, *6*(12), 8115–8120. https://doi.org/10.1021/acscatal.6b01719

Kortlever, R., Peters, I., Koper, S., & Koper, M. T. M. (2015). Electrochemical $CO_2$ Reduction to Formic Acid at Low Overpotential and with High Faradaic Efficiency on Carbon-Supported Bimetallic Pd–Pt Nanoparticles. *ACS Catalysis*, *5*(7), 3916–3923. https://doi.org/10.1021/acscatal.5b00602

Kortlever, R., Shen, J., Schouten, K. J. P., Calle-Vallejo, F., & Koper, M. T. M. (2015). Catalysts and Reaction Pathways for the Electrochemical Reduction of Carbon Dioxide. *The Journal of Physical Chemistry Letters*, *6*(20), 4073–4082. https://doi.org/10.1021/acs.jpclett.5b01559

Lee, J.-C., Kim, J.-Y., Joo, W.-H., Hong, D., Oh, S.-H., Kim, B., Lee, G.-D., Kim, M., Oh, J., & Joo, Y.-C. (2020). Thermodynamically Driven Self-formation of Copper-embedded Nitrogen-doped Carbon Nanofiber Catalysts for a Cascade Electroreduction of Carbon Dioxide to Ethylene. *Journal of Materials Chemistry A*, *8*(23), 11632–11641. https://doi.org/10.1039/D0TA03322G

Li, J., Kuang, Y., Meng, Y., Tian, X., Hung, W. H., Zhang, X., Li, A., Xu, M., Zhou, W., Ku, C. S., Chiang, C. Y., Zhu, G., Guo, J., Sun, X., & Dai, H. (2020). Electroreduction of $CO_2$ to Formate on a Copper-Based Electrocatalyst at High Pressures with High Energy Conversion Efficiency. *Journal of the American Chemical Society*, *142*(16), 7276–7282. https://doi.org/10.1021/JACS.0C00122/SUPPL_FILE/JA0C00122_SI_001.PDF

Li, X., Bi, W., Chen, M., Sun, Y., Ju, H., Yan, W., Zhu, J., Wu, X., Chu, W., Wu, C., & Xie, Y. (2017). Exclusive Ni–$N_4$ Sites Realize Near-Unity CO Selectivity for Electrochemical $CO_2$ Reduction. *Journal of the American Chemical Society*, *139*(42), 14889–14892. https://doi.org/10.1021/jacs.7b09074

Liang, C., Kim, B., Yang, S., Yang Liu, Y. L., Francisco Woellner, C., Li, Z., Vajtai, R., Yang, W., Wu, J., Kenis, P. J. A., & Ajayan, P. M. (2018). High Efficiency Electrochemical Reduction of $CO_2$ Beyond the Two-electron Transfer Pathway on Grain Boundary Rich Ultra-small $SnO_2$ Nanoparticles. *Journal of Materials Chemistry A*, *6*(22), 10313–10319. https://doi.org/10.1039/C8TA01367E

Liu, M., Pang, Y., Zhang, B., De Luna, P., Voznyy, O., Xu, J., Zheng, X., Dinh, C. T., Fan, F., Cao, C., de Arquer, F. P. G., Safaei, T. S., Mepham, A., Klinkova, A., Kumacheva, E., Filleter, T., Sinton, D., Kelley, S. O., & Sargent, E. H. (2016a). Enhanced Electrocatalytic $CO_2$ Reduction Via Field-induced Reagent Concentration. *Nature*, *537*(7620), 382–386. https://doi.org/10.1038/nature19060

Liu, M., Pang, Y., Zhang, B., De Luna, P., Voznyy, O., Xu, J., Zheng, X., Dinh, C. T., Fan, F., Cao, C., de Arquer, F. P. G., Safaei, T. S., Mepham, A., Klinkova, A., Kumacheva, E., Filleter, T., Sinton, D., Kelley, S. O., & Sargent, E. H. (2016b). Enhanced Electrocatalytic $CO_2$ Reduction Via Field-induced Reagent Concentration. *Nature*, *537*(7620), 382–386. https://doi.org/10.1038/nature19060

Lu, L., Sun, X., Ma, J., Yang, D., Wu, H., Zhang, B., Zhang, J., & Han, B. (2018). Highly Efficient Electroreduction of $CO_2$ to Methanol on Palladium–Copper Bimetallic Aerogels. *Angewandte Chemie*, *130*(43), 14345–14349. https://doi.org/10.1002/ange.201808964

Lu, Q., Rosen, J., Zhou, Y., Hutchings, G. S., Kimmel, Y. C., Chen, J. G., & Jiao, F. (2014). A Selective and Efficient Electrocatalyst for Carbon Dioxide Reduction. *Nature Communications*, *5*(1), 3242. https://doi.org/10.1038/ncomms4242

Luo, W., Zhang, Q., Zhang, J., Moioli, E., Zhao, K., & Züttel, A. (2020). Electrochemical Reconstruction of ZnO for Selective Reduction of $CO_2$ to CO. *Applied Catalysis B: Environmental*, *273*, 119060. https://doi.org/10.1016/j.apcatb.2020.119060

Martín, A. J., Larrazábal, G. O., & Pérez-Ramírez, J. (2015). Towards Sustainable Fuels and Chemicals Through the Electrochemical Reduction of $CO_2$: Lessons from Water Electrolysis. *Green Chemistry*, *17*(12), 5114–5130. https://doi.org/10.1039/C5GC01893E

Martin, N. M., Velin, P., Skoglundh, M., Bauer, M., & Carlsson, P.-A. (2017). Catalytic Hydrogenation of $CO_2$ to Methane Over Supported Pd, Rh and Ni Catalysts. *Catalysis Science & Technology*, 7(5), 1086–1094. https://doi.org/10.1039/C6CY02536F

Mun, Y., Kim, K., Kim, S., Lee, S., Lee, S., Kim, S., Choi, W., Kim, S. K., Han, J. W., & Lee, J. (2018). A Novel Strategy to Develop Non-noble Metal Catalyst for $CO_2$ Electroreduction: Hybridization of Metal-organic Polymer. *Applied Catalysis B: Environmental*, *236*, 154–161. https://doi.org/10.1016/J.APCATB.2018.05.025

Nakagawa, S., Kudo, A., Azuma, M., & Sakata, T. (1991). Effect of Pressure on the Electrochemical Reduction of $CO_2$ on Group VIII Metal Electrodes. *Journal of Electroanalytical Chemistry and Interfacial Electrochemistry*, *308*(1–2), 339–343. https://doi.org/10.1016/0022-0728(91)85080-9

Nisar, A., Khan, S., Hameed, M., Nisar, A., Ahmad, H., & Mehmood, S. A. (2021). Bio-conversion of $CO_2$ into Biofuels and Other Value-added Chemicals Via Metabolic Engineering. *Microbiological Research*, *251*, 126813. https://doi.org/10.1016/J.MICRES.2021.126813

Nitopi, S. (2019). Understanding the Factors that Govern Activity and Selectivity of the Electrochemical Carbon Dioxide Reduction Reaction on Copper Catalysts: A Dissertation Submitted to the Department of Chemical Engineering. *Stanford University ProQuest Dissertations & Theses., 28158127*. http://purl.stanford.edu/km176nj5548

Oh, Y., Vrubel, H., Guidoux, S., & Hu, X. (2014). Electrochemical Reduction of $CO_2$ in Organic Solvents Catalyzed by $MoO_2$. *Chemical Communications*, *50*(29), 3878–3881. https://doi.org/10.1039/C3CC49262A

Okoye-Chine, C. G., Otun, K., Shiba, N., Rashama, C., Ugwu, S. N., Onyeaka, H., & Okeke, C. T. (2022). Conversion of Carbon Dioxide into Fuels—A Review. *Journal of $CO_2$ Utilization*, *62*, 102099. https://doi.org/10.1016/j.jcou.2022.102099

Ooka, H., Figueiredo, M. C., & Koper, M. T. M. (2017). Competition between Hydrogen Evolution and Carbon Dioxide Reduction on Copper Electrodes in Mildly Acidic Media. *Langmuir*, *33*(37), 9307–9313. https://doi.org/10.1021/ACS.LANGMUIR.7B00696

Ramdin, M., Morrison, A. R. T., De Groen, M., Van Haperen, R., De Kler, R., Van Den Broeke, L. J. P., Martin Trusler, J. P., De Jong, W., & Vlugt, T. J. H. (2019). High Pressure Electrochemical Reduction of $CO_2$ to Formic Acid/Formate: A Comparison between Bipolar Membranes and Cation Exchange Membranes. *Industrial and Engineering Chemistry Research*, *58*(5), 1834–1847. https://doi.org/10.1021/ACS.IECR.8B04944

Ramírez-Valencia, L. D., Bailón-García, E., Carrasco-Marín, F., & Pérez-Cadenas, A. F. (2021). From $CO_2$ to Value-added Products: A Review about Carbon-based Materials for Electro-chemical $CO_2$ Conversion. In *Catalysts* (Vol. 11, Issue 3, pp. 1–70). MDPI. https://doi.org/10.3390/catal11030351

Rasul, S., Anjum, D. H., Jedidi, A., Minenkov, Y., Cavallo, L., & Takanabe, K. (2015). A Highly Selective Copper–Indium Bimetallic Electrocatalyst for the Electrochemical Reduction of Aqueous $CO_2$ to CO. *Angewandte Chemie*, *127*(7), 2174–2178. https://doi.org/10.1002/ange.201410233

Ren, D., Deng, Y., Handoko, A. D., Chen, C. S., Malkhandi, S., & Yeo, B. S. (2015). Selective Electrochemical Reduction of Carbon Dioxide to Ethylene and Ethanol on Copper(I) Oxide Catalysts. *ACS Catalysis*, *5*(5), 2814–2821. https://doi.org/10.1021/cs502128q

Saravanan, A., Senthil Kumar, P., Vo, D.-V. N., Jeevanantham, S., Bhuvaneswari, V., Anantha Narayanan, V., Yaashikaa, P. R., Swetha, S., & Reshma, B. (2021). A Comprehensive Review on Different Approaches for $CO_2$ Utilization and Conversion Pathways. *Chemical Engineering Science*, *236*, 116515. https://doi.org/10.1016/j.ces.2021.116515

Savino, U., Sacco, A., Bejtka, K., Castellino, M., Farkhondehfal, M. A., Chiodoni, A., Pirri, F., & Tresso, E. (2022). Well Performing Fe-$SnO_2$ for $CO_2$ Reduction to HCOOH. *Catalysis Communications*, *163*, 106412. https://doi.org/10.1016/j.catcom.2022.106412

Sun, Z., Ma, T., Tao, H., Fan, Q., & Han, B. (2017). Fundamentals and Challenges of Electrochemical $CO_2$ Reduction Using Two-Dimensional Materials. *Chem*, *3*(4), 560–587. https://doi.org/10.1016/J.CHEMPR.2017.09.009

Tan, D., Cui, C., Shi, J., Luo, Z., Zhang, B., Tan, X., Han, B., Zheng, L., Zhang, J., & Zhang, J. (2019). Nitrogen-carbon Layer Coated Nickel Nanoparticles for Efficient Electrocatalytic Reduction of Carbon Dioxide. *Nano Research*, *12*(5), 1167–1172. https://doi.org/10.1007/s12274-019-2372-1

Tari, V. (2015a). Earth, Man, and Environment. In Pandurang Y. Patil (Ed.), *Definition Principles and Scope of Environment* (pp. 1–20). DPS Publishing House.

Tari, V. (2015b). Physicochemical and Biological Factors in the Environment. In Pandurang Y. Patil (Ed.), *Definition Principles and Scope of Environment* (pp. 21–29). DPS Publishing House.

Tari, V., Gupta, V., Joshi, A., Kurve, P., & Chandra, N. (2024). Biodiversity Loss Due to Forest Diversion. In A. Rathore (Ed.), *Impact of Societal Development and Infrastructure on Biodiversity Decline* (pp. 137–165). IGI Global. https://doi.org/10.4018/979-8-3693-6950-0.ch010

Tari, V., Kannan, K., Koliar, J., Gurushankar, K., & Wahyuni, D. K. (2025). Influence of Zinc Oxide Nanomaterials on Crop Plants: A Prospective Analysis. In D. Jabbrova, D. Sarkar, S. Rakesh, S. Singh, & R. Datta (Eds.), *Smart Nanomaterials in Agroforestry Systems*. Springer. https://doi.org/10.1007/978-981-96-1337-3_8

Tian, J., Cao, C., Ma, D.-D., Han, S.-G., He, Y., Wu, X.-T., & Zhu, Q.-L. (2022). Killing Two Birds with One Stone: Selective Oxidation of Small Organic Molecule as Anodic Reaction to Boost $CO_2$ Electrolysis. *Small Structures*, *3*(5). https://doi.org/10.1002/sstr.202100134

Vasileff, A., Zheng, Y., & Qiao, S. Z. (2017). Carbon Solving Carbon's Problems: Recent Progress of Nanostructured Carbon-Based Catalysts for the Electrochemical Reduction of $CO_2$. *Advanced Energy Materials*, *7*(21), 1700759. https://doi.org/10.1002/AENM.201700759

Vieira, É. D., Andrietta, M. D. G. S., & Andrietta, S. R. (2013). Yeast Biomass Production: A New Approach in Glucose-Limited Feeding Strategy. *Brazilian Journal of Microbiology*, *44*(2), 551–558. https://doi.org/10.1590/S1517-83822013000200035

Wang, Q., Wang, X., Wu, C., Cheng, Y., Sun, Q., Dong, H., & Yu, H. (2017). Electrodeposition of Tin on Nafion-Bonded Carbon Black as an Active Catalyst Layer for Efficient Electroreduction of $CO_2$ to Formic Acid. *Scientific Reports*, *7*(1), 1–11. https://doi.org/10.1038/s41598-017-14233-y

Wang, W., Duan, J., Liu, Y., & Zhai, T. (2022). Structural Reconstruction of Catalysts in Electroreduction Reaction: Identifying, Understanding, and Manipulating. *Advanced Materials*, *34*(52). https://doi.org/10.1002/adma.202110699

Wen, G., Lee, D. U., Ren, B., Hassan, F. M., Jiang, G., Cano, Z. P., Gostick, J., Croiset, E., Bai, Z., Yang, L., & Chen, Z. (2018). Orbital Interactions in Bi-Sn Bimetallic Electrocatalysts for Highly Selective Electrochemical $CO_2$ Reduction toward Formate Production. *Advanced Energy Materials*, *8*(31). https://doi.org/10.1002/aenm.201802427

Wu, S., Yi, F., Ping, D., Huang, S., Zhang, Y., Han, L., Wang, S., Wang, H., Yang, X., Guo, D., Liu, G., & Fang, S. (2022). Constructing Single-atomic Nickel Sites in Carbon Nanotubes for Efficient $CO_2$ Electroreduction. *Carbon*, *196*, 1–9. https://doi.org/10.1016/j.carbon.2022.04.038

Yin, J., Yin, Z., Jin, J., Sun, M., Huang, B., Lin, H., Ma, Z., Muzzio, M., Shen, M., Yu, C., Zhang, H., Peng, Y., Xi, P., Yan, C.-H., & Sun, S. (2021). A New Hexagonal Cobalt Nanosheet Catalyst for Selective $CO_2$ Conversion to Ethanal. *Journal of the American Chemical Society*, *143*(37), 15335–15343. https://doi.org/10.1021/jacs.1c06877

Yuan, C.-Z., Li, H.-B., Jiang, Y.-F., Liang, K., Zhao, S.-J., Fang, X.-X., Ma, L.-B., Zhao, T., Lin, C., & Xu, A.-W. (2019). Tuning the Activity of N-doped Carbon for $CO_2$ Reduction *via in situ* Encapsulation of Nickel Nanoparticles into Nano-hybrid Carbon Substrates. *Journal of Materials Chemistry A*, *7*(12), 6894–6900. https://doi.org/10.1039/C8TA11500A

Zangeneh, F. T., Sahebdelfar, S., & Ravanchi, M. T. (2011). Conversion of Carbon Dioxide to Valuable Petrochemicals: An Approach to Clean Development Mechanism. *Journal of Natural Gas Chemistry*, *20*(3), 219–231. https://doi.org/10.1016/S1003-9953(10)60191-0

Zhang, B., chemistry, J. Z.-J. of energy, & 2017, undefined. (2017). Rational Design of Cu-based Electrocatalysts for Electrochemical Reduction of Carbon Dioxide. *Journal of Energy Chemistry*, *26*, 1050–1066. https://doi.org/10.1016/j.jechem.2017.10.011

Zhang, J., Xia, S., Wang, Y., Wu, J., & Wu, Y. (2024). Recent Advances in Dynamic Reconstruction of Electrocatalysts for Carbon Dioxide Reduction. *IScience*, *27*(6). https://doi.org/10.1016/J.ISCI.2024.110005

Zhang, S., Kang, P., Bakir, M., Lapides, A. M., Dares, C. J., & Meyer, T. J. (2015). Polymer-supported CuPd Nanoalloy as a Synergistic Catalyst for Electrocatalytic Reduction of Carbon Dioxide to Methane. *Proceedings of the National Academy of Sciences*, *112*(52), 15809–15814. https://doi.org/10.1073/pnas.1522496112

Zhang, S., Kang, P., & Meyer, T. J. (2014). Nanostructured Tin Catalysts for Selective Electrochemical Reduction of Carbon Dioxide to Formate. *Journal of the American Chemical Society*, *136*(5), 1734–1737. https://doi.org/10.1021/JA4113885

Zhang, T., Li, Z., Zhang, J., & Wu, J. (2020). Enhance $CO_2$-to-$C_2^+$ Products Yield Through Spatial Management of CO Transport in Cu/ZnO Tandem Electrodes. *Journal of Catalysis*, *387*, 163–169. https://doi.org/10.1016/j.jcat.2020.05.002

Zhang, W., Hu, Y., Ma, L., Zhu, G., Wang, Y., Xue, X., Chen, R., Yang, S., & Jin, Z. (2018). Progress and Perspective of Electrocatalytic $CO_2$ Reduction for Renewable Carbonaceous Fuels and Chemicals. *Advanced Science*, *5*(1), 1700275. https://doi.org/10.1002/ADVS.201700275

Zhou, Y., Che, F., Liu, M., Zou, C., Liang, Z., De Luna, P., Yuan, H., Li, J., Wang, Z., Xie, H., Li, H., Chen, P., Bladt, E., Quintero-Bermudez, R., Sham, T.-K., Bals,

S., Hofkens, J., Sinton, D., Chen, G., & Sargent, E. H. (2018). Dopant-induced Electron Localization Drives $CO_2$ Reduction to $C_2$ Hydrocarbons. *Nature Chemistry*, *10*(9), 974–980. https://doi.org/10.1038/s41557-018-0092-x

Zhu, W., Michalsky, R., Metin, Ö., Lv, H., Guo, S., Wright, C. J., Sun, X., Peterson, A. A., & Sun, S. (2013). Monodisperse Au Nanoparticles for Selective Electrocatalytic Reduction of $CO_2$ to CO. *Journal of the American Chemical Society*, *135*(45), 16833–16836. https://doi.org/10.1021/ja409445p

Zhu, W., Zhang, Y.-J., Zhang, H., Lv, H., Li, Q., Michalsky, R., Peterson, A. A., & Sun, S. (2014). Active and Selective Conversion of $CO_2$ to CO on Ultrathin Au Nanowires. *Journal of the American Chemical Society*, *136*(46), 16132–16135. https://doi.org/10.1021/ja5095099

Chapter 11

# Green hydrogen production from proton exchange membrane and anion exchange membrane-based water electrolysis

*Kamana K. Mishra, Uma Sharma, Karthik Kannan, Yong-Song Chen, and Prabhakar Singh*

## 11.1 INTRODUCTION

Initially, our research was focused on developing energy alternatives to non-renewable energy sources to address their limited availability and sustainability concerns. However, with the increasing awareness of greenhouse gas (GHG) emissions and their detrimental impact on the environment, our focus has now shifted toward producing green energy. The urgency to mitigate climate change and reduce carbon footprints has driven us to explore cleaner and more sustainable energy solutions (Tari et al., 2024). Renewable energy sources like solar cells and wind turbines provide sustainable power, but their variability and intermittency require efficient energy management solutions. Especially during periods of low wind or insufficient solar radiation, hydrogen serves as an outstanding energy storage medium due to its ability to be generated through clean and sustainable processes using renewable energy. It can be stored and later converted back into electricity when energy is needed. Such techniques help in promoting a greener future by achieving carbon neutrality.

Hydrogen is primarily produced through three main methods: steam methane reforming, coal gasification, and water electrolysis. Although the first two methods currently dominate global hydrogen production, they are highly dependent on fossil fuels and result in significant $CO_2$ emissions. Hydrogen generated through carbon-intensive methods is referred to as "gray hydrogen. $CO_2$ emissions can be controlled through carbon capture and storage (CCS) or carbon capture, utilization, and storage (CCUS); however, these processes increase production costs and require strict methane emission control, producing what is called "blue hydrogen." The most environmentally friendly method is water electrolysis powered by renewable energy sources such as solar, hydro, wind, nuclear energy, or geothermal. This process generates "green hydrogen," which is entirely carbon-free and contributes to sustainable energy solutions (Hou and Yang, n.d.).

Water electrolysis is a process that uses electricity to split water into hydrogen ($H_2$) and oxygen ($O_2$). And this takes place in water electrolyzers, devices that contain an anode, a cathode, and an electrolyte (Kannan and

DOI: 10.1201/9781003584742-11

Tari, 2025). Water electrolysis involves two processes: the oxygen evolution reaction (OER) at the anode and the hydrogen evolution reaction (HER) at the cathode.

When electricity is applied to the electrodes, at the anode, water molecules lose electrons and form oxygen gas and protons or hydroxide ions, depending on the system. Oxygen gas is released as a byproduct (Chen et al., 2023).

$$\text{Anode}: H_2O \rightarrow 2H^+ + \tfrac{1}{2}O_2 + 2e^-$$

Whereas at the cathode, the protons or hydroxide ions travel through the electrolyte to the cathode. Electrons from the external circuit combine with protons to form hydrogen gas.

$$\text{Cathode}: 2H^+ + 2e^- \rightarrow H_2$$

Hydrogen is collected at the cathode and used as clean fuel, and oxygen is collected at the anode, either vented or utilized in other processes.

$$\text{Overall reaction}: H_2O \rightarrow H_2 + \tfrac{1}{2}O_2$$

The history of water electrolysis is a fascinating tale of scientific curiosity and discovery that dates back over two centuries. The journey of water electrolysis began in 1789, when Dutch chemists Jan Rudolph Deiman and Adriaan Paets van Troostwijk first used an electrostatic generator to split water into hydrogen and oxygen, marking the earliest exploration of electricity's ability to drive chemical change. The field of electrolysis advanced significantly in 1800 when Alessandro Volta invented the first true battery, the voltaic pile, capable of providing a steady electric current. Their groundbreaking work showed that electricity could decompose water into its elemental gases, a revelation that laid the foundation for electrochemistry. The next major milestone came with Michael Faraday's pioneering work in the 1830s. Faraday established the fundamental principles of electrolysis, including his two laws of electrolysis, which quantitatively related the amount of electric current passed through a substance to the amount of chemical change produced. In 1888, Dmitry Lachinov developed a technique for synthesizing hydrogen and oxygen industrially through water electrolysis. By 1902, over 400 industrial water electrolyzers were already in service. The first large water electrolysis plant with a capacity of 10,000 $Nm^3$ $H_2$/h came into service in 1939. Over time, water electrolysis has become a cornerstone of clean energy research, particularly for producing green hydrogen. Today, water electrolysis is at the heart of global efforts to transition to sustainable energy systems (Santos et al., 2013). Modern electrolyzers use advanced catalysts, membranes, and energy sources like renewable electricity to achieve higher efficiencies. From its humble beginnings with Deiman and Troostwijk to its central role in the green hydrogen economy, water electrolysis has evolved through the contributions of many great scientists and engineers. It remains a vibrant and essential area of research, driving progress toward a sustainable energy future.

Currently, there are four different methods to generate green hydrogen from the water electrolysis technique. These methods differ by the electrolyte used to conduct ions (or charged atoms) in the system. Some methods use hydroxide ions ($OH^-$), others use hydrogen ions ($H^+$), and others use oxygen ions ($O^{2-}$). They are (i) alkaline water electrolysis (AWE), (ii) solid oxide water electrolysis (SOEC), (iii) proton exchange membrane water electrolysis (PEMWE), and (iv) anion-exchange membrane water electrolysis (AEMWE). Detailed working of these electrolyzer systems is shown in Figure 11.1.

## 11.2 PROTON EXCHANGE MEMBRANE WATER ELECTROLYZERS (PEMWE)

Among the four electrolysis techniques, the most commercially mature method of generating hydrogen are AWE and proton exchange membrane (PEM) water electrolysis. While both of these electrolysis techniques are being considered to produce high-purity hydrogen, we will focus on the PEM systems. PEM electrolysis originated when Grubb and Neidrach developed the technology at General Electric in the 1950s, and later it was sold to United Technologies. Initially, it was used primarily for oxygen generation and providing life support in closed environments such as space and submarines. PEM electrolysis was first used on a small scale for lab work and filling weather balloons. Later, larger systems were developed for cooling power plants and treating metals and glass.

At first, the system itself was the main cost factor, and customers had to pay extra for hydrogen with the same purity and quantity. So, there was little motivation to improve the electrolyzer stack. As the technology grew and could produce hundreds of kilograms of hydrogen per day, the stack became more important for reducing costs. Electricity costs also became a major factor. For large-scale hydrogen production, cost targets are stricter because it has to compete with cheaper hydrogen from steam methane reforming. Since PEM electrolysis and PEM fuel cells have developed differently, manufacturers are only now starting to apply fuel cell knowledge to electrolyzers. This creates opportunities to cut both capital and operating costs significantly.

PEM water electrolysis, with its numerous advantages such as high energy efficiency, high HER kinetics, compact design, and safe operation by eliminating the alkaline electrolyte, is considered the most efficient and environmentally friendly technology for the production of high-purity green

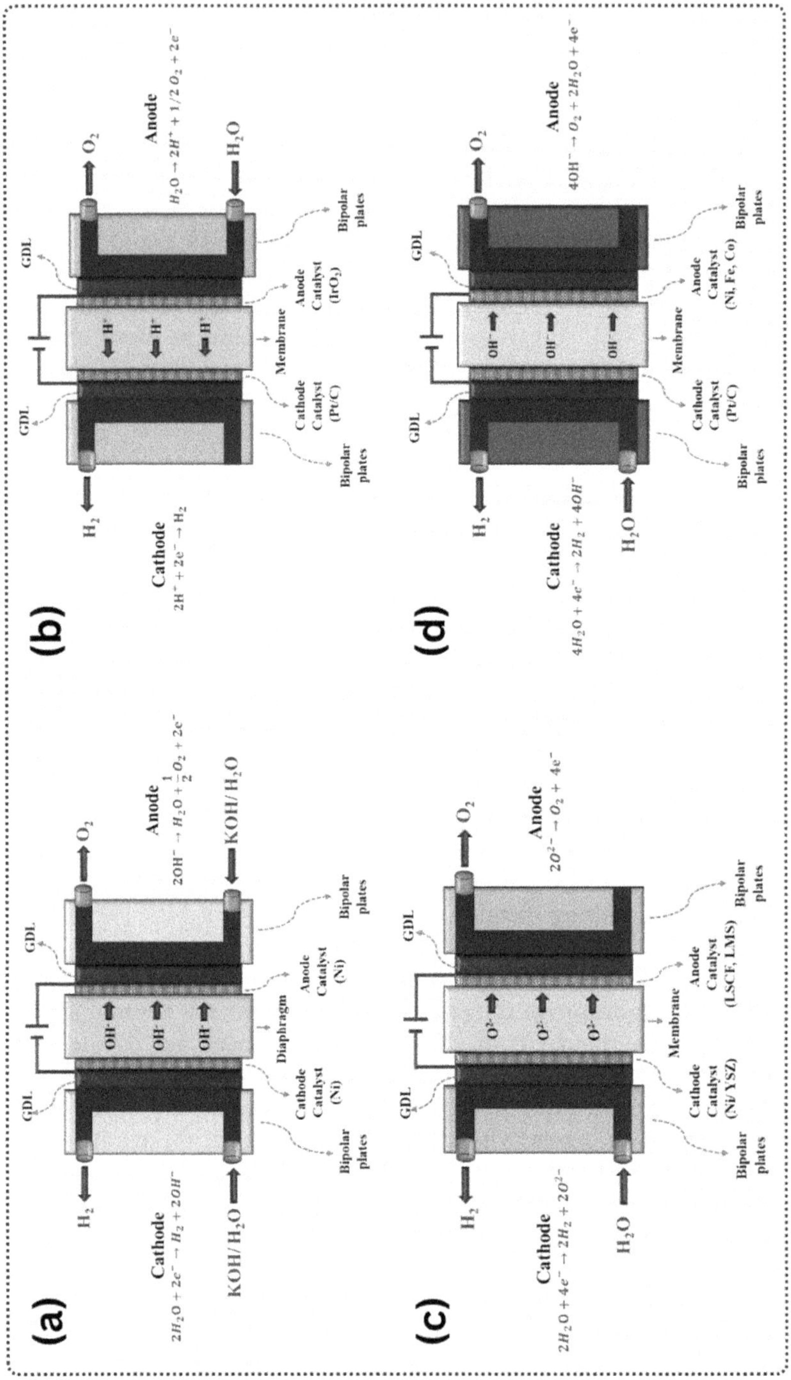


*Figure 11.1* Schematic configuration of water electrolyzers (a) AWE, (b) PEMWE, (c) SOEC, (d) AEMWE.

hydrogen. Additionally, PEMWE offers extended cycle life as its high ionic conductivity and current density prevent the formation of harmful substances such as carbonates. Due to all these attractive features, PEMWE stands as a promising technology toward achieving green energy aims. Therefore, several research organizations and industries globally are currently working on the development of enhanced PEMWEs and scale-up technologies.

### 11.2.1 Operation of PEMWE

A PEMWE cell comprising bipolar plates (BPs) having flow channels and a membrane electrode assembly (MEA) operates according to the principle illustrated in Figure 11.2. In PEMWE, the electrochemical water splitting involves two half-cell reactions: anodic HER and cathodic OER, resulting in the production of hydrogen and oxygen gases. Water is fed at the anode side of the cell, where it flows through the channels of the separator plates and PTLs, ultimately reaching the anodic catalyst layer. In the process, the application of a direct current voltage leads to the decomposition of water molecules into oxygen gas, electrons, and protons. Protons migrate from the anode side to the cathode side with the help of the proton-conducting and electron-insulating membrane (PEM). Whereas electrons are transported to the cathodic surface via the external power circuit, where protons and electrons combine to form hydrogen gas. The hydrogen produced at the cathode and the oxygen produced at the anode eventually leave the electrolysis cell through the flow channels of the BPs after passing through the electrocatalytic layers.

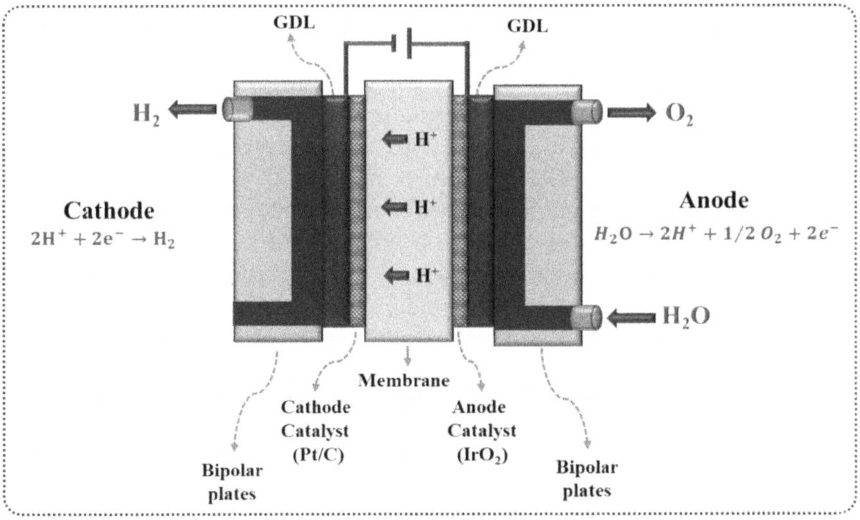


*Figure 11.2* Schematic illustration of the working principle of PEM water electrolysis.

## 11.3 KEY COMPONENTS IN PEMWE

### 11.3.1 Bipolar plates

BPs, also known as separator plates, are one of the most essential components of PEMWEs, serving multiple functions such as conducting electricity between adjacent cells in a stack, providing structural support to electrodes and membranes, distributing water evenly to reaction sites, and facilitating the removal of produced hydrogen and oxygen and managing heat generated during the electrolysis process through flow channels (Tijani et al., 2019). To achieve optimal performance in an electrolysis system, BPs must meet several key requirements. These characteristics include a lightweight structure and a low interfacial contact resistance (ICR) of less than 10 mΩ-cm$^2$, excellent corrosion resistance, high electrical conductivity, and gas impermeability. Additionally, BPs should be cost-effective and should offer robust mechanical strength (Tijani et al., 2019). Materials having such properties enhance the performance and reliability of PEM electrolytic cells by enabling the BPs to function effectively. Traditionally, titanium is the most widely used material due to its lightweight, strength, and corrosion resistance in acidic environments. However, titanium easily undergoes oxidation in the acidic environment of the cell, resulting in the formation of $TiO_2$ coatings and, eventually, leading to a decline in the BPs' conductivity and efficiency of water electrolysis (Babic et al., 2017; Tijani et al., 2019). Nevertheless, the high cost of Ti-based BPs, particularly those with Pt and Au coatings, which account for approximately 53% of the total electrolyzer stack cost, hinders their commercial use, making cost reduction a key area of research (Lee et al., 2019). Austenitic stainless steel (SS316 and SS403) is a viable substitute for titanium because of its excellent mechanical strength and affordable cost. However, because the anode operates under highly corrosive conditions, it is essential to apply a corrosion-resistant protective coating, such as due to Au, TiN, Au/Ni, SnO2, Ti/Pt and CrN on stainless-steel and titanium plates to ensure long-term stability (Babic et al., 2017; Cai et al., 2022; Tijani et al., 2019). Coating titanium and stainless-steel BPs using noble metals like Au and Pt enhances surface stability and prevents oxidation. Studies on gold-coated stainless-steel BPs, fabricated using selective laser melting and electroplating, showed low ICR (6.4 mΩ-cm$^2$) and good durability under conditions of 1.45 MPa pressure and a cell voltage of 1.71 V at a current density of 2 A/cm$^2$ (Babic et al., 2017; Tijani et al., 2019). Similarly, Pt/Ti coatings applied through vacuum plasma spraying and magnetron sputtering effectively protected stainless steel substrates and lowered ICR. However, in the absence of an intermediary Ti layer, pitting corrosion occurred (Cai et al., 2022). Despite their effectiveness, the high cost of noble metal coatings limits commercial viability, prompting research into cost-effective alternatives using non-noble materials (Cai et al., 2022). To develop an inexpensive alternative to commercially available Pt-coated Ti BPs, Hiroaki Wakayama et al. used $Ti_4O_7$-coated Ti, which exhibited a low contact resistance of 4–5 mΩ-cm$^2$ (Cai et al., 2022; Lee et al., 2019). Another approach studied by Laedre S. et al. involved a Ta–ITO (Tantalum-Indium Tin Oxide) bi-layer coating, which showed ICR values of 503 mΩ-cm$^2$ at 2.6 $V_{RHE}$, improving corrosion resistance while maintaining good electrical conductivity (Lee et al., 2019). Other coatings such as TiN, Nb, NiTiP, Ti–Ag, Ti–Ag–N, and (Ti, Zr) N have also been explored to enhance performance (Cai et al., 2022; Lee et al., 2019).

Furthermore, graphite-based materials are also used as BPs in small-scale PEM electrolyzers operating below 1.8 V. However, at higher voltages, they oxidize, producing $CO_2$ and reducing conductivity and electrocatalytic activity (Cai et al., 2022; Lee et al., 2019). Alternative materials like aluminum, stainless steel, and copper have been explored due to their low cost, but they are at risk of corrosion in oxidative anodic conditions (Lee et al., 2019). In addition to material selections, the design of flow channels in BPs plays an important role in maintaining the efficiency and overall cost of the electrolyzer stack. Flow channels, typically engraved on the surface of BPs, serve as a uniform distributor of reactant water and heat. Also, flow channels help to collect the produced gases from the electrodes. Various flow field designs, such as parallel, serpentine, point, porous, and others, have been developed to enhance the performance of PEMWEs (Babic et al., 2017; Lee et al., 2019). To investigate the impact of these designs, Toghyani et al. conducted a 3D numerical analysis using a 25 cm$^2$ active area to compare different configurations, including single-path, dual-path, triple-path, and quadruple-path serpentine designs, as well as straight parallel flow fields. Among these, the single-path serpentine design demonstrated the best performance in PEM water electrolysis systems (Figure 11.3b) (Lee et al., 2019; Toghyani et al., 2018).

## 11.4 GAS DIFFUSION LAYER

The gas diffusion layer (GDL), also known as the porous transport layer (PTL) or liquid-gas diffusion layer (LGDL), is a vital component in PEMWEs located between the catalyst layer (CL) and the BPs. It plays a crucial role in facilitating the transport of reactant water and product gases and ensuring electrical conductivity by connecting the electrode to current collectors. Along with this, it also provides mechanical support to the MEA, promotes uniform current distribution, and reduces contact resistance. Due to the mixed phase and harsh environment during water electrolysis, the GDL must exhibit high mechanical stability along with high electrical and thermal conductivity, excellent corrosion resistance, and optimized porosity (Gabbasa et al., 2014). Carbon-based materials like carbon felt, carbon cloth, and carbon paper, commonly used on the cathode side, are generally considered unsuitable for the anode side because they tend to undergo

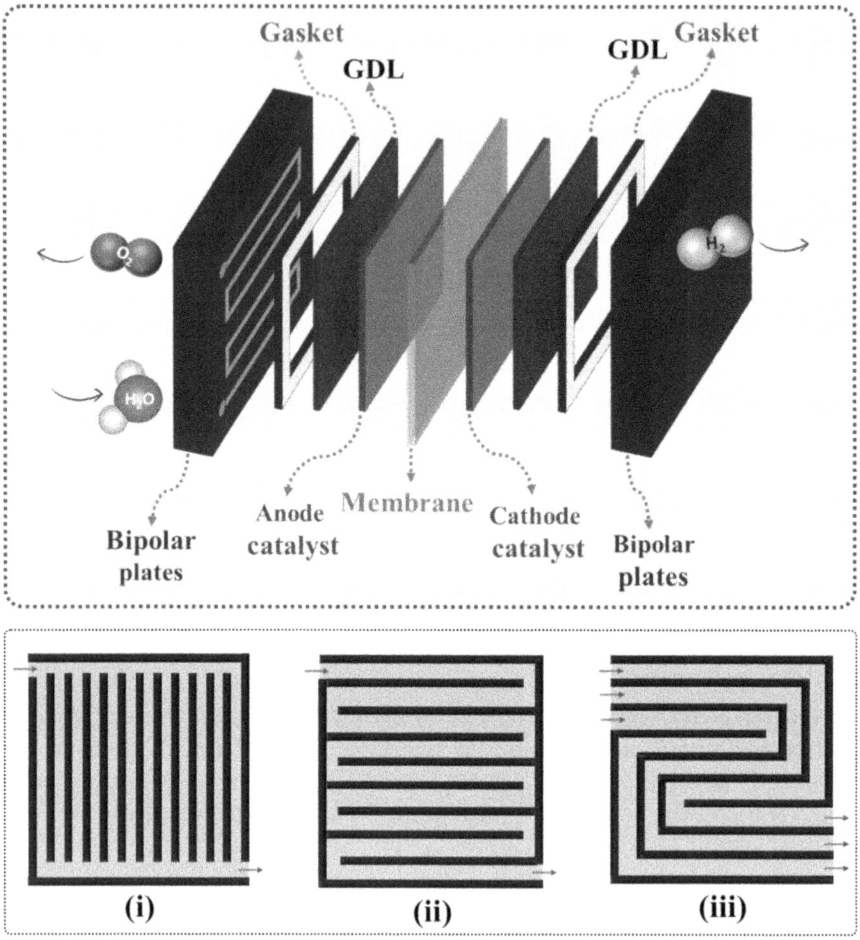


*Figure 11.3* (a) Schematic view of water electrolyzer cell layers (b) different flow field designs of bipolar plates (BPs) (i) parallel (ii) serpentine (iii) multiple channel serpentine.

corrosion and degradation (Beall et al., 2021; Doan et al., 2021). Therefore, due to high corrosion resistance and good mechanical and chemical stability, titanium-based materials like titanium mesh, felt, foam, and sintered powders are typically used on the anode side of the GDL (Beall et al., 2021; Doan et al., 2021; Schmidt et al., 2020). Moreover, to improve durability, GDLs are often coated with noble metals (e.g., Pt, Ir, Au) or treated via electrochemical nitridation or chemical etching (e.g., $TiN/TiO_x$, $TiH_x$) to suppress passivation and enhance conductivity (Z. Kang et al., 2017; Liu et al., 2021; Shapiro et al., 2005). Apart from the chosen material, the performance of a GDL largely depends on its structural parameters such as pore size, porosity, and thickness. Several articles reported that to provide a good balance between gas and water transport while minimizing mass transport and ohmic losses, optimal pore diameters should be in the range of 10–13 μm and porosities above 22–30% (Lala & Shahgaldi, 2025; Majasan et al., n.d.). Thickness also plays a crucial role in influencing both mechanical stability and mass transport. It has been found that thinner layers (e.g., 25 μm) can reduce ohmic resistance and enhance hydrogen production. However, excessively thin GDLs may lack durability and lead to poor mechanical stability. On the other hand, thicker layers (e.g., >300 μm) can manage gas transport better but may introduce higher resistance and reduced performance. Experimental studies have shown that optimizing the microstructure and morphology of GDLs can significantly improve overall electrolyzer efficiency and durability, especially under high current densities. Design advancements such as microporous layers (MPLs) or multilayer designs (ML-PTLs) have shown a significant effect on regulating reactant flow, improving long-term stability, and reducing pressure losses (Kang, Yang, Mo, Yu, et al., 2018; Kulkarni et al., 2022; Schuler et al., 2020). Tailored multilayer GDLs with gradient porosity and reduced surface roughness can enhance interfacial contact, limit membrane damage, and facilitate more uniform pressure distribution (Kulkarni et al., 2022). To further enhance the GDL's electrochemical stability and interfacial contact with catalyst layers, various surface engineering techniques such as coatings of noble metals like platinum (Pt), gold, and iridium (Ir) are employed (Z. Kang et al., 2017; Liu et al., 2021). Although these techniques are effective in suppressing titanium passivation and reducing interfacial resistance, they significantly increase the system's cost. Alternative approaches to improve resistance to corrosion and conductivity include the use of titanium nitrides and oxides formed via electrochemical nitridation or plasma sputtering. Passivation of titanium in the form of $TiO_2$ remains a significant concern, as it increases ICR and diminishes performance over time (Q. Feng et al., 2017; Rakousky et al., 2016; Shapiro et al., 2005).

GDLs' contribution to the overall cost of the PEMWE stack accounts for up to 17%, which has prompted the exploration of low-cost alternatives, including stainless steel substrates coated with corrosion-resistant layers like Nb/Ti or thin films of precious metals (Beall et al., 2021; Chinnaiah, Kannan, Palko, Grishina, Chen, et al., 2024b). Innovations in manufacturing techniques, such as tape casting and diffusion bonding, are also helping scale up the production of advanced GDLs with custom porosity and performance characteristics. Overall, the GDL is integral to ensuring efficient mass and charge transport, maintaining long-term operational stability, and achieving optimal performance in PEMWEs.

## 11.5 MEMBRANE

The proton exchange membrane (PEM) acts as a solid electrolyte in PEMWEs, and allows selective proton transport while blocking the crossover of hydrogen and oxygen gases. To ensure efficient, durable, and safe operation of PEMWEs, PEMs should possess a balanced set of physical, chemical, and electrochemical properties such as high proton conductivity, low gas crossover, excellent chemical and thermal stability, mechanical robustness under acidic and high-pressure operating conditions, and should be economically viable for commercial-scale deployment (Shin et al., 2017). Considering these essential qualities, perfluorosulfonic acid (PFSA)-based membranes, including Nafion®, Aciplex®, Flemion®, and Aquivion®, have emerged as the most widely adopted materials among the various PEMs developed to date (Salehmin et al., 2022; Shin et al., 2017). PFSA is a linear fluorocarbon polymer synthesized from tetrafluoroethylene and perfluoro-alkyl vinyl ethers as its primary monomers. The side chains terminate with sulfonic acid groups, which facilitate efficient proton conduction through the membrane when adequately hydrated (Kusoglu & Weber, 2017; Ren et al., 2024). While Nafion demonstrates great electrochemical qualities, it also has some drawbacks, including high cost, low heat stability over 90 °C, relatively high hydrogen permeability, and limited durability in long-term operation (Goñi-Urtiaga et al., 2012; Wei et al., 2010).

To address these limitations and to improve proton conductivity, reduce gas crossover, and enhance overall durability, various investigations have been conducted, which include the incorporation of inorganic fillers like $SiO_2$, $TiO_2$, and sulfated $ZrO_2$ to enhance water retention and mechanical integrity (Antonucci et al., 2008; Baglio et al., 2009; Siracusano et al., 2012, 2017). Composite or hybrid membranes can also be made with materials like bismuth oxide clusters, boron nitride, or graphene oxide functionalized with proton carriers (Liu et al., 2021; H. Wang, Sun, et al., 2021). Furthermore, biaxially stretched Nafion membranes have shown improved performance by forming oriented hydrophilic channels, while hybrid Nafion membranes with fillers like sulfated titania or $CeO_2$ have demonstrated enhanced proton transport and radical scavenging capabilities (Hrbek et al., 2020; Siracusano et al., 2017).

Additionally, hydrocarbon polymers are also considered as a potential membrane material due to their lower cost, tunable chemical structures, and environmental advantages (e.g., fluorine-free) (Deimede et al., 2014; Klose et al., 2020; Siracusano et al., 2013). Sulfonated hydrocarbon-based membranes exhibit favorable thermal and mechanical properties along with good proton conductivities due to the presence of sulfonic acid groups, which create hydrophilic domains that promote ion conduction (Daryaei et al., 2017; Smith et al., 2013). Excessive sulfonation, on the other hand, can cause unfavorable swelling, diminished mechanical integrity, and quick degradation over time, especially at high temperatures. Poly (ether ether ketone) (sPEEK), poly (arylene ether sulfone) (sPAES), polybenzimidazole (sPBI), and sulfonated poly (phenylene sulfone) (sPPS) are examples of sulfonated hydrocarbon-based membranes (Waribam et al., 2023; Wei et al., 2010).

To overcome the compromise in chemical and dimensional stability due to a higher degree of sulfonation (DS) value, composite membranes blending hydrocarbon polymers with inorganic or organic fillers have been developed.

For example, sPEEK blended with poly (ether sulfone) or embedded with MXene-$Cu_2O$ has shown improved antioxidant capability and mechanical strength (Waribam et al., 2023). Similarly, sulfonated poly (phenylene ether sulfone) reinforced with $CeO_2$ nanoparticles has demonstrated superior stability due to its radical scavenging function and interfacial toughness (Choi et al., 2022). With their reinforced architecture and reduced equivalent weight, short-side chain PFSA membranes such as Aquivion® have demonstrated encouraging gains in thermal stability and conductivity (Siracusano et al., 2014; Skulimowska et al., 2014). PTFE fiber-reinforced Aquivion® membranes demonstrated acceptable voltage degradation rates and high current densities in long-term durability testing (T. Li et al., 2020). However, their stability and performance still require validation under real PEMWE operating conditions in comparison to conventional Nafion membranes. In practical applications, the membrane must effectively prevent hydrogen permeation under varying conditions to avoid risks of flammability and performance loss. To avoid hydrogen permeation without compromising conductivity, different methods, including structural reinforcement and surface treatments, can be performed (Lee et al., 2022; Trinke et al., 2017). In addition, membranes must have low high-frequency resistance (HFR), good interfacial bonding with catalytic layers, and less gas crossover at high current densities and temperatures (Klose et al., 2020; Park et al., 2021).

Considering all of the aforementioned membrane characteristics, Nafion remains the benchmark for PEMs in electrolyzers. However, ongoing research into composite and hydrocarbon-based membranes highlights the potential for developing cost-effective, durable, and high-performance alternatives. For these membranes to be successfully used in commercial PEMWE systems, they need to have low hydrogen permeability, chemical resistance, mechanical and thermal durability, and strong proton conductivity. Therefore, future efforts should focus on optimizing membrane structure, sulfonation processes, and hybrid formulations, as well as long-term durability studies under realistic electrolysis conditions to advance the next generation of PEMs (Albert et al., 2015; Wei et al., 2010).

## 11.6 ELECTROCATALYSTS

Electrocatalysts are crucial components in determining the efficiency, cost, and long-term durability of PEMWEs. As discussed earlier, the process of water splitting involves two critical half-cell reactions: the OER at the anode and the HER at the cathode. The HER is a two-electron process, whereas the OER is a complex four-electron process and therefore poses challenges due to its sluggish kinetics and high overpotential requirements, limiting large-scale hydrogen production. A valid approach for advancing PEMWE technology is developing advanced electrocatalysts with enhanced kinetics and reduced overpotentials. A wide range of catalysts has been suggested for use in PEMWE, as listed in Tables 11.1 and 11.2.

*Table 11.1* Electrocatalysts suitable for OER

| *Sample* | *Electrolyte* | $\eta_{10}$ (*V* **vs.** *RHE)* | *Tafel slope (mV dec* $-^1$*)* | *Chronopotentiometry tests* | *Ref.* |
|---|---|---|---|---|---|
| **NiIrRuAl nanowires** | 0.1 M $HClO_4$ | 237 | 50 | 16 h at 5 mA $cm^{-2}$ | N. Liu et al. (2020) |
| **Ir- tungsten** | 0.1 M $HClO_4$ | | 42 | 100 h at 10.0 mA/$cm^2$ | Gao et al. (2020) |
| **Laminar Ir** | 0.1 M $HClO_4$ | 240 | 40.8 | 8 h at 2.5 mA $cm^{-2}$ | Pi et al. (2016) |
| **Co-RuIr** | 0.1 M $HClO_4$ | 235 | 66.9 | 25 h at 10.0 mA $cm^{-2}$ | Shan et al. (2019) |
| **Ir/g-$C_3N_4$/NG** | 0.5 M $H_2SO_4$ | 287 | 72.8 | | B. Jiang et al. (2018) |
| **Au@AuIr$_2$** | 0.5 M $H_2SO_4$ | 261 | 58.3 | Above 40 h at 10.0 mA $cm^{-2}$ | H. Wang, Chen, et al. (2021) |
| **IrCoNi/C** | 0.5 M $H_2SO_4$ | 303 | 53.8 | 1000 cycles at 2 mA $cm^{-2}$ | J. Feng et al. (2017) |
| **$RuO_2$-WC NPs** | 0.5 M $H_2SO_4$ | 347 | 88.5 | 10 h at 10.0 mA $cm^{-2}$ | S. C. Sun et al. (2022) |
| **Ir@WOxNR** | 0.5 M $H_2SO_4$ | 330 | 46.8–52.3 | 40 h from 10 to 100 mA $cm^{-2}$ | G. Jiang et al. (2021) |
| ***np*-IrAl** | 0.5 M $H_2SO_4$ | 240 | 40 | 30 h at 100.0 mA $cm^{-2}$ | Q. Li et al. (2020) |
| **RuIr$O_x$** | 0.5 M $H_2SO_4$ | 233 | 42 | 3000 CV cycles | Zhuang et al. (2019) |
| **Ir–Ni–Co oxide** | 0.1 M $HClO_4$ | 285 | 53 | 5.56 h at 10.0 mA $cm^{-2}$ | Zaman et al. (2017) |
| **$IrO_2$@Ir$O_x$** | 0.1 M $HClO_4$ | 309 | 45 | 6 h at 10.0 mA $cm^{-2}$ | Baik et al. (2021) |
| **$RuO_2$@Ir$O_x$** | 0.5 M $H_2SO_4$ | 215 | 72 | 300 h at 1000.0 mA $cm^{-2}$ | Lv et al. (2020) |
| **IrRu@Te** | 0.5 M $H_2SO_4$ | 220 | 35 | 20 h at 10.0 mA $cm^{-2}$ | J. Xu et al. (2020) |
| **$Ni_2$Ta** | 0.50 M $H_2SO_4$ | 570 | | 66 h at 10.0 mA $cm^{-2}$ | Mondschein et al. (2018) |
| **RuCu NSs/C-350 1C** | 0.50 M $H_2SO_4$ | 236 | 40.4 | 13.5 h at 5.0 mA $cm^{-2}$ | Yao et al. (2019) |
| **La-$RuO_2$** | 0.5 M $H_2SO_4$ | 208 | 57.4 | 28 h at 10.0 mA $cm^{-2}$ | Wu et al. (2022) |
| **Pt-$RuO_2$ HNSs** | 0.5 M $H_2SO_4$ | 228 | 51 | 100 h at 10.0 mA $cm^{-2}$ | J. Wang et al. (2022) |

| | | | | | |
|---|---|---|---|---|---|
| **Ru-SA/$Ti_3C_2$Tx** | 0.1 M $HClO_4$ | 290 | 37.9 | 32 h at 10.0 mA $cm^{-2}$ | Peng et al. (2020) |
| **$Ce_x$-$IrO_2$** | 0.5 M $H_2SO_4$ | 224 | 55.9 | 100 h at 10.0 mA $cm^{-2}$ | Y. Wang et al. (2020) |
| **$Fe_2O_3$ film** | 0.5 M $H_2SO_4$ | 650 | 56 | 24 h at 10.0 Ma $cm^{-2}$ | Kwong et al. (2018) |
| **Codoped $SrIrO_3$(6H)** | 0.1 M $HClO_4$ | 235 | 51.8 | 20 h at 10.0 mA $cm^{-2}$ | Yang et al. (2019) |
| **Cr0.6Ru0.4O2** | 0.5 M $H_2SO_4$ | 178 | 58 | 10 h at 10.0 mA $cm^{-2}$ | Runge et al. (2019) |
| **$RuO_2$/$Co_3O_4$–RuCo@NC** | 0.5 M $H_2SO_4$ | 247 | 89 | 8 h at 10.0 mA $cm^{-2}$ | Fan et al. (2019) |
| **Ni0.5Mn0.5Sb1.7Oy** | 1.0 M $H_2SO_4$ | 672 ± 9 | 60 ± 3 | 168 h at 10.0 mA $cm^{-2}$ | Moreno-Hernandez et al. (2017) |
| **IrCo NPs** | 0.5 M $H_2SO_4$ | 270 | 71.8 | 2 h at 1.58 V | X. Sun et al. (2019) |
| **$RuO_2$/C** | 0.5 M $H_2SO_4$ | 267 | 52.6 | 30 h at 1.5 V | Cao et al. (2019) |
| **Ir nanowires** | 0.5 M $HClO_4$ | 270 | 43.6 | 500 cycles at 10.0 mA $cm^{-2}$ | Fu et al. (2018) |
| **SrIrO** | 0.5 M $H_2SO_4$ | 263 | 49.7 | 2000 cycles at 10.0 mA $cm^{-2}$ | Zhang et al. (2021) |

*Table 11.2* Electrocatalysts suitable for HER

| *Sample* | *Electrolyte* | $\eta_{10}$ *(V vs. RHE)* | *Tafel slope (mV dec* $-^{1}$*)* | *Chronopotentiometry tests* | *Ref.* |
|---|---|---|---|---|---|
| **NiIrRuAl nanowires** | 0.1 M $HClO_4$ | 14 | 20 | 28 h at 5 mA $cm^{-2}$ | N. Liu et al. (2020) |
| **Co-RuIr** | 0.1 M $HClO_4$ | 14 | 31.1 | 25 h at 10 mA $cm^{-2}$ | Shan et al. (2019) |
| **Ir/g-$C_3N_4$/NG** | 0.5 M $H_2SO_4$ | 58 | 66 | 10 h at 10.0 mA $cm^{-2}$ | |
| **Ir nanowires** | 0.1 M $HClO_4$ | 24.6 | 29.8 | — | Fu et al. (2018) |
| **Ir/g-$C_3N_4$/NG** | 0.5 M $H_2SO_4$ | 22 | 22 | 2000 cycles at 10.0 mA $cm^{-2}$ | B. Jiang et al. (2018) |
| **Ir nanoparticles/C** | 0.5 M $H_2SO_4$ | 28 | 23.5 | — | F. Luo et al. (2020) |
| **WC@NC** | 0.5 M $H_2SO_4$ | 127 | 56.3 | 20 h at 200.0 mA $cm^{-2}$ | Q. Feng, Xiong, et al. (2019) |
| **$RuO_2$/$Co_3O_4$–RuCo@NC** | 0.5 M $H_2SO_4$ | 141 | 63 | 14 h at 10.0 mA $cm^{-2}$ | Fan et al. (2019) |
| **IrCo NPs** | 0.5 M $H_2SO_4$ | 23.9 | 25.7 | 5.56 h at -0.28 V | X. Sun et al. (2019) |
| **$IrP_2$-rGO** | 0.5 M $H_2SO_4$ | 8 | 30 | 20 h at 10.0 mA $cm^{-2}$ | Pu et al. (2020) |
| **$Co_3Mo_3N$** | 0.5 M $H_2SO_4$ | 108±8 | 79.8 | — | Y. Sun et al. (2022) |
| **$Fe_3P$** | 0.5 M $H_2SO_4$ | 49 | 57 | 20 h at an overpotential of 120 mV | Schipper et al. (2018) |
| **$Mo_3S_{13}$-NCNT** | 0.5 M $H_2SO_4$ | 188 | 40 | 100 h at 1000.0 mA $cm^{-2}$ | Holzapfel et al. (2020) |
| **V0.1Mo0.89S2—yNy** | 0.5 M $H_2SO_4$ | 51 | 36.44 | 10,000 cycles at 10 mV $s^{-1}$ | Bolar et al. (2021) |
| **$Ni_2P$/CNT** | 0.5 M $H_2SO_4$ | 124 | 53 | 500 cycles at 100 mV $s^{-1}$ | Pan et al. (2015) |

Electrocatalysts based on Pt group metals (PGMs), including Pt, Ru, Pd, and Ir, are widely employed for the HER process in PEMWEs, with common examples being Pt/C and Ir/C. In acidic conditions, Pt remains the benchmark catalyst due to its excellent catalytic activity and chemical stability. Pt-based catalysts, such as Pt/C, exhibit favorable kinetics and overpotentials as low as 25 mV at current densities of 10 mA $cm^{-2}$, with Tafel slopes of around 31 mV $dec^{-1}$ (Niu et al., 2022). Despite having exceptional catalytic activity and stability, Pt is not feasible for large-scale industrial applications due to its high cost and scarcity. Therefore, significant work has been done over the last few decades to lower operating costs by increasing Pt's durability and utilization efficiency. The use of single-atom Pt catalysts, Pt-based alloys (such as PtRu, PtNi), and advanced support materials like graphitic carbon nanofibers and reduced graphene oxide have been investigated to lower the Pt loading while preserving high performance (Cheng et al., 2020; Muntean et al., 2020; Rakočević et al., 2021). The goal of these modifications is to enhance charge transfer and increase the electrochemical surface area. Cobalt phosphide (CoP), molybdenum sulfides ($MoS_2$, $Mo_3S_{13}$), nickel-molybdenum alloys, and transition metal dichalcogenides are a few Pt-free HER catalysts that have shown better performance in efforts to move away from Pt. For example, CoP has demonstrated stable performance in both lab-scale and commercial-scale PEMWE cells with a current density of 1.86 A $cm^{-2}$ at 2.05 V over 1700 hours (King et al., 2019). Similarly, $Mo_3S_{13}$ supported on nitrogen-doped carbon nanotubes achieved 4 A $cm^{-2}$ at a cell voltage of 2.36 V, highlighting its high activity and long-term stability (Holzapfel et al., 2020).

The OER, on the other hand, is significantly more challenging due to its four-electron mechanism and sluggish kinetics. Therefore, to enhance the OER performance in PEM water electrolyzers, it is essential to use highly active catalysts that can lower the overpotential and accelerate reaction kinetics. Among the most effective catalyst materials currently available, Ir-based catalysts, particularly $IrO_2$ and amorphous $IrO_x$, are the most widely employed OER catalysts in PEMWE because of their stability in acidic environments (Spöri et al., 2017; Wang et al., 2019). These materials facilitate more efficient oxygen evolution by promoting faster electron transfer and reducing the energy required to drive the reaction. However, Ir's high cost and scarcity limit its usability and therefore lead to numerous strategies for reducing its loading while maintaining or enhancing performance. Among the many investigated methods, doping, alloying, core-shell nanostructures, and the application of conductive supports have all proven effective in enhancing the OER mechanism. For example, $IrO_x$ coatings on titanium supports and Ir nanoparticles supported on graphitic carbon nitride or nitrogen-doped carbon have demonstrated increased mass activities and decreased overpotentials (B. Jiang et al., 2018; Wang et al., 2020). Ru-Ir alloying has also been successful, with $Ir_{0.7}Ru_{0.3}O_x$ nanostructures outperforming pure $IrO_2$ in terms of durability and performance (Siracusano et al., 2015). Other materials, like perovskites and pyrochlores, have also drawn interest for OER catalysis. Ir-doped perovskites (e.g., Ir-STO, Ir-CCTO) have shown higher intrinsic activity and lower Ir content while maintaining structural stability (Alia et al., 2018; Liang et al., 2019). Pyrochlore-type oxides such as $Y_2Ru_2O_7$ and its doped derivatives have demonstrated fast charge transfer and high oxygen defect density, resulting in enhanced OER kinetics and stability in acidic conditions (Q. Feng, Wang, et al., 2019). Another class of advanced catalysts with high atom utilization and superior catalytic activity is single-atom catalysts (SACs), especially Ir atoms scattered on spinel supports (Xi et al., 2020).

Despite these advancements, one of the main challenges of the currently available catalysts is durability and cost. Electrocatalysts must withstand extremely acidic and oxidative conditions without losing their structural integrity over thousands of hours of operation. Therefore, to address this issue, Ir-based catalysts supported on robust structures such as $TiO_2$, $WO_x$, and MXenes have been explored in recent studies, and it was found that these catalysts enhance both performance and durability (Peng et al., 2020). Meanwhile, non-PGM substitutes such as $MnO_x$, $FeO_x$, $CoO_x$, and $NiO_x$ are still being studied for their potential in acidic OER, even though they still have issues with stability and surface reconstruction under operating circumstances (Chauhan et al., 2024; Chinnaiah, Kannan, Krishnamoorthi, et al., 2024a; Li et al., 2018; Mishra et al., 2025; Sharma et al., 2025).

In summary, Pt and Ir remain the most effective electrocatalysts for HER and OER, respectively, in PEMWEs. However, extensive efforts are being made to develop low-loading strategies and alternative materials with comparable activity and stability to reduce reliance on these scarce and expensive metals. Therefore, for PEMWE technology to achieve high efficiency, long-term durability, and cost-effectiveness at a commercial scale in the future, a continued innovation and optimization of these electrocatalysts are required.

## 11.7 MEMBRANE ELECTRODE ASSEMBLY

The MEA serves as the core component of PEMWEs. The MEA consists of the PEM, catalyst layers (CLs), and PTLs. In PEMWEs, electrochemical reactions take place at the MEA, and it also governs the efficiency, stability, and cost-effectiveness of the electrolyzer system. For MEA, two different types of configurations are available: the catalyst-coated membrane (CCM) and the porous transport electrode (PTE) or gas diffusion electrode (GDE) type. In the CCM configuration, catalyst inks prepared by adding ionomers are directly applied to both sides of the membrane using methods like spraying, decal transfer, or ultrasonic coating. This method contributes to reduced ohmic losses and improved durability by ensuring a small, low-resistance contact and strong mechanical stability (Su et al., 2013; Xie et al., 2021; Zhirov et al., 2012). Whereas in the PTE configuration, catalyst layers are directly applied onto PTLs, which are then laminated with the membrane through hot pressing. This method simplifies manufacturing and avoids PEM deformation. However, it often suffers from weak cohesion between the catalyst and membrane, leading to reduced proton transfer efficiency and mechanical stability (Bühler et al., 2019; Choe et al., 2018). Nevertheless, electrodeposition of $IrO_2$ layers and sputtering of Pt coatings have demonstrated significant improvements in durability and catalytic activity (Kim et al., 2021).

Apart from the configuration, ionomer content within the catalyst layer also plays a critical role as excessive ionomer leads to agglomeration and increased resistance, while too little results in poor dispersion and inhomogeneity (Bühler et al., 2019; Li et al., 2019). Another crucial factor that must be considered is catalyst loading, because overly thick CLs from high loadings can reduce catalyst utilization due to agglomeration and rough surfaces that hinder effective contact and increase mass transport losses (Z. Kang, Yang, Mo, Li, et al., 2018).

Furthermore, nanostructured catalysts such as nanowires and nanoarrays with underwater superaerophobicity have shown significant improvement in HER kinetics by enhancing bubble removal and increasing active site exposure (Darband et al., 2019). Experimental results show that smaller pore sizes and higher porosity (e.g., 50 μm with 40–60% porosity) of the PTL, particularly in the PTE configuration, lead to better mass activity and gas management (Yu et al., 2022).

MEAs in PEMWEs also face several degradation mechanisms due to catalyst agglomeration, PEM contamination by metal ions (e.g., $Na^{+}$, $Fe^{3+}$, $Cu^{2+}$), and gas-induced mechanical stresses (Khatib et al., 2019; Li et al., 2019; Rakousky et al., 2018). Recovery techniques like acid treatment can partially restore MEA performance, though they incur additional costs (Rakousky et al., 2018). Selecting appropriate ionomers, optimizing electrode structure, and employing protective coatings such as Ir or TiC layers can also improve MEA's performance (Kúš et al., 2016).

Ultimately, the performance of PEMWEs is closely based on the design and integrity of the MEA. A well-designed MEA must exhibit low proton transfer resistance, high catalyst utilization, strong cohesion between PEM and CLs, and a structure that supports abundant triple-phase boundaries (TPBs). It should also resist contamination, mechanical stress, and electrochemical degradation under extended operational periods. As the PEMWE market grows, continued development of MEAs using nanostructured catalysts, advanced deposition methods, and integrated manufacturing processes will be key to achieving durable, efficient, and cost-effective hydrogen production.

## 11.8 CHALLENGES AND PERSPECTIVES

Despite having several advantages for green hydrogen production, such as high energy efficiency, compact design, and fast response to intermittent renewable power, PEMWEs face significant challenges related to cost, durability, efficiency, and manufacturability, which hinder their commercialization and large-scale deployment. The cost of green hydrogen produced via PEMWE remains significantly higher than that produced via conventional methods due to expensive materials, complex system integration, and limited operational lifetimes. One of the primary causes of the high manufacturing costs in PEMWE is the high cost of the components, especially the catalysts, PEMs, PTLs, and BPs. Electrocatalyst materials like Ir and Pt, commonly used for OER and HER, respectively, are scarce and costly. Previous research has focused on replacing these materials with earth-abundant alternatives. However, these alternative electrocatalysts often lead to trade-offs in stability and efficiency. Therefore, the current research focuses on reducing noble metal loading, improving catalyst utilization, and developing advanced catalyst structures, such as atomically dispersed Ir/Ru sites, superstructured catalysts, and corrosion-resistant composite supports. Furthermore, computational screening and artificial intelligence (AI) are now employed to discover and optimize catalytic materials with high intrinsic activity and durability. In addition to the expensive electrocatalyst materials, the PEM is a major contributor to the high cost of PEMWE. And to reduce the cost, PEM requires significant optimization to enhance chemical, mechanical, and thermal stability under harsh electrochemical environments. Membranes must operate at elevated temperatures and differential pressures without compromising performance (Feng et al., 2022; Stoerzinger et al., 2014). Long-term degradation in PEMWE is contributed by issues such as radical attack, membrane thinning, and pinhole formation. The most widely accepted PEM is Nafion. However, to investigate its cost-effective replacement alternative membranes, such as hydrocarbon-based or composite PEMs with enhanced conductivity and thermal tolerance, are under investigation.

Furthermore, BPs and PTLs that are responsible for fluid management, gas transport, and current conduction also contribute significantly to the high cost of PEMWE. Titanium plates coated with noble metals are currently used to prevent corrosion and hydrogen embrittlement, but they contribute up to 53% of the total stack cost (Feng et al., 2022). Therefore, research is undergoing to replace titanium with corrosion-resistant, cost-effective alternatives such as 3D-printed non-metallic flow plates with optimized flow-field designs. For PTLs, enhancements in porosity, conductivity, and microstructural stability are necessary to improve mass and charge transport, reduce mechanical losses, and enable long-term operation.

In addition to cost-effective operation, PEMWE must demonstrate stable operation over 50,000–100,000 hours to meet industrial targets. Metal ion dissolution, MEA delamination, and catalyst aggregation are examples of degradation processes that need to be fully comprehended and controlled. It has been suggested that in-situ/operando diagnostics and advanced stress testing (AST) be used to forecast and prolong component lifetimes. Models that relate degradation rates to electrochemical performance, including those that take Ir dissolution or membrane thinning into consideration, are essential for calculating stack lifespans in real-world working scenarios.

To further reduce the cost of green hydrogen, innovations at the system level are required. New cell architectures, such as bubble-free electrode operation through capillary-induced water transport, have shown promise in reducing energy losses and simplifying the system balance of plant.

In conclusion, a complex strategy involving automated production, predictive modeling, system integration, and advanced material research is required for the commercial implementation of PEMWE. In addition to optimizing stability and efficiency at the material level, design and process improvements must be used to lower component and system costs. PEMWE technology has the potential to significantly contribute to global decarbonization initiatives by facilitating the production of green hydrogen at a reasonable cost and scale, provided that it continues to advance and research investments are made strategically.

## 11.9 ANION EXCHANGE MEMBRANE-BASED WATER ELECTROLYSIS

The acidic nature of PEMWE restricts the choice of electrocatalysts because only PGM is acceptable for use as electrodes in PEMWE. This makes PEMWE expensive and impedes its commercialization and large-scale deployment. These barriers can be addressed with anion-exchange membrane water electrolyzer (AEMWE), which has emerged as a next-generation technology. The anion exchange membrane-based water electrolysis has a combined effect of alkaline and PEM water electrolysis. Furthermore, an anion exchange membrane (AEM) cell uses non-noble catalyst materials, which lowers the cost of producing hydrogen. Due to the utilization of non-noble catalysts, AEM water electrolysis technology has become the current focus of many researchers. Figure 11.4 illustrates how an anion exchange membrane (AEM) separates $H_2$ and $O_2$ by conducting $OH^-$ ions.

The low-temperature AEMWE system operates through two half-cell reactions: HER at the cathode and OER at the anode. At the cathode, water molecules are reduced to produce hydrogen gas and hydroxide ions ($OH^-$), which then migrate through the AEM toward the anode under the influence of the positively charged anode. At the anode, these hydroxide ions are oxidized to generate oxygen gas, water, and electrons. The electrons released from this oxidation are transferred through an external circuit back to the cathode, completing the circuit. This setup enables the separate collection of hydrogen at the cathode and oxygen at the anode, preventing gas crossover within the reaction cell. To facilitate efficient ion transport and catalyst particle cohesion, particle-based catalytic layers are fabricated using anion

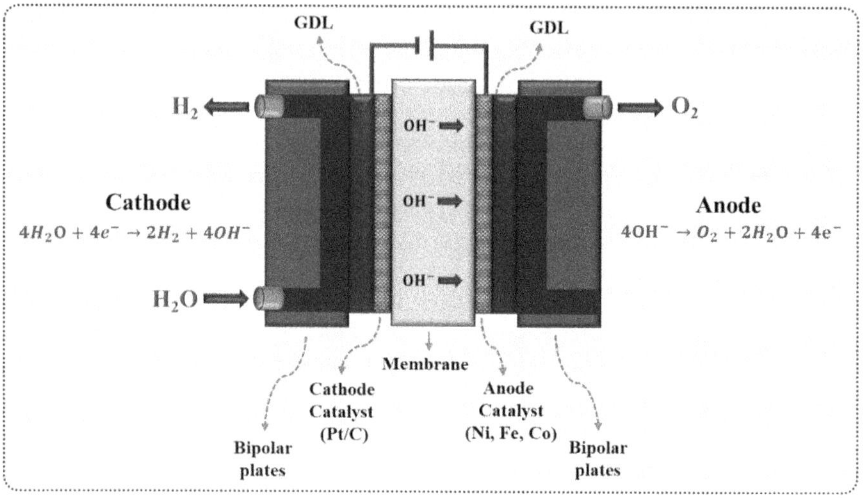


*Figure 11.4* Representation of anion exchange membrane water electrolysis (AEMWE).

exchange ionomers (AEI), which bind the catalyst particles and conduct $OH^-$ ions through the catalytic layer. The overall working principle of AEMWE is typically illustrated with a schematic representation, such as Figure 11.4.

The primary factor that limits the water electrolysis process is the sluggishness of the OER mechanism, which requires significant effort to overcome. Along with slow kinetics for OER or HER, stability, and high overpotential concerns, AEM must possess several properties such as high ionic conductivity (more than 100 mS/cm at the operating temperature), excellent alkaline stability, and good thermal and dimensional stability to develop highly efficient electrocatalysts (Kang et al., 2022).

## 11.10 KEY COMPONENTS IN THE AEMWE

### 11.10.1 Anion exchange membranes (AEMs)

The performance of AEMWE depends extensively on the AEM; thus, it should have strong ionic conductivity, chemical and electrochemical stability, thermal and mechanical robustness, and resistance to alkaline degradation while being cost-effective and sustainable (Dekel, 2018; Zakaria & Kamarudin, 2021). AEMs incorporate positively charged functional groups, such as quaternary ammonium, imidazolium, piperidinium, and phosphonium, attached to polymer backbones like poly (phenylene oxide), polysulfone, polybenzimidazole, and polystyrene for the transportation of $OH^-$ ions (Liu et al., 2019; F. Xu et al., 2021). Hydroxide conductivity values for modern AEMs have reached up to 300 mS $cm^{-1}$, surpassing the proton conductivity of PEMs (Zhegur-Khais et al., 2020). However, a major challenge for AEM durability is the degradation of these cationic groups through mechanisms such as Hofmann elimination and SN2 reactions, particularly under highly alkaline conditions.

There are several AEMs, including Fumatech Fumasep®, Tokuyama A201, Ionomer Aemion™, and Dioxide materials Sustainion®, which have shown promise and are commercially available. However, newer research-grade membranes, such as PiperION from Versogen, are being developed to improve conductivity and stability. But their commercial use is hindered by cost and scalability (Henkensmeier et al., 2021; Versogen, n.d.; Flatsome, n.d.).

Additionally, properties like gas and water permeability are critical, as excessive transport can lead to hydrogen crossover or cathode flooding. For example, FAA3 and Tokuyama A201 exhibit hydrogen permeabilities of $7.3 \times 10^{-17}$ and $5.3 \times 10^{-17}$ mol. $cm^{-1}.S^{-1}.Pa^{-1}$, respectively, under 1 M KOH at 50 °C (Gunasekara et al., 2012; Höfner, n.d.). These characteristics, along with mechanical traits like tensile strength and elongation at break, define the membrane's suitability for durable, high-performance AEMWE systems (Konovalova et al., 2018; Luo et al., 2018). Ultimately, optimizing the balance between conductivity, stability, and mechanical integrity and the development of efficient AEMs will pave the way for large-scale, low-cost electrolyzers.

## 11.11 CATALYSTS

The alkaline environment permits the use of low-cost transition metal catalysts. The development and use of non-PGMs catalysts are essential for lowering the cost of AEMWE systems. The main challenge for this catalyst is the optimization of its chemical composition, catalytic activity, and stability. These non-PGM catalysts have lower mass-specific activity than PGMs, which need high catalyst loading, resulting in a higher ohmic loss. From the literature, it is also known that using a PGM catalyst, the HER kinetics are sluggish in alkaline conditions compared to acidic conditions (Ferriday et al., 2021). Here we have discussed the PGM and non-PGMs electrocatalyst AEMWE (Choe et al., 2014).

### 11.11.1 PGM electrocatalyst for HER/OER in AEMWE

In recent years, significant research has explored the impact of PGM catalysts—such as Pt, palladium (Pd), and ruthenium (Ru)—on the HER in alkaline exchange membrane water electrolysis (AEMWE) systems. Pt remains the benchmark for HER due to its outstanding catalytic activity and low overpotential requirements. Many studies have explained high current densities and long-term durability using Pt-based catalysts, often paired with $IrO_2$ for the OER, achieving stable performance at higher temperatures (Park et al., 2023; Wang et al., 2024).

Efforts to reduce costs and dependence on Pt have led to investigations into Pd- and Ru-based catalysts. Pd supported on $CeO_2$ has shown better HER performance compared to commercial Pt/C due to strong metal-metal oxide interactions. Ru-based catalysts, including Ru/$TiO_2$ and Ru-Ni alloys, have also demonstrated promising activity and stability, with some achieving overpotentials as low as 15 mV for 10 mA-$cm^{-2}$ current density. Despite their promising activity, the high cost and varying abundance of These Pd and Ru have promising activity but they are costly, limiting their scalability. To address these limitations, strategies such as nanoparticle engineering, porous structure development, and single-atom catalysts have been employed to maximize PGM utilization while maintaining high activity. Alloying PGMs with earth-abundant metals is another promising approach. However, the search for truly cost-effective and sustainable alternatives continues, with ongoing research into PGM-free catalysts as the ultimate goal for scalable green hydrogen production (Wang et al., 2024).

On the other hand, for OER performance, the use of PGMs such as Pt, Ru, and Ir will increase the performance due to their catalytic behavior. The most used catalysts in PGMs, especially Ir, are widely used for their superior OER catalytic performance in alkaline media, significantly reducing overpotentials and enhancing efficiency. In much research, it is discussed that Ir-based catalysts show high current densities and good compatibility with various AEMs, especially at high temperatures and KOH (alkaline) concentrations. However, the limited availability and high cost of Ir have attracted research toward other cost-effective alternatives like Ru. Ru-based catalysts, including Ru-alloys and Ru supported on carbon or metal oxides, are promising due to their high activity and lower cost. Alloying Ru with transition metals like nickel (Ni), cobalt (Co), or iron (Fe) has improved both stability and performance. Structural optimization strategies such as core-shell architectures and nanoparticle design have also been utilized to enhance PGM durability and reduce usage. Despite progress, challenges remain in catalyst degradation and economic viability, but PGMs, especially Ir and Ru, continue to be the most effective OER catalysts in AEMWE for commercial hydrogen production (Wang et al., 2024).

### 11.11.2 Non-PGM electrocatalyst for HER/OER in AEMWE

The development of non-PGM catalysts for the OER in AEMWEs is pivotal for a cost-effective and scalable hydrogen production. Conventional catalysts like Pt and Ru offer excellent activity and durability, but their widespread use is hindered by high costs and scarcity. As a result, research has shifted toward non-PGM catalysts, especially transition metal-based catalysts such as Ni, Fe, and Co, which offer a more affordable solution along with promising catalytic performance in alkaline environments. These aspects are discussed in detail below:

Ni-Based Catalysts: High-surface-area Ni and Ni-alloy catalysts have been central to traditional AWE. Ni alloys like NiFe and NiCoFe have demonstrated notable OER activity, particularly when freshly synthesized. However, those incorporating aluminum (Al) and molybdenum (Mo) encounter stability challenges under high-potential, high-pH conditions (Wang et al., 2024).

Nano-Structured Catalysts: Ni-based catalysts with nanostructured features have exhibited remarkably high OER mass activities, largely attributed to their enhanced surface area. For example, a catalyst composed of a ~50 nm Fe core coated with a Ni shell of 5 nm exhibited a remarkable mass activity of 1795 A/g was achieved at an overpotential of 0.35 V in 1 M KOH (Chinnaiah, Kannan, Palko, Grishina, Guganathan, et al., 2024c; Xiao et al., 2021). FeNi Catalysts: Non-PGM FeNi catalysts have shown improved electrolysis performance in AEMWE due to improved catalyst utilization and more efficient mass transport (Tong & Chen, 2024).

CoCu-Based Catalysts: Cobalt-copper (CoCu)-based catalysts have garnered interest due to their improved OER activity in AEMWE systems. Cu-doped $Co_3O_4$ catalysts have demonstrated promising cell voltages and current densities, although the mechanism behind the improvement remains under investigation (Chauhan et al., 2017; Li & Li, 2023).

High Entropy Alloys (HEAs): Amorphous and crystalline FeCoNiSiB-based HEAs have been developed as OER electrodes, with the amorphous variants offering higher OER activity due to the diversity of crystalline structures.

As discussed above, PGMs like Pt, Ir, Ru, and Pb are expensive as well as deficient, and can be replaced with earth-abundant transition metals such as Ni, Fe, and Co as a potential solution.

Ni stands out for its good HER activity in alkaline conditions. However, due to its limited intrinsic catalytic activity, it is often alloyed with elements like Fe, Mo, P, or V, forming compounds such as NiFe, NiMo, Ni-P, and Ni-V, which improve activity, stability, and corrosion resistance.

Several studies have explored various electrocatalysts, such as Vanadium-doped $Ni_3N$ nanosheets, which have enhanced HER by promoting $H_2O$ adsorption and facilitating heterolytic cleavage. Ni@Ni core-shell structures modified with CoP or Pt improved HER performance, particularly in saline electrolytes. $FeNi_3/FeNiO_x$ nanoparticles demonstrated better OER activity than $RuO_2$ but still lagged behind Pt/C-$RuO_2$ for HER in terms of overpotential. NiMo films on various substrates (e.g., Ni foam, carbon paper, stainless steel mesh) were evaluated, but Pt/C still showed superior current density at lower overpotentials. Additional research into metal phosphides, sulfides, and carbides has shown potential for cost-effective HER catalysts (Malaj et al., 2024; Varadarajan et al., 2024).

Despite these advances, non-PGM catalysts face challenges, particularly in terms of long-term stability and performance under commercial-scale conditions. Ongoing research efforts are aimed at bridging the performance gap between the PGM-based and non-PGM-based electrocatalysts for AEMWE applications.

## 11.12 ELECTROLYTES

Electrolytes play a crucial role in AEMWE. They enable efficient $OH^-$ ion transport and maintain the pH balance with minimum membrane degradation. The selection of electrolytes impacts the performance of the system, durability, and efficiency. Studies have shown that KOH is the most effective alkaline electrolyte, especially in dry-cathode configurations. It offers low overpotential, high conductivity, and reduced high-frequency resistance. DI water and other alkali hydroxides like NaOH and LiOH show variable performance. However, LiOH often increases overpotential due to $OH^-$ stabilization. The use of near-neutral pH electrolytes can reduce corrosion, but faces limitations in conductivity and stability. Saline solutions like $NaClO_4$ and $KNO_3$ can enhance performance but require optimized concentrations. Additionally, $K_2CO_3$ and $KHCO_3$ allow moderate performance but lag behind KOH (Rossi et al., 2023; Tricker et al., 2023).

The development of stable, non-precious metal catalysts and advanced characterization tools for AEMWE is always a challenge.

## 11.13 CHALLENGES AND PERSPECTIVES

AEMWE is considered a promising technology for $H_2$ production, as it combines the benefits of both AWE and PEMWE, enabling a more environmentally friendly synthesis process and reduced operating costs. Despite its potential, AEMWE remains in the early stages of development and needs further exploration. AEMWE technology currently faces two key challenges: enhancing the stability and durability of AEMs and developing high-performance, non-PGM catalysts for both HER and OER. Addressing these drawbacks will help them to withstand high temperatures and pressure conditions while maintaining strong and high catalytic performance over extended periods.

To enhance the performance of AEMWE, it is crucial to optimize the non-PGM catalyst for both HER and OER. Additionally, improving AEM stability requires a deeper understanding of interactions between hydroxide ion conductivity, polymer backbone structures, and functional groups, especially under high temperature and pressure conditions that often degrade membranes. Organic-inorganic composite AEMs offer superior stability and conductivity but remain underexplored beyond single-cell tests. Preventing membrane plasticization through hydrophobic, rigid polymer structures and incorporating small amounts of 2D nanomaterials can improve mechanical strength and performance. Additionally, transition metal doping can overcome slow HER kinetics, and Li-based alkaline catholytes enhance the system's durability and efficiency.

## 11.14 CONCLUSION

For a sustainable energy future, development and deployment of green hydrogen technologies are essential. PEMWE emerges as one of the most promising solutions, which offers distinct advantages such as high hydrogen purity, rapid response to dynamic power inputs from renewables, compact system design, and superior energy efficiency. Despite these advantages, several challenges, particularly related to cost, durability, and scalability, remain a concerning point. PEMWEs' high performance and high cost are caused by essential parts such as electrocatalysts, PEMs, GDLs, and BPs. Although coated titanium components and catalysts based on noble metals offer superior conductivity and stability, their cost and limited availability hinder large-scale adoption. Therefore, ongoing research is focused on finding low-cost alternatives, including non-precious metal catalysts, composite membranes, and coated stainless steel materials. Furthermore, some other methods like microstructural designing, advanced coatings, and additive manufacturing are also helping reduce interfacial resistance and enhance long-term performance.

Simultaneously, anion exchange membrane water electrolysis (AEMWE), a next-generation option that promises cheaper capital costs by using earth-abundant materials and non-corrosive electrolytes, is also gaining attention. AEMWE still has durability problems, though, especially with membrane stability at high pH levels.

Overall, while PEMWE remains the most promising option for immediate commercialization due to its technological maturity and reliability, both PEMWE and AEMWE need to make concentrated efforts in material innovation, system integration, and cost reduction strategies to fulfill their potential in the green hydrogen economy. With continued interdisciplinary research, suitable material selection, and process optimization, water electrolysis can become a cornerstone of decarbonized global energy systems.

## REFERENCES

Albert, A., Barnett, A. O., Thomassen, M. S., Schmidt, T. J., & Gubler, L. (2015). Radiation-grafted polymer electrolyte membranes for water electrolysis cells: Evaluation of key membrane properties. *ACS Applied Materials and Interfaces*, 7(40), 22203–22212. https://doi.org/10.1021/acsami.5b04618

Alia, S. M., Shulda, S., Ngo, C., Pylypenko, S., & Pivovar, B. S. (2018). Iridium-based nanowires as highly active, oxygen evolution reaction electrocatalysts. *ACS Catalysis*, *8*(3), 2111–2120. https://doi.org/10.1021/acscatal.7b03787

Antonucci, V., Di Blasi, A., Baglio, V., Ornelas, R., Matteucci, F., Ledesma-Garcia, J., Arriaga, L. G., & Aricò, A. S. (2008). High temperature operation of a composite membrane-based solid polymer electrolyte water electrolyser. *Electrochimica Acta*, *53*(24), 7350–7356. https://doi.org/10.1016/j.electacta.2008.04.009

Babic, U., Suermann, M., Büchi, F. N., Gubler, L., & Schmidt, T. J. (2017). Critical review—Identifying critical gaps for polymer electrolyte water electrolysis development. *Journal of The Electrochemical Society*, *164*(4), F387–F399. https://doi.org/10.1149/2.1441704jes

Baglio, V., Ornelas, R., Matteucci, F., Martina, F., Ciccarella, G., Zama, I., Arriaga, L. G., Antonucci, V., & Aricó, A. S. (2009). Solid polymer electrolyte water electrolyser based on Nafion-$TiO_2$ composite membrane for high temperature operation. *Fuel Cells*, *9*(3), 247–252. https://doi.org/10.1002/fuce.200800063

Baik, C., Lee, S. W., & Pak, C. (2021). Glycine-induced ultrahigh-surface-area $IrO_2$@IrOx catalyst with balanced activity and stability for efficient water splitting. *Electrochimica Acta*, *390*. https://doi.org/10.1016/j.electacta.2021.138885

Beall, C. E., Fabbri, E., & Schmidt, T. J. (2021). Perovskite oxide based electrodes for the oxygen reduction and evolution reactions: The underlying mechanism. *ACS Catalysis*, *11*(5), 3094–3114. https://doi.org/10.1021/acscatal.0c04473

Bolar, S., Shit, S., Murmu, N. C., Samanta, P., & Kuila, T. (2021). Activation strategy of $MoS_2$ as her electrocatalyst through doping-induced lattice strain, band gap engineering, and active crystal plane design. *ACS Applied Materials and Interfaces*, *13*(1), 765–780. https://doi.org/10.1021/acsami.0c20500

Bühler, M., Hegge, F., Holzapfel, P., Bierling, M., Suermann, M., Vierrath, S., & Thiele, S. (2019). Optimization of anodic porous transport electrodes for proton exchange membrane water electrolyzers. *Journal of Materials Chemistry A*, *7*(47), 26984–26995. https://doi.org/10.1039/c9ta08396k

Cai, Z., Wang, P., Zhang, J., Chen, A., Zhang, J., Yan, Y., & Wang, X. (2022). Reinforced layered double hydroxide oxygen-evolution electrocatalysts: A polyoxometallic acid wet-etching approach and synergistic mechanism. *Advanced Materials*, *34*(26). https://doi.org/10.1002/adma.202110696

Cao, L., Luo, Q., Chen, J., Wang, L., Lin, Y., Wang, H., Liu, X., Shen, X., Zhang, W., Liu, W., Qi, Z., Jiang, Z., Yang, J., & Yao, T. (2019). Dynamic oxygen adsorption on single-atomic Ruthenium catalyst with high performance for acidic oxygen evolution reaction. *Nature Communications*, *10*(1). https://doi.org/10.1038/s41467-019-12886-z

Chauhan, M., Bangwal, A. S., Borodianskiy, K., Mishra, K. K., & Singh, P. (2024). Electrochemical behavior of nickel oxide nanoelectrocatalyst in diverse electrolytic environments and its application as an effective electrode for supercapacitors. *Electrochimica Acta*, *473*. https://doi.org/10.1016/j.electacta.2023.143496

Chauhan, M., Reddy, K. P., Gopinath, C. S., & Deka, S. (2017). Copper Cobalt Sulfide nanosheets realizing a promising electrocatalytic oxygen evolution reaction. *ACS Catalysis*, *7*(9), 5871–5879. https://doi.org/10.1021/acscatal.7b01831

Chen, Y., Liu, C., Xu, J., Xia, C., Wang, P., Xia, B. Y., Yan, Y., & Wang, X. (2023). Key components and design strategy for a proton exchange membrane water electrolyzer. In *Small Structures* (Vol. 4, Issue 6). John Wiley and Sons Inc. https://doi.org/10.1002/sstr.202200130

Cheng, Q., Hu, C., Wang, G., Zou, Z., Yang, H., & Dai, L. (2020). Carbon-defect-driven electroless deposition of Pt atomic clusters for highly efficient hydrogen evolution. *Journal of the American Chemical Society*, *142*(12), 5594–5601. https://doi.org/10.1021/jacs.9b11524

Chinnaiah, K., Kannan, K., Krishnamoorthi, R., Palko, N., & Gurushankar, K. (2024a). Nanostructured Ag/NiO composites for supercapacitor and antibacterial applications, and in-silico theoretical investigation. *Journal of Physics and Chemistry of Solids*, *184*. https://doi.org/10.1016/j.jpcs.2023.111730

Chinnaiah, K., Kannan, K., Palko, N., Grishina, M., Chen, Y. S., & Gurushankar, K. (2024b). Exploring the potential of Withania somnifera-mediated Ag/$Mn_3O_4$ nanocomposites as electrode material for high-performance supercapattery device. *Journal of the Taiwan Institute of Chemical Engineers*, *157*. https://doi.org/10.1016/j.jtice.2024.105441

Chinnaiah, K., Kannan, K., Palko, N., Grishina, M., Guganathan, L., & Gurushankar, K. (2024c). Bioengineered Ag/NiO nanocomposites as advanced battery-supercapacitor electrodes for highly efficient symmetric hybrid devices. *Ionics*, *30*(3), 1691–1707. https://doi.org/10.1007/s11581-023-05361-9

Choe, S., Lee, B. S., Cho, M. K., Kim, H. J., Henkensmeier, D., Yoo, S. J., Kim, J. Y., Lee, S. Y., Park, H. S., & Jang, J. H. (2018). Electrodeposited $IrO_2$/Ti electrodes as durable and cost-effective anodes in high-temperature polymer-membrane-electrolyte water electrolyzers. *Applied Catalysis B: Environmental*, *226*, 289–294. https://doi.org/10.1016/j.apcatb.2017.12.037

Choe, Y. K., Fujimoto, C., Lee, K. S., Dalton, L. T., Ayers, K., Henson, N. J., & Kim, Y. S. (2014). Alkaline stability of benzyl trimethyl ammonium functionalized polyaromatics: A computational and experimental study. *Chemistry of Materials*, *26*(19), 5675–5682. https://doi.org/10.1021/cm502422h

Choi, S., Shin, S. H., Lee, D. H., Doo, G., Lee, D. W., Hyun, J., Lee, J. Y., & Kim, H. T. (2022). Enhancing the durability of hydrocarbon-membrane-based polymer electrolyte water electrolysis using a radical scavenger-embedded interlocking interfacial layer. *Journal of Materials Chemistry A*, *10*(2), 789–798. https://doi.org/10.1039/d1ta08222a

Darband, G. B., Aliofkhazraei, M., & Shanmugam, S. (2019). Recent advances in methods and technologies for enhancing bubble detachment during electrochemical water splitting. In *Renewable and Sustainable Energy Reviews* (Vol. 114). Elsevier Ltd. https://doi.org/10.1016/j.rser.2019.109300

Daryaei, A., Miller, G. C., Willey, J., Choudhury, S. R., Vondrasek, B., Kazerooni, D., Burtner, M. R., Mittelsteadt, C., Lesko, J. J., Riffle, J. S., & McGrath, J. E. (2017). Synthesis and membrane properties of Sulfonated Poly(arylene ether sulfone) statistical copolymers for electrolysis of water: Influence of meta- and para-substituted comonomers. *ACS Applied Materials and Interfaces*, *9*(23), 20067–20075. https://doi.org/10.1021/acsami.7b02401

Deimede, V., Labou, D., & Neophytides, S. G. (2014). Polymer electrolyte membranes based on blends of sulfonated polysulfone and PEO-grafted polyethersulfone for low temperature water electrolysis. *Journal of Applied Polymer Science*, *131*(4). https://doi.org/10.1002/app.39922

Dekel, D. R. (2018). Review of cell performance in anion exchange membrane fuel cells. *Journal of Power Sources*, *375*, 158–169. https://doi.org/10.1016/j.jpowsour.2017.07.117

Doan, T. L., Lee, H. E., Shah, S. S. H., Kim, M. J., Kim, C. H., Cho, H. S., & Kim, T. (2021). A review of the porous transport layer in polymer electrolyte membrane water electrolysis. In *International Journal of Energy Research* (Vol. 45, Issue 10, pp. 14207–14220). John Wiley & Sons Ltd. https://doi.org/10.1002/er.6739

Fan, Z., Jiang, J., Ai, L., Shao, Z., & Liu, S. (2019). Rational design of ruthenium and cobalt-based composites with rich metal-insulator interfaces for efficient and stable overall water splitting in acidic electrolyte. *ACS Applied Materials and Interfaces*, *11*(51), 47894–47903. https://doi.org/10.1021/acsami.9b15844

Feng, J., Lv, F., Zhang, W., Li, P., Wang, K., Yang, C., Wang, B., Yang, Y., Zhou, J., Lin, F., Wang, G. C., & Guo, S. (2017). Iridium-based multimetallic porous hollow nanocrystals for efficient overall-water-splitting catalysis. *Advanced Materials*, *29*(47). https://doi.org/10.1002/adma.201703798

Feng, Q., Wang, Q., Zhang, Z., Xiong, Y., Li, H., Yao, Y., Yuan, X. Z., Williams, M. C., Gu, M., Chen, H., Li, H., & Wang, H. (2019). Highly active and stable ruthenate pyrochlore for enhanced oxygen evolution reaction in acidic medium electrolysis. *Applied Catalysis B: Environmental*, *244*, 494–501. https://doi.org/10.1016/j.apcatb.2018.11.071

Feng, Q., Xiong, Y., Xie, L., Zhang, Z., Lu, X., Wang, Y., Yuan, X. Z., Fan, J., Li, H., & Wang, H. (2019). Tungsten carbide encapsulated in grape-like N-doped carbon nanospheres: One-step facile synthesis for low-cost and highly active electrocatalysts in proton exchange membrane water electrolyzers. *ACS Applied Materials and Interfaces*, *11*(28), 25123–25132. https://doi.org/10.1021/acsami.9b04725

Feng, Q., Yuan, X. Z., Liu, G., Wei, B., Zhang, Z., Li, H., & Wang, H. (2017). A review of proton exchange membrane water electrolysis on degradation mechanisms and mitigation strategies. In *Journal of Power Sources* (Vol. 366, pp. 33–55). Elsevier B.V. https://doi.org/10.1016/j.jpowsour.2017.09.006

Feng, Y. Y., Si, S., Deng, G., Xu, Z. X., Pu, Z., Hu, H. S., & Wang, C. Bin. (2022). Copper-doped ruthenium oxide as highly efficient electrocatalysts for the evolution of oxygen in acidic media. *Journal of Alloys and Compounds*, *892*. https://doi.org/10.1016/j.jallcom.2021.162113

Ferriday, T. B., Middleton, P. H., & Kolhe, M. L. (2021). Review of the hydrogen evolution reaction—A basic approach. In *Energies* (Vol. 14, Issue 24). MDPI. https://doi.org/10.3390/en14248535

Flatsome. (n.d.). Hello world. https://www.xergyincstore.com/anionic-membranes/

Fu, L., Yang, F., Cheng, G., & Luo, W. (2018). Ultrathin Ir nanowires as high-performance electrocatalysts for efficient water splitting in acidic media. *Nanoscale*, *10*(4), 1892–1897. https://doi.org/10.1039/c7nr09377b

Gabbasa, M., Sopian, K., Fudholi, A., & Asim, N. (2014). A review of unitized regenerative fuel cell stack: Material, design and research achievements. *International Journal of Hydrogen Energy*, *39*(31), 17765–17778. https://doi.org/10.1016/j.ijhydene.2014.08.121

Gao, J., Huang, X., Cai, W., Wang, Q., Jia, C., & Liu, B. (2020). Rational design of an Iridium-Tungsten composite with an Iridium-rich surface for acidic water oxidation. *ACS Applied Materials and Interfaces*, *12*(23), 25991–26001. https://doi.org/10.1021/acsami.0c05906

Goñi-Urtiaga, A., Presvytes, D., & Scott, K. (2012). Solid acids as electrolyte materials for proton exchange membrane (PEM) electrolysis: Review. In *International Journal of Hydrogen Energy* (Vol. 37, Issue 4, pp. 3358–3372). Elsevier Ltd. https://doi.org/10.1016/j.ijhydene.2011.09.152

Gunasekara, I., Lee, M., Abbott, D., & Mukerjee, S. (2012). Mass transport and oxygen reduction kinetics at an anion exchange membrane interface: Microelectrode studies on effect of carbonate exchange. *ECS Electrochemistry Letters*, *1*(2). https://doi.org/10.1149/2.007202eel

Henkensmeier, D., Najibah, M., Harms, C., Žitka, J., Hnát, J., & Bouzek, K. (2021). Overview: State-of-the art commercial membranes for anion exchange membrane water electrolysis. *Journal of Electrochemical Energy Conversion and Storage*, *18*(2). https://doi.org/10.1115/1.4047963

Höfner, T. (n.d.). *Entwicklung von Membran-Elektroden-Einheiten für die Anionenaustauschmembran-Wasserelektrolyse Development of Membrane Electrode Assemblies for the Anion Exchange Membrane Water Electrolysis.*

Holzapfel, P. K. R., Bühler, M., Escalera-López, D., Bierling, M., Speck, F. D., Mayrhofer, K. J. J., Cherevko, S., Pham, C. V., & Thiele, S. (2020). Fabrication of a Robust PEM water electrolyzer based on non-noble metal cathode catalyst: $[Mo_3S_{13}]2-$ Clusters anchored to N-doped carbon nanotubes. *Small*, *16*(37). https://doi.org/10.1002/smll.202003161

Hou, J., & Yang, M. (n.d.). *Green Hydrogen Production by Water Electrolysis.*

Hrbek, T., Kúš, P., Yakovlev, Y., Nováková, J., Lobko, Y., Khalakhan, I., Matolín, V., & Matolínová, I. (2020). Sputter-etching treatment of proton-exchange membranes: Completely dry thin-film approach to low-loading catalyst-coated membranes for water electrolysis. *International Journal of Hydrogen Energy*, *45*(41), 20776–20786. https://doi.org/10.1016/j.ijhydene.2020.05.245

Jiang, B., Wang, T., Cheng, Y., Liao, F., Wu, K., & Shao, M. (2018). Ir/g-$C_3N_4$/Nitrogen-doped graphene nanocomposites as bifunctional electrocatalysts for overall water splitting in acidic electrolytes. *ACS Applied Materials and Interfaces*, *10*(45), 39161–39167. https://doi.org/10.1021/acsami.8b11970

Jiang, G., Yu, H., Li, Y., Yao, D., Chi, J., Sun, S., & Shao, Z. (2021). Low-loading and highly stable membrane electrode based on an Ir@WOxNR ordered array for PEM water electrolysis. *ACS Applied Materials and Interfaces*, *13*(13), 15073–15082. https://doi.org/10.1021/acsami.0c20791

Kang, S. Y., Park, J. E., Jang, G. Y., Kim, O. H., Kwon, O. J., Cho, Y. H., & Sung, Y. E. (2022). High-performance and durable water electrolysis using a highly conductive and stable anion-exchange membrane. *International Journal of Hydrogen Energy*, *47*(15), 9115–9126. https://doi.org/10.1016/j.ijhydene.2022.01.002

Kang, Z., Mo, J., Yang, G., Li, Y., Talley, D. A., Retterer, S. T., Cullen, D. A., Toops, T. J., Brady, M. P., Bender, G., Pivovar, B. S., Green, J. B., & Zhang, F. Y. (2017). Thin film surface modifications of thin/tunable liquid/gas diffusion layers for high-efficiency proton exchange membrane electrolyzer cells. *Applied Energy*, *206*, 983–990. https://doi.org/10.1016/j.apenergy.2017.09.004

Kang, Z., Yang, G., Mo, J., Li, Y., Yu, S., Cullen, D. A., Retterer, S. T., Toops, T. J., Bender, G., Pivovar, B. S., Green, J. B., & Zhang, F. Y. (2018). Novel thin/tunable gas diffusion electrodes with ultra-low catalyst loading for hydrogen evolution reactions in proton exchange membrane electrolyzer cells. *Nano Energy*, *47*, 434–441. https://doi.org/10.1016/j.nanoen.2018.03.015

Kang, Z., Yang, G., Mo, J., Yu, S., Cullen, D. A., Retterer, S. T., Toops, T. J., Brady, M. P., Bender, G., Pivovar, B. S., Green, J. B., & Zhang, F. Y. (2018). Developing titanium micro/nano porous layers on planar thin/tunable LGDLs for high-efficiency hydrogen production. *International Journal of Hydrogen Energy*, *43*(31), 14618–14628. https://doi.org/10.1016/j.ijhydene.2018.05.139

Kannan, K., & Tari, V. (Eds.). (2025). *Novel Energy Storage and Conversion Technologies for Two-Dimensional MXenes and MBenes*. https://www.amazon.com/dp/9798369395462

Khatib, F. N., Wilberforce, T., Ijaodola, O., Ogungbemi, E., El-Hassan, Z., Durrant, A., Thompson, J., & Olabi, A. G. (2019). Material degradation of components in polymer electrolyte membrane (PEM)electrolytic cell and mitigation mechanisms: A review. *Renewable and Sustainable Energy Reviews*, *111*, 1–14. https://doi.org/10.1016/j.rser.2019.05.007

Kim, H., Kim, J., Kim, J., Han, G. H., Guo, W., Hong, S., Park, H. S., Jang, H. W., Kim, S. Y., & Ahn, S. H. (2021). Dendritic gold-supported iridium/iridium oxide ultra-low loading electrodes for high-performance proton exchange membrane water electrolyzer. *Applied Catalysis B: Environmental*, *283*. https://doi.org/10.1016/j.apcatb.2020.119596

King, L. A., Hubert, M. K. A., Capuano, C., Manco, J., Danilovic, N., Valle, E., Hellstern, T. R., Ayers, K., & Jaramillo, T. F. (2019). A non-precious metal hydrogen catalyst in a commercial polymer electrolyte membrane electrolyser. *Nature Nanotechnology*, *14*(11), 1071–1074. https://doi.org/10.1038/s41565-019-0550-7

Klose, C., Saatkamp, T., Münchinger, A., Bohn, L., Titvinidze, G., Breitwieser, M., Kreuer, K. D., & Vierrath, S. (2020). All-hydrocarbon MEA for PEM water electrolysis combining low hydrogen crossover and high efficiency. *Advanced Energy Materials*, *10*(14). https://doi.org/10.1002/aenm.201903995

Konovalova, A., Kim, H., Kim, S., Lim, A., Park, H. S., Kraglund, M. R., Aili, D., Jang, J. H., Kim, H. J., & Henkensmeier, D. (2018). Blend membranes of polybenzimidazole and an anion exchange ionomer (FAA3) for alkaline water electrolysis: Improved alkaline stability and conductivity. *Journal of Membrane Science*, *564*, 653–662. https://doi.org/10.1016/j.memsci.2018.07.074

Kulkarni, D., Huynh, A., Satjaritanun, P., O'Brien, M., Shimpalee, S., Parkinson, D., Shevchenko, P., DeCarlo, F., Danilovic, N., Ayers, K. E., Capuano, C., & Zenyuk, I. V. (2022). Elucidating effects of catalyst loadings and porous transport layer morphologies on operation of proton exchange membrane water electrolyzers. *Applied Catalysis B: Environmental*, *308*. https://doi.org/10.1016/j.apcatb.2022.121213

Kúš, P., Ostroverkh, A., Ševčíková, K., Khalakhan, I., Fiala, R., Skála, T., Tsud, N., & Matolin, V. (2016). Magnetron sputtered Ir thin film on TiC-based support sublayer as low-loading anode catalyst for proton exchange membrane water electrolysis. *International Journal of Hydrogen Energy*, *41*(34), 15124–15132. https://doi.org/10.1016/j.ijhydene.2016.06.248

Kusoglu, A., & Weber, A. Z. (2017). New insights into perfluorinated sulfonic-acid ionomers. *Chemical Reviews*, *117*(3), 987–1104. https://doi.org/10.1021/acs.chemrev.6b00159

Kwong, W. L., Lee, C. C., Shchukarev, A., Björn, E., & Messinger, J. (2018). High-performance iron (III) oxide electrocatalyst for water oxidation in strongly acidic media. *Journal of Catalysis*, *365*, 29–35. https://doi.org/10.1016/j.jcat.2018.06.018

Lala, S. R. F., & Shahgaldi, S. (2025). Mass and charge transport phenomena in porous transport layer for proton exchange membrane water electrolyzers: A review. In *Energy Reports* (Vol. 13, pp. 162–183). Elsevier Ltd. https://doi.org/10.1016/j.egyr.2024.11.089

Lee, B., Lee, H., Cho, H. S., Cho, W. C., Kim, C. H., & Lim, H. (2019). Projected economic outlook and scenario analysis for $H_2$ production by alkaline water electrolysis on the basis of the unit electricity price, the learning rate, and the automation level. *Sustainable Energy and Fuels*, *3*(7), 1799–1807. https://doi.org/10.1039/c9se00148d

Lee, C. J., Song, J., Yoon, K. S., Rho, Y., Yu, D. M., Oh, K. H., Lee, J. Y., Kim, T. H., Hong, Y. T., Kim, H. J., Yoon, S. J., & So, S. (2022). Controlling hydrophilic channel alignment of perfluorinated sulfonic acid membranes via biaxial drawing for high performance and durable polymer electrolyte membrane water electrolysis. *Journal of Power Sources*, *518*. https://doi.org/10.1016/j.jpowsour.2021.230772

Li, A., Sun, Y., Yao, T., & Han, H. (2018). Earth-abundant transition-metal-based electrocatalysts for water electrolysis to produce renewable hydrogen. In *Chemistry - A European Journal* (Vol. 24, Issue 69, pp. 18334–18355). Wiley-VCH Verlag. https://doi.org/10.1002/chem.201803749

Li, H., & Li, G. (2023). Novel palladium-based nanomaterials for multifunctional ORR/OER/HER electrocatalysis. In *Journal of Materials Chemistry A* (Vol. 11, Issue 17, pp. 9383–9400). Royal Society of Chemistry. https://doi.org/10.1039/d3ta01059g

Li, N., Araya, S. S., & Kær, S. K. (2019). Long-term contamination effect of iron ions on cell performance degradation of proton exchange membrane water electrolyser. *Journal of Power Sources*, *434*. https://doi.org/10.1016/j.jpowsour.2019.226755

Li, Q., Li, J., Xu, J., Zhang, N., Li, Y., Liu, L., Pan, D., Wang, Z., & Deepak, F. L. (2020). Ultrafine-grained porous Ir-based catalysts for high-performance overall water splitting in acidic media. *ACS Applied Energy Materials*, *3*(4), 3736–3744. https://doi.org/10.1021/acsaem.0c00201

Li, T., Shen, J., Chen, G., Guo, S., & Xie, G. (2020). Performance comparison of proton exchange membrane fuel cells with Nafion and Aquivion Perfluorosulfonic acids with different equivalent weights as the electrode binders. *ACS Omega*, *5*(28), 17628–17636. https://doi.org/10.1021/acsomega.0c02110

Liang, X., Shi, L., Liu, Y., Chen, H., Si, R., Yan, W., Zhang, Q., Li, G., Yang, L., & Zou, X. (2019). Activating inert, nonprecious perovskites with Iridium dopants for efficient oxygen evolution reaction under acidic conditions. *Angewandte Chemie*, *131*(23), 7713–7717. https://doi.org/10.1002/ange.201900796

Liu, B., Hu, B., Du, J., Cheng, D., Zang, H. Y., Ge, X., Tan, H., Wang, Y., Duan, X., Jin, Z., Zhang, W., Li, Y., & Su, Z. (2021). Precise molecular-level modification of Nafion with Bismuth oxide clusters for high-performance proton-exchange membranes. *Angewandte Chemie - International Edition*, *60*(11), 6076–6085. https://doi.org/10.1002/anie.202012079

Liu, C., Shviro, M., Gago, A. S., Zaccarine, S. F., Bender, G., Gazdzicki, P., Morawietz, T., Biswas, I., Rasinski, M., Everwand, A., Schierholz, R., Pfeilsticker, J., Müller, M., Lopes, P. P., Eichel, R. A., Pivovar, B., Pylypenko, S., Friedrich, K. A., Lehnert, W., & Carmo, M. (2021). Exploring the interface of skin-layered titanium fibers for electrochemical water splitting. *Advanced Energy Materials*, *11*(8). https://doi.org/10.1002/aenm.202002926

Liu, N., Zhang, Y., Ma, W., Zhang, Z., Yin, K., Si, C., & Kou, T. (2020). Hierarchically porous nickel-iridium-ruthenium-aluminum alloys with tunable compositions and electrocatalytic activities towards the oxygen/hydrogen evolution reaction in acid electrolyte. *Journal of Materials Chemistry A*, *8*(13), 6245–6255. https://doi.org/10.1039/d0ta00445f

Liu, X., Li, Y., Xue, J., Zhu, W., Zhang, J., Yin, Y., Qin, Y., Jiao, K., Du, Q., Cheng, B., Zhuang, X., Li, J., & Guiver, M. D. (2019). Magnetic field alignment of stable proton-conducting channels in an electrolyte membrane. *Nature Communications*, *10*(1). https://doi.org/10.1038/s41467-019-08622-2

Luo, F., Guo, L., Xie, Y., Xu, J., Qu, K., & Yang, Z. (2020). Iridium nanorods as a robust and stable bifunctional electrocatalyst for pH-universal water splitting. *Applied Catalysis B: Environmental*, *279*. https://doi.org/10.1016/j.apcatb.2020.119394

Luo, X., Wright, A., Weissbach, T., & Holdcroft, S. (2018). Water permeation through anion exchange membranes. *Journal of Power Sources*, *375*, 442–451. https://doi.org/10.1016/j.jpowsour.2017.05.030

Lv, H., Wang, S., Li, J., Shao, C., Zhou, W., Shen, X., Xue, M., & Zhang, C. (2020). Self-assembled $RuO_2$@IrOx core-shell nanocomposite as high efficient anode catalyst for PEM water electrolyzer. *Applied Surface Science*, *514*. https://doi.org/10.1016/j.apsusc.2020.145943

Majasan, J. O., Iacoviello, F., Cho, J. I. S., Maier, M., Lu, X., Neville, T. P., Dedigama, I., Shearing, P. R., & Brett, D. J. L. (n.d.). *Correlative Study of Microstructure and Performance for Porous Transport Layers in Polymer Electrolyte Membrane Water Electrolysers by X-ray Computed Tomography and Electrochemical Characterization.*

Malaj, F., Tampucci, A., Lentini, D., Brogi, L., Berretti, E., Coletti, C., Forti, S., Rossi, A., & Santoro, C. (2024). One-pot synthesis of $FeNi_3$/FeNiOx nanoparticles for PGM-free anion exchange membrane water electrolysis. *Electrochimica Acta*, *507*. https://doi.org/10.1016/j.electacta.2024.145109

Mishra, K. K., Chauhan, M., Singh, S., & Singh, P. (2025). Catalytic performance of hydrothermally processed NiO nanoparticles in environmentally sustainable electrolytic medium. *Fuel*, *396*. https://doi.org/10.1016/j.fuel.2025.135313

Mondschein, J. S., Kumar, K., Holder, C. F., Seth, K., Kim, H., & Schaak, R. E. (2018). Intermetallic $Ni_2Ta$ electrocatalyst for the oxygen evolution reaction in highly acidic electrolytes. *Inorganic Chemistry*, *57*(10), 6010–6015. https://doi.org/10.1021/acs.inorgchem.8b00503

Moreno-Hernandez, I. A., Macfarland, C. A., Read, C. G., Papadantonakis, K. M., Brunschwig, B. S., & Lewis, N. S. (2017). Crystalline nickel manganese antimonate as a stable water-oxidation catalyst in aqueous 1.0 M $H_2SO_4$. *Energy and Environmental Science*, *10*(10), 2103–2108. https://doi.org/10.1039/c7ee01486d

Muntean, R., Pascal, D. T., Rost, U., Mărginean, G., Brodmann, M., & Vaszilcsin, N. (2020). Synthesis and characterisation of platinum–cobalt–manganese ternary alloy catalysts supported on carbon nanofibers: An alternative catalyst for hydrogen evolution reaction. *International Journal of Hydrogen Energy*, *45*(49), 26217–26225. https://doi.org/10.1016/j.ijhydene.2020.02.041

Niu, H., Xia, C., Huang, L., Zaman, S., Maiyalagan, T., Guo, W., You, B., & Xia, B. Y. (2022). Rational design and synthesis of one-dimensional platinum-based nanostructures for oxygen-reduction electrocatalysis. In *Chinese Journal of Catalysis* (Vol. 43, Issue 6, pp. 1459–1472). Science Press. https://doi.org/10.1016/S1872-2067(21)63862-7

Pan, Y., Hu, W., Liu, D., Liu, Y., & Liu, C. (2015). Carbon nanotubes decorated with nickel phosphide nanoparticles as efficient nanohybrid electrocatalysts for the hydrogen evolution reaction. *Journal of Materials Chemistry A*, *3*(24), 13087–13094. https://doi.org/10.1039/c5ta02128f

Park, J. E., Kim, J., Han, J., Kim, K., Park, S. Bin, Kim, S., Park, H. S., Cho, Y. H., Lee, J. C., & Sung, Y. E. (2021). High-performance proton-exchange membrane water electrolysis using a sulfonated poly(arylene ether sulfone) membrane and ionomer. *Journal of Membrane Science*, *620*. https://doi.org/10.1016/j.memsci.2020.118871

Park, S. Bin, Park, J. E., Na, G., Choi, C., Cho, Y. H., & Sung, Y. E. (2023). Low-cost and high-performance anion-exchange membrane water electrolysis stack using non-noble metal-based materials. *ACS Applied Energy Materials*, *6*(17), 8738–8748. https://doi.org/10.1021/acsaem.3c01215

Peng, X., Zhao, S., Mi, Y., Han, L., Liu, X., Qi, D., Sun, J., Liu, Y., Bao, H., Zhuo, L., Xin, H. L., Luo, J., & Sun, X. (2020). Trifunctional single-atomic Ru sites enable efficient overall water splitting and oxygen reduction in acidic media. *Small*, *16*(33). https://doi.org/10.1002/smll.202002888

Pi, Y., Zhang, N., Guo, S., Guo, J., & Huang, X. (2016). Ultrathin laminar Ir superstructure as highly efficient oxygen evolution electrocatalyst in broad pH range. *Nano Letters*, *16*(7), 4424–4430. https://doi.org/10.1021/acs.nanolett.6b01554

Pu, Z., Liu, T., Zhao, W., Shi, X., Liu, Y., Zhang, G., Hu, W., Sun, S., & Liao, S. (2020). Versatile route to fabricate precious-metal phosphide electrocatalyst for acid-stable hydrogen oxidation and evolution reactions. *ACS Applied Materials and Interfaces*, *12*(10), 11737–11744. https://doi.org/10.1021/acsami.9b23426

Rakočević, L., Srejić, I., Maksić, A., Golubović, J., & Štrbac, S. (2021). Hydrogen evolution on reduced graphene oxide-supported pdau nanoparticles. *Catalysts*, *11*(4). https://doi.org/10.3390/catal11040481

Rakousky, C., Keeley, G. P., Wippermann, K., Carmo, M., & Stolten, D. (2018). The stability challenge on the pathway to high-current-density polymer electrolyte membrane water electrolyzers. *Electrochimica Acta*, *278*, 324–331. https://doi.org/10.1016/j.electacta.2018.04.154

Rakousky, C., Reimer, U., Wippermann, K., Carmo, M., Lueke, W., & Stolten, D. (2016). An analysis of degradation phenomena in polymer electrolyte membrane water electrolysis. *Journal of Power Sources*, *326*, 120–128. https://doi.org/10.1016/j.jpowsour.2016.06.082

Ren, K., Liu, X., & Rao, Z. (2024). Effect of temperature and external electric field on proton transport in ordered and amorphous proton exchange membranes: A molecular dynamics study. *International Journal of Hydrogen Energy*, *54*, 1422–1435. https://doi.org/10.1016/j.ijhydene.2023.11.152

Rossi, R., Taylor, R., & Logan, B. E. (2023). Increasing the electrolyte salinity to improve the performance of anion exchange membrane water electrolyzers. *ACS Sustainable Chemistry and Engineering*, *11*(23), 8573–8579. https://doi.org/10.1021/acssuschemeng.3c01245

Runge, J., Bathiany, S., Bollt, E., Camps-Valls, G., Coumou, D., Deyle, E., Glymour, C., Kretschmer, M., Mahecha, M. D., Muñoz-Marí, J., van Nes, E. H., Peters, J., Quax, R., Reichstein, M., Scheffer, M., Schölkopf, B., Spirtes, P., Sugihara, G., Sun, J., … Zscheischler, J. (2019). Inferring causation from time series in Earth system sciences. *Nature Communications*, *10*(1). https://doi.org/10.1038/s41467-019-10105-3

Salehmin, M. N. I., Husaini, T., Goh, J., & Sulong, A. B. (2022). High-pressure PEM water electrolyser: A review on challenges and mitigation strategies towards green and low-cost hydrogen production. In *Energy Conversion and Management* (Vol. 268). Elsevier Ltd. https://doi.org/10.1016/j.enconman.2022.115985

Santos, D. M. F., Sequeira, C. A. C., & Figueiredo, J. L. (2013). Hydrogen production by alkaline water electrolysis. In *Quim. Nova* (Vol. 36, Issue 8).

Schipper, D. E., Zhao, Z., Thirumalai, H., Leitner, A. P., Donaldson, S. L., Kumar, A., Qin, F., Wang, Z., Grabow, L. C., Bao, J., & Whitmire, K. H. (2018). Effects of catalyst phase on the hydrogen evolution reaction of water splitting: Preparation of phase-pure films of FeP, $Fe_2P$, and $Fe_3P$ and their relative catalytic activities. *Chemistry of Materials*, *30*(10), 3588–3598. https://doi.org/10.1021/acs.chemmater.8b01624

Schmidt, G., Suermann, M., Bensmann, B., Hanke-Rauschenbach, R., & Neuweiler, I. (2020). Modeling overpotentials related to mass transport through porous transport layers of PEM water electrolysis cells. *Journal of The Electrochemical Society*, *167*(11), 114511. https://doi.org/10.1149/1945-7111/aba5d4

Schuler, T., Ciccone, J. M., Krentscher, B., Marone, F., Peter, C., Schmidt, T. J., & Büchi, F. N. (2020). Hierarchically structured porous transport layers for polymer electrolyte water electrolysis. *Advanced Energy Materials*, *10*(2). https://doi.org/10.1002/aenm.201903216

Shan, J., Ling, T., Davey, K., Zheng, Y., & Qiao, S. Z. (2019). Transition-metal-doped ruir bifunctional nanocrystals for overall water splitting in acidic environments. *Advanced Materials*, *31*(17). https://doi.org/10.1002/adma.201900510

Shapiro, D., Duffy, J., Kimble, M., & Pien, M. (2005). Solar-powered regenerative PEM electrolyzer/fuel cell system. *Solar Energy*, *79*(5), 544–550. https://doi.org/10.1016/j.solener.2004.10.013

Sharma, M., Mishra, K. K., Singh, S., Singh, S. K., & Singh, P. (2025). Lattice strain engineering in $Ni_3TeO_6$ nanostructure for enhanced oxygen evolution reaction in alkaline medium. *International Journal of Hydrogen Energy*, *132*, 116–129. https://doi.org/10.1016/j.ijhydene.2025.04.424

Shin, D. W., Guiver, M. D., & Lee, Y. M. (2017). Hydrocarbon-based polymer electrolyte membranes: Importance of morphology on ion transport and membrane stability. In *Chemical Reviews* (Vol. 117, Issue 6, pp. 4759–4805). American Chemical Society. https://doi.org/10.1021/acs.chemrev.6b00586

Siracusano, S., Baglio, V., Lufrano, F., Staiti, P., & Aricò, A. S. (2013). Electrochemical characterization of a PEM water electrolyzer based on a sulfonated polysulfone membrane. *Journal of Membrane Science*, *448*, 209–214. https://doi.org/10.1016/j.memsci.2013.07.058

Siracusano, S., Baglio, V., Navarra, M. A., Panero, S., Antonucci, V., & Aricò, A. S. (2012). Investigation of composite Nafion/Sulfated Zirconia membrane for solid polymer electrolyte electrolyzer applications. *The International Journal of Electrochemical Science*, 7. www.electrochemsci.org

Siracusano, S., Baglio, V., Nicotera, I., Mazzapioda, L., Aricò, A. S., Panero, S., & Navarra, M. A. (2017). Sulfated titania as additive in Nafion membranes for water electrolysis applications. *International Journal of Hydrogen Energy*, *42*(46), 27851–27858. https://doi.org/10.1016/j.ijhydene.2017.05.136

Siracusano, S., Baglio, V., Stassi, A., Merlo, L., Moukheiber, E., & Arico, A. S. (2014). Performance analysis of short-side-chain Aquivion® perfluorosulfonic acid polymer for proton exchange membrane water electrolysis. *Journal of Membrane Science*, *466*, 1–7. https://doi.org/10.1016/j.memsci.2014.04.030

Siracusano, S., Van Dijk, N., Payne-Johnson, E., Baglio, V., & Aricò, A. S. (2015). Nanosized IrOx and IrRuOx electrocatalysts for the $O_2$ evolution reaction in PEM water electrolysers. In *Applied Catalysis B: Environmental* (Vol. 164, pp. 488–495). Elsevier. https://doi.org/10.1016/j.apcatb.2014.09.005

Skulimowska, A., Dupont, M., Zaton, M., Sunde, S., Merlo, L., Jones, D. J., & Rozière, J. (2014). Proton exchange membrane water electrolysis with short-side-chain Aquivion® membrane and $IrO_2$ anode catalyst. *International Journal of Hydrogen Energy*, *39*(12), 6307–6316. https://doi.org/10.1016/j.ijhydene.2014.02.082

Smith, D. W., Oladoyinbo, F. O., Mortimore, W. A., Colquhoun, H. M., Thomassen, M. S., Ødegård, A., Guillet, N., Mayousse, E., Klicpera, T., & Hayes, W. (2013). A microblock ionomer in proton exchange membrane electrolysis for the production of high purity hydrogen. *Macromolecules*, *46*(4), 1504–1511. https://doi.org/10.1021/ma3026145

Spöri, C., Kwan, J. T. H., Bonakdarpour, A., Wilkinson, D. P., & Strasser, P. (2017). Stabilitätsanforderungen von Elektrokatalysatoren für die Sauerstoffentwicklung: der Weg zu einem grundlegenden Verständnis und zur Minimierung der Katalysatordegradation. *Angewandte Chemie*, *129*(22), 6088–6117. https://doi.org/10.1002/ange.201608601

Stoerzinger, K. A., Qiao, L., Biegalski, M. D., & Shao-Horn, Y. (2014). Orientation-dependent oxygen evolution activities of rutile $IrO_2$ and $RuO_2$. *Journal of Physical Chemistry Letters*, *5*(10), 1636–1641. https://doi.org/10.1021/jz500610u

Su, H., Linkov, V., & Bladergroen, B. J. (2013). 1 noble metal loadings for hydrogen production from solid polymer electrolyte water electrolysis. *International Journal of Hydrogen Energy*, *38*(23), 9601–9608. https://doi.org/10.1016/j.ijhydene.2013.05.099

Sun, S. C., Jiang, H., Chen, Z. Y., Chen, Q., Ma, M. Y., Zhen, L., Song, B., & Xu, C. Y. (2022). Bifunctional WC-supported $RuO_2$ nanoparticles for robust water splitting in acidic media. *Angewandte Chemie - International Edition*, *61*(21). https://doi.org/10.1002/anie.202202519

Sun, X., Liu, F., Chen, X., Li, C., Yu, J., & Pan, M. (2019). Iridium-doped ZIFs-derived porous carbon-coated IrCo alloy as competent bifunctional catalyst for overall water splitting in acid medium. *Electrochimica Acta*, *307*, 206–213. https://doi.org/10.1016/j.electacta.2019.03.179

Sun, Y., Wang, L., Guselnikova, O., Semyonov, O., Fraser, J., Zhou, Y., López, N., & Ganin, A. Y. (2022). Revealing the activity of $Co_3Mo_3N$ and $Co_3Mo_3N_{0.5}$ as electrocatalysts for the hydrogen evolution reaction. *Journal of Materials Chemistry A*, *10*(2), 855–861. https://doi.org/10.1039/d1ta08389a

Tari, V., Gupta, V., Joshi, A., Kurve, P., & Chandra, N. (2024). Biodiversity loss due to forest diversion. In A. Rathore (Ed.), *Impact of Societal Development and Infrastructure on Biodiversity Decline* (pp. 137–165). IGI Global. https://doi.org/10.4018/979-8-3693-6950-0.ch010

Tijani, A. S., Ghani, M. F. A., Rahim, A. H. A., Muritala, I. K., & Binti Mazlan, F. A. (2019). Electrochemical characteristics of (PEM) electrolyzer under influence of charge transfer coefficient. *International Journal of Hydrogen Energy*, *44*(50), 27177–27189. https://doi.org/10.1016/j.ijhydene.2019.08.188

Toghyani, S., Afshari, E., Baniasadi, E., & Atyabi, S. A. (2018). Thermal and electrochemical analysis of different flow field patterns in a PEM electrolyzer. *Electrochimica Acta*, *267*, 234–245. https://doi.org/10.1016/j.electacta.2018.02.078

Tong, Y., & Chen, P. (2024). Nanostructure engineering of ruthenium-modified electrocatalysts for efficient electrocatalytic water splitting. In *Journal of Materials Chemistry A* (Vol. 12, Issue 7, pp. 3844–3878). Royal Society of Chemistry. https://doi.org/10.1039/d3ta07230d

Tricker, A. W., Lee, J. K., Shin, J. R., Danilovic, N., Weber, A. Z., & Peng, X. (2023). Design and operating principles for high-performing anion exchange membrane water electrolyzers. *Journal of Power Sources*, *567*. https://doi.org/10.1016/j.jpowsour.2023.232967

Trinke, P., Bensmann, B., & Hanke-Rauschenbach, R. (2017). Current density effect on hydrogen permeation in PEM water electrolyzers. *International Journal of Hydrogen Energy*, *42*(21), 14355–14366. https://doi.org/10.1016/j.ijhydene.2017.03.231

Varadarajan, S., Kavitha, A., Selvaraju, P., Muthu, S. E., Gurushankar, K., Shanmugan, S., & Kannan, K. (2024). Enhanced photocatalytic hydrogen evolution by $TiO_2$: A synergistic approach with defect-rich $SnS_2$ and $Ti3C_2$ MXene cocatalysts. *Hydrogen*, *5*(4), 940–957. https://doi.org/10.3390/hydrogen5040050

Versogen. (n.d.). Our proprietary membranes are the most advanced and cost-effective in the world. Here's why https://versogen.com/revolutionary-anion-exchange-membranes/

Wang, C., Lan, F., He, Z., Xie, X., Zhao, Y., Hou, H., Guo, L., Murugadoss, V., Liu, H., Shao, Q., Gao, Q., Ding, T., Wei, R., & Guo, Z. (2019). Iridium-based catalysts for solid polymer electrolyte electrocatalytic water splitting. In *ChemSusChem* (Vol. 12, Issue 8, pp. 1576–1590). Wiley-VCH Verlag. https://doi.org/10.1002/cssc.201802873

Wang, H., Chen, Z. N., Wu, D., Cao, M., Sun, F., Zhang, H., You, H., Zhuang, W., & Cao, R. (2021). Significantly enhanced overall water splitting performance by partial oxidation of Ir through Au modification in core-shell alloy structure. *Journal of the American Chemical Society*, *143*(12), 4639–4645. https://doi.org/10.1021/jacs.0c12740

Wang, H., Sun, N., Xu, X., Wang, S., Kang, W., Zhuang, X., Yin, Y., & Cheng, B. (2021). Adenosine triphosphate@graphene oxide proton channels for proton exchange membranes constructed via electrostatic layer-by-layer deposition. *Journal of Membrane Science*, *620*. https://doi.org/10.1016/j.memsci.2020.118880

Wang, J., Yang, H., Li, F., Li, L., Wu, J., Liu, S., Cheng, T., Xu, Y., Shao, Q., & Huang, X. (2022). Single-site Pt-doped $RuO_2$ hollow nanospheres with interstitial C for high-performance acidic overall water splitting. *Science Advances*, *8*, eabl9271. https://www.science.org

Wang, X., Jiang, Z., Ma, Y., Su, X., Zhao, X., Zhu, A., & Zhang, Q. (2024). Ni/Fe based electrocatalyst for highly-efficient anion exchange membrane water electrolysis. *Journal of Power Sources*, *591*. https://doi.org/10.1016/j.jpowsour.2023.233819

Wang, Y., Hao, S., Liu, X., Wang, Q., Su, Z., Lei, L., & Zhang, X. (2020). Ce-doped $IrO_2$ electrocatalysts with enhanced performance for water oxidation in acidic media. *ACS Applied Materials and Interfaces*, *12*(33), 37006–37012. https://doi.org/10.1021/acsami.0c00389

Waribam, P., Jaiyen, K., Samart, C., Ogawa, M., Guan, G., & Kongparakul, S. (2023). MXene-copper oxide/sulfonated polyether ether ketone as a hybrid composite proton exchange membrane in electrochemical water electrolysis. *Catalysis Today*, *407*, 96–106. https://doi.org/10.1016/j.cattod.2022.02.007

Wei, G., Xu, L., Huang, C., & Wang, Y. (2010). SPE water electrolysis with SPEEK/PES blend membrane. *International Journal of Hydrogen Energy*, *35*(15), 7778–7783. https://doi.org/10.1016/j.ijhydene.2010.05.041

Wu, Y., Yao, R., Zhao, Q., Li, J., & Liu, G. (2022). La-$RuO_2$ nanocrystals with efficient electrocatalytic activity for overall water splitting in acidic media: Synergistic effect of La doping and oxygen vacancy. *Chemical Engineering Journal, 439*. https://doi.org/10.1016/j.cej.2022.135699

Xi, P., Huang, B., Yin, J., Jin, J., Lu, M., Zhang, H., Peng, Y., & Yan, C. H. (2020). Iridium single atoms coupling with oxygen vacancies boosts oxygen evolution reaction in acid media. *Journal of the American Chemical Society, 142*(43), 18378–18386. https://doi.org/10.1021/jacs.0c05050

Xiao, J., Oliveira, A. M., Wang, L., Zhao, Y., Wang, T., Wang, J., Setzler, B. P., & Yan, Y. (2021). Water-fed hydroxide exchange membrane electrolyzer enabled by a fluoride-incorporated nickel-iron oxyhydroxide oxygen evolution electrode. *ACS Catalysis, 11*(1), 264–270. https://doi.org/10.1021/acscatal.0c04200

Xie, Z., Yu, S., Yang, G., Li, K., Ding, L., Wang, W., & Zhang, F. Y. (2021). Optimization of catalyst-coated membranes for enhancing performance in proton exchange membrane electrolyzer cells. *International Journal of Hydrogen Energy, 46*(1), 1155–1162. https://doi.org/10.1016/j.ijhydene.2020.09.239

Xu, F., Qiu, K., Li, J., Lin, B., & Ding, J. (2021). Cross-linked anion exchange membranes with flexible, long-chain, bis-imidazolium cation cross-linker. *International Journal of Hydrogen Energy, 46*(74), 37039–37048. https://doi.org/10.1016/j.ijhydene.2021.08.181

Xu, J., Lian, Z., Wei, B., Li, Y., Bondarchuk, O., Zhang, N., Yu, Z., Araujo, A., Amorim, I., Wang, Z., Li, B., & Liu, L. (2020). Strong electronic coupling between ultrafine Iridium-Ruthenium nanoclusters and conductive, acid-stable Tellurium nanoparticle support for efficient and durable oxygen evolution in acidic and neutral media. *ACS Catalysis, 10*(6), 3571–3579. https://doi.org/10.1021/acscatal.9b05611

Yang, L., Chen, H., Shi, L., Li, X., Chu, X., Chen, W., Li, N., & Zou, X. (2019). Enhanced Iridium mass activity of 6H-phase, Ir-based perovskite with nonprecious incorporation for acidic oxygen evolution electrocatalysis. *ACS Applied Materials and Interfaces, 11*(45), 42006–42013. https://doi.org/10.1021/acsami.9b11287

Yao, Q., Huang, B., Zhang, N., Sun, M., Shao, Q., & Huang, X. (2019). Channel-rich RuCu nanosheets for pH-universal overall water splitting electrocatalysis. *Angewandte Chemie - International Edition, 58*(39), 13983–13988. https://doi.org/10.1002/anie.201908092

Yu, S., Li, K., Wang, W., Xie, Z., Ding, L., Kang, Z., Wrubel, J., Ma, Z., Bender, G., Yu, H., Baxter, J., Cullen, D. A., Keane, A., Ayers, K., Capuano, C. B., & Zhang, F. Y. (2022). Tuning catalyst activation and utilization via controlled electrode patterning for low-loading and high-efficiency water electrolyzers. *Small, 18*(14). https://doi.org/10.1002/smll.202107745

Zakaria, Z., & Kamarudin, S. K. (2021). A review of alkaline solid polymer membrane in the application of AEM electrolyzer: Materials and characterization. In *International Journal of Energy Research* (Vol. 45, Issue 13, pp. 18337–18354). John Wiley & Sons Ltd. https://doi.org/10.1002/er.6983

Zaman, W. Q., Wang, Z., Sun, W., Zhou, Z., Tariq, M., Cao, L., Gong, X. Q., & Yang, J. (2017). Ni-Co codoping breaks the limitation of single-metal-doped $IrO_2$ with higher oxygen evolution reaction performance and less Iridium. *ACS Energy Letters, 2*(12), 2786–2793. https://doi.org/10.1021/acsenergylett.7b01032

Zhang, L., Jang, H., Li, Z., Liu, H., Kim, M. G., Liu, X., & Cho, J. (2021). $SrIrO_3$ modified with laminar $Sr_2IrO_4$ as a robust bifunctional electrocatalyst for overall water splitting in acidic media. *Chemical Engineering Journal, 419*. https://doi.org/10.1016/j.cej.2021.129604

Zhegur-Khais, A., Kubannek, F., Krewer, U., & Dekel, D. R. (2020). Measuring the true hydroxide conductivity of anion exchange membranes. *Journal of Membrane Science*, *612*. https://doi.org/10.1016/j.memsci.2020.118461

Zhirov, A. V., Komarov, A. O., Danilov, V. V., & Shorokhov, S. A. (2012). Effect of glycerine concentration on dissolution and oxidation of mild steel during anodic cementation. *Surface Engineering and Applied Electrochemistry*, *48*(3), 289–291. https://doi.org/10.3103/S1068375512030143

Zhuang, Z., Wang, Y., Xu, C. Q., Liu, S., Chen, C., Peng, Q., Zhuang, Z., Xiao, H., Pan, Y., Lu, S., Yu, R., Cheong, W. C., Cao, X., Wu, K., Sun, K., Wang, Y., Wang, D., Li, J., & Li, Y. (2019). Three-dimensional open nano-netcage electrocatalysts for efficient pH-universal overall water splitting. *Nature Communications*, *10*(1). https://doi.org/10.1038/s41467-019-12885-0

Chapter 12

# Advances in the green hydrogen generation and novel hydrogen powered resources

## Approach toward environmentally friendly energy solutions

*Kaushlendra Agrahari, Vivek Kumar Nautiyal, Teerthraj Verma, R. R. Awasthi, and Pankaj Kumar Tripathi*

## 12.1 INTRODUCTION

Energy is an essential part of our daily life and a fundamental requirement for nearly all human activities. However, we rely on various sources of energy, while the demand for it continues to grow daily. Several natural and non-renewable energy sources such as coal, natural gas, and petroleum (Figure 12.1), account for approximately 80% of the energy consumed worldwide. The growing global population and the relentless drive to improve living standards continue to increase energy demand (Tari et al., 2024a). At present, a large number of countries rely predominantly on fossil fuels as their primary energy source (Tari, 2015). It is estimated that the existing fossil fuel reserves could sustain for a maximum of 40 to 150 years. The world has already witnessed a steady rise in petroleum and natural gas prices over time. At the current rate of fossil fuel consumption, prices are expected to continue climbing significantly. If fossil fuels continue to be consumed at the same rate, their reserves will be depleted much sooner than anticipated. Consequently, the rising energy consumption and the dwindling fossil fuel reserves pose significant threats to global energy and economic security. Additionally, fossil fuels produce greenhouse gases and other large-scale pollutants, contributing to environmental degradation and global warming (Wani et al., 2025; Tari et al., 2021, 2024b; Siddiqui et al., 2022). These pollutants include hazardous air contaminants that lead to chronic health issues (Lelieveld et al., 2019; Kumar et al., 2021; York, 2012; Holechek et al., 2022; Chiari & Zecca, 2011). To address these critical challenges, it is essential to reassess our current dependence on fossil fuels and seek sustainable alternatives. Many researchers and engineers are focusing on developing resources that are environmentally friendly, cost-effective, and more sustainable to meet energy demands and replace the existing fossil fuel systems. After extensive evaluation, hydrogen has emerged as a promising

DOI: 10.1201/9781003584742-12

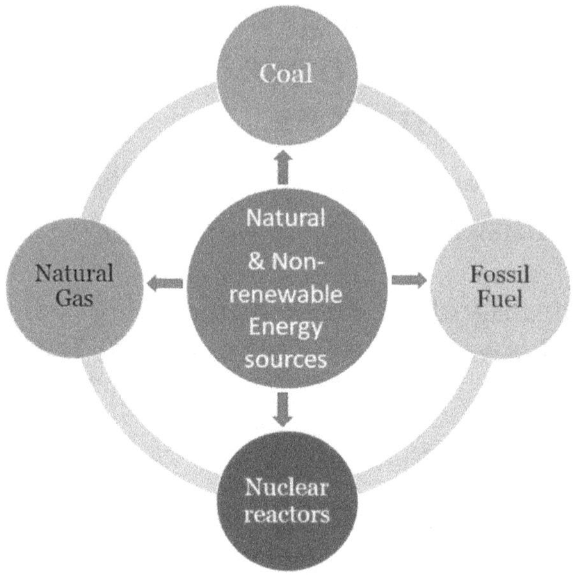


*Figure 12.1* Some common natural and non-renewable energy sources.

alternative energy source due to its exceptional properties. The fundamental idea of utilizing hydrogen as an energy carrier holds the potential to address global energy crisis challenges effectively.

Hydrogen has long been a subject of research as a potential energy source for future markets. However, storing hydrogen remains a significant challenge due to its extremely low density. Hydrogen is typically stored as a cryogenic liquid, a pressurized gas, or chemically bonded to suitable solid-state nanomaterials (Bhattacharyya & Mohan, 2015). Among the various storage methods, solid-state systems show great promise due to their larger gravimetric hydrogen storage capacity (Schlapbach & Zuttel, 2001). Solid-state metal hydride storage materials are relatively safe, efficient, and reversible, making them increasingly popular for hydrogen storage applications (Billur et al., 2007). Despite these advancements, no metal hydride has fully satisfied all the key requirements for the hydrogen market, primarily due to issues such as limited capacity for hydrogen storage, slow kinetics, and challenging absorption/desorption temperatures. This chapter focuses on hydrogen storage, hydrogen energy, the hydrogen market, and the current state of solid-state hydrogen storage using various chemical and physical storage materials (Schlapbach, 2002; Zuttel, 2003). Furthermore, it aims to develop novel metal hydride nanomaterials, offering a potential breakthrough for practical hydrogen storage applications and advancing the hydrogen economy (Prabhukhot et al., 2016; Ren et al., 2017).

## 12.2 HYDROGEN ENERGY

Hydrogen is widely considered as a promising candidate for low-cost, clean, and green fuel to meet future energy needs. It possesses unique characteristics, being the most abundant element found in the universe, the lightest element with the lowest molecular weight, combined with the highest energy content of any fuel source. Hydrogen is sustainable, innocuous, and environmentally friendly, unlike coal, petroleum, or natural gas. Hydrogen energy can be derived from both renewable and non-renewable energy sources. Renewable energy sources include wave, hydro, geothermal, wind, solar, and biomass energy, while non-renewable sources include nuclear energy, coal, and natural gas (Jain, 2009; Mazloomi & Gomes, 2012). These energy sources can be stored as fuel and used for power generation and transportation systems, as shown in Figure 12.2. Hydrogen also possesses an exceptional energy storage capacity. For instance, 1 kilogram of hydrogen holds around 120 MJ of energy, which is nearly double that of most traditional fuels (Schneemann et al., 2018; Gujral et al., 2022).

Hydrogen energy, as a sustainable energy solution for the future, offers significant benefits. It promotes sustainability via the utilization of renewable energy sources, reduces hazardous gas pollution, and improves air quality by enabling near-zero emissions of carbon and greenhouse gases. For these

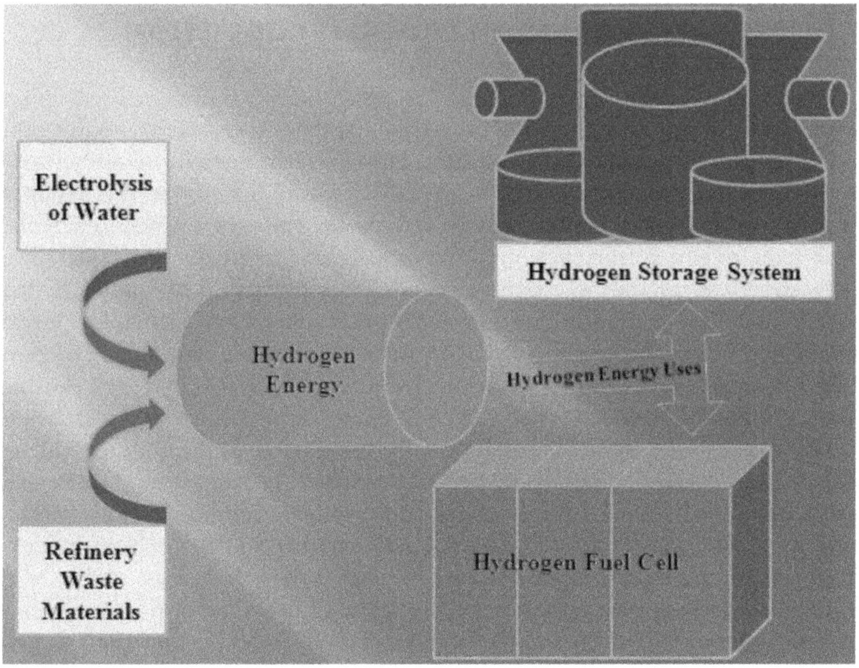


*Figure 12.2* Hydrogen energy power system.

critical reasons, hydrogen energy is regarded as a vital solution to the global energy challenges and a pivotal contributor to environmental safety.

The main objective of a hydrogen economy is to produce hydrogen on a large scale as an energy source capable of replacing the fossil fuels presently used in the transportation, industrial, and vehicle sectors (Langer et al., 2007). The hydrogen economy is envisioned as a sophisticated and sustainable solution to the interconnected challenges facing the planet, including food shortages, deterioration of natural sources, and growing global population. Transitioning to a hydrogen economy offers significant advantages, such as economic growth, environmental benefits, enhanced energy security, and better outcomes for end users. However, hydrogen's extremely low density poses significant challenges for storage, which is a critical issue for commercial applications. Additionally, while hydrogen is already used in chemical and refinery industries, its production, storage, and transportation costs remain high, limiting its broader applications. Despite these challenges, the hydrogen economy holds immense potential, attracting global interest from governments as they explore innovative energy solutions. For instance, the European Commission (EU) has set an ambitious target that 35% of newly manufactured vehicles in the EU should run on hydrogen energy by 2040 (Hwang & Varma, 2014; Rosen & Fayegh, 2016; Pudukudy et al., 2014).

## 12.3 INVESTIGATION OF GREEN CHEMISTRY PRINCIPLES-BASED HYDROGEN PRODUCTION METHODS

Hydrogen energy, as a clean and green fuel for the future, is continually gaining attention as a sustainable solution. However, the current manufacturing of hydrogen is not always environmentally friendly. Therefore, it is essential to improve production processes to reduce dependency on non-renewable resources, enhance efficiency, minimize waste, and make the process greener. The present investigation adopts a systematic approach to advancing hydrogen production techniques in line with green chemistry principles. Green chemistry focuses on generating chemical products while minimizing or eliminating hazardous substances. Various techniques exist for hydrogen production, utilizing diverse raw materials and energy sources. Key approaches to synthesizing green hydrogen include improving energy efficiency, producing safer products and auxiliaries, utilizing renewable resources, suppressing derivatives, employing catalysis, ensuring biodegradability, and preventing pollution. The focus on hydrogen as an alternative fuel stems from its environmentally friendly properties, non-toxicity, safety, lightness, and ease of transport, making it suitable for a wide range of potential applications.

There are various chemical methods such as electrolysis, thermolysis, artificial photosynthesis, etc. for hydrogen production which is based on green chemistry principles (Jiao et al., 2015; Sapountzi et al., 2017).

### 12.3.1 Electrolysis method

The most effective electrochemical water-splitting technique is the well-known electrolysis method for hydrogen production and is widely used in industry. In this process, an electric current is applied to the electrodes, facilitating electron movement. The chemical reaction of this process is given as follows:

$$2H_2O \xrightarrow{\text{electricity}} 2H_2 \uparrow + O_2 \uparrow \tag{12.1}$$

The characteristics of electrolysis are based on alkaline electrolysis, which is extensively used for hydrogen production through the chemical electrolysis process.

### 12.3.2 Thermolysis method

The thermal water dissociation technique generally relies on a single splitting system. When thermal energy, in the form of extremely high temperatures, is applied to water molecules, they begin to dissociate into hydrogen and oxygen atoms. The thermolysis chemical reaction can be represented as follows:

$$2H_2O \xrightarrow{\text{Heat}} 2H_2 \uparrow + O_2 \uparrow \tag{12.2}$$

Thermal water-splitting techniques, including steam reforming and gasification, have been extensively investigated as viable techniques for hydrogen production.

### 12.3.3 Artificial photosynthesis

Artificial photosynthesis is an important method for hydrogen production and is considered a biomimetic process. However, it is primarily focused on the generation of hydrogen energy. This process is based on the capture of carbon dioxide ($CO_2$) from the air using a photosensitive synthetic pigment. It involves the formation of oxygen and carbohydrates through the chemical reaction of $CO_2$ with water, followed by the conversion of carbohydrates into hydrogen and carbon dioxide. The process can be represented by the following reaction:

$$6CO_2 + 6H_2O \xrightarrow{\text{Light+artifical pigment}} C_6H_{12}O_6 + 6O_2 \uparrow \tag{12.3}$$

$$C_6H_{12}O_6 + 6H_2O \xrightarrow{\text{Light+artifical pigment}} 6CO_2 + 12H_2 \tag{12.4}$$

The general chemical reaction involves the decomposition of water in the presence of light using pigments. Artificial photosynthesis is an eco-friendly approach within green chemistry, offering a low-cost and scalable method for mass-producing hydrogen energy.

## 12.4 METHODS FOR MEASUREMENTS OF HYDROGEN INTERACTION WITH MATTER

The interaction of hydrogen with matter in bulk and thin films can be measured using different methods. For bulk materials, classical volumetric and gravimetric techniques are commonly employed. Hydrogen concentration in bulk is determined by measuring either the pressure drop or the increase in the sample's weight. However, applying these methods to thin films is difficult because of the limited amount of hydrogen content. In thin films, hydrogen concentration can be accurately measured using techniques such as elastic recoil detection study, electrochemical loading, neutron reflectivity, and nuclear reaction analysis. In electrochemical loading, the metal film serves as the cathode. This process involves the electrochemical reduction of water, with one electron corresponding to each absorbed $H_2$ atom. Consequently, the thin film becomes hydrogenated through this reaction.

### 12.4.1 Hydrogen storage system

In the current scenario, hydrogen storage has become a critical technology, essential for the advancement of progressive nations worldwide. The primary motivation behind adopting hydrogen as a green energy source is its potential to address global energy crises and mitigate climate change. For the hydrogen economy to thrive, portable and stationary hydrogen storage systems are needed (Niaz et al., 2015). Mobile storage systems are expected to be the largest consumers of hydrogen in the upcoming hydrogen economy. These systems utilize mechanisms that transform chemical energy into heat energy, and subsequently into mechanical energy, thereby reducing greenhouse gas emissions, improving sustainable energy systems, and decreasing dependence on fossil fuels (David, 2005). Stationary hydrogen storage systems typically require large spaces and operate under high temperatures and high pressure conditions. To meet the demand for stationary applications, advancements are needed to develop systems capable of storing large quantities of hydrogen efficiently. In contrast, mobile hydrogen storage demands compact systems optimized for transportation. Key performance metrics for hydrogen storage systems include gravimetric capacity, defined as the amount of hydrogen stored per unit weight, and volumetric capacity, defined as the amount of hydrogen stored per unit volume. Achieving high values in both is important for effective storage of hydrogen. Excessively large storage

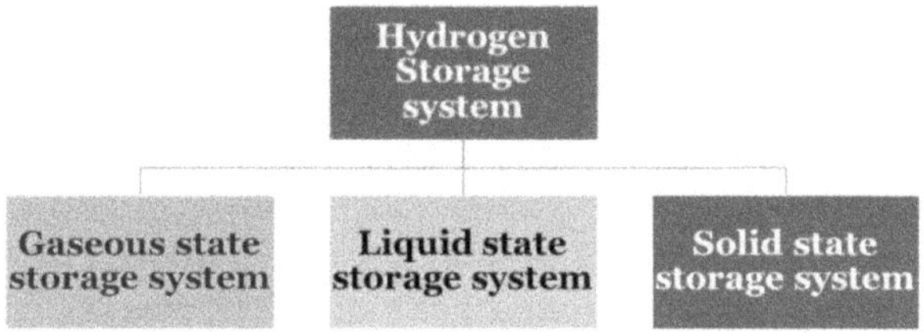


*Figure 12.3* Primary methods of hydrogen storage system.

systems can limit vehicle range and reduce available baggage space, highlighting the need for balanced design considerations. Other critical aspects of hydrogen storage include low operating temperatures and pressures, rapid hydrogen uptake and discharge kinetics, minimal heat formation, and reduced energy requirements for hydrogen release. Additionally, desirable features include trivial thermal losses during exothermic hydride formation, low-cost recycling and recharging infrastructure, minimal energy dissipation during charge–discharge cycles, high operational safety, multi-cycle reversibility of hydrogen absorption and desorption, and long-term resistance to oxygen and moisture. Hydrogen can be stored using three primary methods, each with its strengths and weaknesses: gaseous, liquid, and solid-state storage systems as depicted in Figure 12.3. These storage types are tailored to diverse applications and storage requirements (Becherif et al., 2015; Jorgensen, 2011).

## 12.4.2 Gaseous state storage system

Physical storage systems are the most conventional methods for storing hydrogen, typically in the form of compressed hydrogen gas. Due to hydrogen's very low density of 0.089 kg/m$^3$, high pressures or extremely low temperatures are required for effective storage. However, this low density also increases the likelihood of hydrogen leakage from enclosed containers under high pressure (Schlapbach, 2002; Trahey et al., 2020).

Commercial hydrogen storage often utilizes steel or aluminum containers. More advanced options include carbon-fiber-reinforced plastic composite containers, which provide sufficient strength and impact resilience, thereby enhancing collision safety. These composite containers are significantly lighter than metal alternatives but are also much more expensive. To address these challenges, researchers and engineers are investigating more cost-effective and practical alternatives. For storing large quantities of hydrogen, compressed hydrogen can be stored in large-scale physical formations such as salt caverns. Salt caverns are highly impermeable to hydrogen, can withstand elevated pressures, and have excellent sealing properties, making them a promising option for large-scale storage.

### 12.4.3 Liquid state hydrogen storage system

The storage of hydrogen in cryogenic liquid form is referred to as the liquid-state storage system. Storing hydrogen as a liquid provides a higher density, with cryogenic liquid hydrogen achieving approximately 71 g/L at its normal boiling point of 20 K. This density is about two times greater than that of pressurized hydrogen gas. However, due to its low boiling point, liquid hydrogen requires advanced cooling technologies that consume approximately 30% of its inherent energy (Valenti, 2016). To minimize heat leakage, specially designed double-walled containers with efficient insulation systems are essential. Cryogenic pressure containers, which are both denser and lighter, offer better protection and performance compared to standard compressed hydrogen containers.

### 12.4.4 Solid-state hydrogen storage system

Hydrogen storage systems must be safe, compact, efficient, inexpensive, and lightweight. As mentioned earlier, storing pressurized hydrogen gas and cryogenic liquid hydrogen entails large space, is very costly, and raises safety concerns for a hydrogen economy. It is clear that hydrogen storage space needs significant scientific advancement, potentially offering more feasible alternatives compared to pressurized hydrogen gas and cryogenic liquid hydrogen.

In particular, fuel cell vehicles face the critical challenge of storing sufficient hydrogen, which remains a significant obstacle. A hydrogen storage system with high volumetric and gravimetric capacity would be ideal for hydrogen-based vehicular applications. Solid-state materials are among the most promising candidates, as they have the capacity to reversibly uptake and desorb hydrogen (Lim et al., 2010; Rusman & Dahari, 2016; Kannan & Tari, 2025; Xu et al., 2024). The advantages and disadvantages of the different storage systems (gaseous, liquid, solid) are summarized in Table 12.1.

In solid-state systems, hydrogen is stored through either physical adsorption or chemical adsorption. In the physical adsorption process, molecular hydrogen adheres to solid surfaces via van der Waals forces. Organic-based

*Table 12.1* Advantages and disadvantages of the different storage systems

| *System* | *Advantage* | *Disadvantage* |
|---|---|---|
| Solid | High volumetric density, safe and compact, Low-pressure operation | Slow kinetics, expensive material. |
| Liquid | No need for high pressure tanks, High density | Cryogenic, High liquefaction cost |
| Gaseous | Simple and mature technology, Rapid filling and release | High-pressure safety risks. low volume density, heavy tanks |

materials, such as polymers, carbon nanotubes, carbon nanowalls, carbon nanorods, fibers, and fullerenes, are commonly used for this purpose (Rauch et al., 2017; Erdogan et al., 2023; Rochat et al., 2017; Munir et al., 2021). Hydrogen can be released, when necessary, by thermal stimulation or other appropriate techniques. While these materials offer good reversibility and fast kinetics, they currently have low hydrogen storage capacity under ambient conditions, limiting their practical application. In the chemical adsorption process, atomic hydrogen is bonded with a host material by strong chemisorptive binding, and the atomic hydrogen is stored in the bulk of the material. Notably, dehydrogenation remains a significant challenge because of various irreversible storage methods associated with chemical compounds. In this chapter, we have discussed physical and chemical materials employed for hydrogen storage.

### 12.4.5 Metal hydrides

Metal hydrides, initially considered among the foremost materials for chemical hydrogen storage, result from the interaction between hydrogen atoms and metals, with their structural features playing a pivotal role in defining their properties. In addition to binary metal hydrides, ternary hydrides, complex metal hydrides, and hydrides derived from ternary alloys such as $Mg_2NiH_4$, $Mg_3La$, and $Mg_3LaNi_{0.1}$ have been actively investigated (Reilly Jr & Wiswall Jr, 1968; Ouyang et al., 2006). Due to its extensive scope and diversity, providing a comprehensive review of metal hydrides and allied materials is impractical here. Instead, we have focused on the key features of metal hydrides and highlighted the latest developments in the field.

In recent years, various metal hydrides, such as magnesium hydride ($MgH_2$), have attracted noteworthy attention as green energy hydrogen storage materials owing to their high gravimetric density (~7.6 wt%), reversibility in hydrogen storage, and cost-effectiveness. Metal hydrides provide a convenient and safe technology for hydrogen storage. The primary focus of research has been on enhancing hydrogen sorption kinetics and lowering the high dissociation temperature (~573 K) (Dornheim et al., 2007; Jia et al., 2015). Notably, Mg has some drawbacks, including its large negative enthalpy of formation and slow hydriding and dehydriding processes under ambient conditions. In view of this, several research groups have explored various catalysts/additives such as Ni, Ti, and Fe at 12 atm of $H_2$ pressure, leading to the formation of nano $MgH_2$-$Ti_5Fe_5Ni_5$ that has improved the hydrogen storage properties of $MgH_2$, and the dehydrogenation temperatures of these newly developed composites are much lower compared to $MgH_2$ (Shahi et al., 2013).

Several synthesis techniques are employed for the fabrication of metal hydride materials. A metal hydride is formed through a chemical reaction, similar to a physical storage method. In metal hydride tanks, hydrogen molecules chemically bond with the metal at moderate temperatures and pressures, typically between 3 and 30 atmospheres, which is much lower than the pressures used in compressed gas tanks. This contributes to the system's safety since hydrogen gas leakage or tank explosions are less likely. Mechanical ball milling technique is used to achieve nanostructures and reduce the grain size to ~10 nm, induce defects, increase the surface area-to-volume ratio, and create grain boundaries, all of which weaken the bond strength of the M-H bond. Moreover, mechanical ball milling with diverse catalysts has been shown to significantly improve hydrogen sorption kinetics. Various synthesis methods have enabled the optimization of Mg-based alloys, resulting in materials with superior kinetics and high hydrogen storage capacity (Iosub et al., 2009).

Although metal hydride tanks tend to be relatively heavy, this weight is not a major concern for stationary and certain niche applications, specifically in small-scale storage. Once lighter materials are discovered to address this issue, their use in mobile applications will become more feasible. Furthermore, metal hydrides offer relatively high volumetric energy densities, excellent storage capabilities, and the ability to absorb and release hydrogen with slight pressure changes, making them a highly promising hydrogen storage solution.

The primary objective of this study is to introduce the fabrication of large-scale hydrogen storage materials for use in transportation. Thin film processing is a promising alternative method, as it allows for the preparation of nanostructured materials with tailored compositions, precise microstructures, and controlled crystal dimensions. Consequently, this provides a more practical strategy for addressing challenges in hydrogen technology development. By reducing the grain size to the nanoscale, the surface topography or interface can be modified to improve hydrogenation thermodynamics. Additionally, phenomena such as elastic interactions in the interfacial region can be explored via the creation of multilayer thin films, helping to minimize kinetic restrictions and resulting in novel nanomaterials with unique properties.

Thin films of metal hydrides and their alloys are valuable for identifying stoichiometric compositions and understanding the optimal mechanisms for hydrogen adsorption. Thin film nanomaterials have several advantages over bulk materials, including reduced susceptibility to embrittlement and degradation, which allows for more efficient cyclic absorption and desorption studies. When hydrogen is loaded into the film, buckling is induced when the film's elastic energy surpasses its adhesion energy. However, weak film-substrate adhesion leads to stress relaxation through a buckle-and-crack complex configuration. If adhesion energy is adequate to maintain film attachment despite high mechanical stress, stress relaxation proceeds via dislocation deformations, resulting in increased elastic out-of-plane expansion. This results in increased elastic out-of-plane expansion, strengthening the clamped thin film and preventing changes in the surface morphology.

It was observed that metal hydrides significantly enhance hydrogen absorption kinetics, reducing the nucleation barrier for hydride phase formation (Vajo, 2011). Hydrogenation of the metal hydride alloy under mild conditions can improve the storage capacity. The crystal structure and electrochemical hydrogen storage properties of the thin film fabrications, catalyzed solution, hydrogen plasma metal reaction, etc. were investigated and found to be enhanced (Sadhasivam et al., 2017; Shao et al., 2012).

### 12.4.6 Carbon-based materials

Carbon is widely recognized as one of the most abundant elements in both living and non-living systems and is renowned for its gas adsorption capabilities, making it extensively used for detoxification and purification purposes. It is well known that carbon is one of the most abundant elements in both living and non-living matter, and it is an eminent material capable of adsorbing gases and has been widely utilized as a detoxifier and purifier. Carbon-based materials have garnered significant attention as propitious candidates for hydrogen storage due to their numerous advantageous properties, including a larger surface area, diverse pore structures with high porosity, excellent chemical stability, low weight, and affordability. Various carbon-based materials like activated nanoporous carbon, single-walled carbon nano-tubes (CNTs), multi-walled carbon nanotubes, carbon nanofibers, and carbon nitrides (CNs), have been recognized as promising candidates for hydrogen storage (Cheng et al., 2001; Zhang et al., 2017; Zhao et al., 2017; Sharma, 2020). Several of these materials exhibit excellent performance in the physical adsorption and desorption of hydrogen. AC is a highly porous, modified synthetic carbon composed of both crystalline graphite and amorphous carbon. The loading of noble metal, such as platinum, onto AC enhances adsorption capacity due to the combination of chemisorption on the platinum surface and physisorption on the carbon surface. In the aforementioned materials, the porous structure plays a crucial role in regulating hydrogen adsorption, where the alignment of pore size with the kinetic diameter of hydrogen molecules underpins their physical hydrogen storage capabilities. Table 12.2 summarizes the hydrogen storage capacity of different carbon-based materials.

### 12.4.7 Metal–organic frameworks (MOFs)

Metal–organic frameworks (MOFs) are crystalline materials formed by the coordination of metal ions or clusters with organic linkers. MOFs have emerged as outstanding candidates and have garnered significant attention for clean energy applications, particularly as a medium for storing gases like methane and hydrogen. Their exceptional properties, including high crystallinity, vast internal surface areas reaching up to 6000 $m^2/g$, and

*Table 12.2* Hydrogen storage capacities of various carbon-based materials

| *Material* | *Temperature (K)* | *Pressure (bar)* | *Capacity (wt.%)* | *Reference* |
|---|---|---|---|---|
| CHCPB-K-700PC | 77.3 | 1.13 | 3.25 | Zhang et al. (2017) |
| SWCNTs | 77 | 100 | 1.73 | Zhao et al. (2017) |
| MWCNT/Pt | 77 | 25 | 0.55 | Sharma (2020) |
| 5 % Pd/C | 298 | 60 | 7.16 | Zhou and Szpunar (2016) |
| $Pd_3Co\text{-}gC_3N_4$ | 298 | 30 | 5.0 | Nair and Sundara (2016) |
| Activated carbon | 77 | 10 | 2.02 | Chang et al. (2015) |
| Reduced GO | 293 | 1.06 | 0.13 | Wei and Mao (2016) |

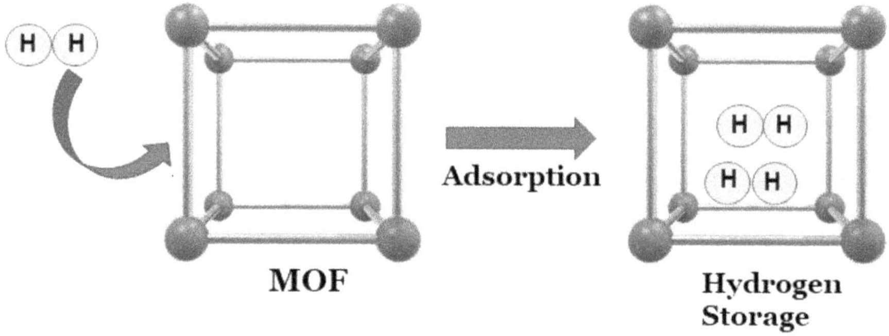


*Figure 12.4* Schematic mechanism of MOFs' hydrogen storage.

ultrahigh porosity (up to 90% empty volume), make them highly suitable for hydrogen storage (Zhou et al., 2012). The primary hydrogen storage mechanism in MOFs is similar to that of other porous materials, where hydrogen molecules occupy and are retained within the material's pores as shown in Figure 12.4. Researchers have reported that MOF materials demonstrate effective hydrogen storage capacity at relatively lower temperatures. Therefore, hydrogen storage at ambient temperatures still poses a significant challenge, and researchers are exploring various approaches to enhance the capacity of MOF materials at elevated temperatures. Hydrogen storage capacity of various MOFs is listed in Table 12.3.

*Table 12.3* Hydrogen storage capacities of various MOFs

| *Material* | *Temperature (K)* | *Pressure (bar)* | *Capacity (wt.%)* | *Ref.* |
|---|---|---|---|---|
| NPF-200 | 77 | 100 | 13.1 | Zhang et al. (2020) |
| Pd CNMS/MOF-5 | 298 | 100 | 1.8 | Naeem et al. (2016) |
| $Zn_2$ (dobpdc) MOFs | 298 | 100 | 1.3 | Gómez-Gualdrón et al. (2016) |
| CB/Pt/MOF-5 | 298 | 100 | 0.62 | Chen et al. (2021) |
| MFU-4/Li | 77-160 | 100-5 | 9.4 | Kim et al. (2015) |
| NU-1501-Al | 77-160 | 100-5 | 14 | Gygi et al. (2016) |
| UBMOF-31 | 77 | 60 | 4.9 | Viditha et al. (2016) |
| BP-COF-5 | 77 | 100 | 8.57 | Bian et al. (2022) |
| Cu-MOF-399 | 77 | 50 | 8.3 | Srivastava et al. (2022) |
| Zn BTC $Rb^+$ | 77 | 1 | 0.15 | Almáši et al. (2021) |

## 12.5 CONCLUSION

In this chapter, we have briefly reviewed different hydrogen production methods and some chemical and physical materials for hydrogen storage. Hydrogen production, storage, and distribution represent the three primary challenges, with safe and reversible storage identified as the most critical obstacle. The quest for a hydrogen storage material with high gravimetric and volumetric densities has prompted the exploration of a wide range of materials, from bulk metal alloys to nanomaterials. In comparison to physical storage materials, chemical storage materials, like liquid organic hydrogen carriers or metal hydrides, are relatively easy to deploy in practical applications as they can be used at ambient temperatures. Materials-based solid hydrogen storage holds great promise and is a competitive alternative for the extensive use of hydrogen as a future fuel source for vehicular energy demand. The production of hydrogen, its storage, and conversion have attained a significant technological level, though there is still ample room for improvements and new discoveries to overcome the challenges associated with it.

## REFERENCES

Almáši, M., Sharma, A., & Zelenka, T. (2021). Anionic zinc(II) metal-organic framework post-synthetically modified by alkali-ion exchange: Synthesis, characterization and hydrogen adsorption properties. *Inorganica Chimica Acta*, *526*, 120505. https://doi.org/10.1016/j.ica.2021.120505

Becherif, M., Ramadan, H.S., Cabaret, K., Picard, F., Simoncini, N., & Bethoux, O. (2015). Hydrogen energy storage: New techno-economic emergence solution analysis. *Energy Procedia*, *74*, 371–380. https://doi.org/10.1016/j.egypro.2015.07.629

Bhattacharyya, R., & Mohan, S. (2015). Solid state storage of hydrogen and its isotopes: An engineering overview. *Renewable and Sustainable Energy Reviews*, *41*, 872–883. https://doi.org/10.1016/j.rser.2014.09.004

Bian, L.Y., Li, X.D., Huang, X.Y., Yang, P.H., Wang, Y.D., Liu, X.Y., & Chen, Z. (2022). Molecular simulation on hydrogen storage properties of five novel covalent organic frameworks with the higher valency. *International Journal of Hydrogen Energy*, *47*, 29390–29398. https://doi.org/10.1016/j.ijhydene.2022.06.285

Billur, S., Darkrimb, F.L., & Hirscherc, M. (2007). Metal hydride materials for solid hydrogen storage: A review. *International Journal of Hydrogen Energy*, *32*, 1121–1140. https://doi.org/10.1016/j.ijhydene.2006.11.022

Chang, Y.M., Tsai, W.T., & Li, M.H. (2015). Characterization of activated carbon prepared from Chlorella-based algal residue. *Bioresource Technology*, *184*, 344–348. https://doi.org/10.1016/j.biortech.2014.09.131

Chen, Z., Mian, M.R., Lee, S.-J., Chen, H., Zhang, X., Kirlikovali, K.O., Shulda, S., Melix, P., Rosen, A.S., & Parilla, P.A. (2021). Fine-tuning a Robust metal-organic framework toward enhanced clean energy gas storage. *Journal of the American Chemical Society*, *143*, 18838–18843. https://doi.org/10.1021/jacs.1c08749

Cheng, H.M., Yang, Q.H., & Liu, C. (2001). Hydrogen storage in carbon nanotubes. *Carbon*, *39*(10), 1447–1454. https://doi.org/10.1016/S0008-6223(00)00306-7

Chiari, L., & Zecca, A. (2011). Constraints of fossil fuels depletion on global warming projections. *Energy Policy*, *39*, 5026–5034. https://doi.org/10.1016/j.enpol.2011.06.011

David, E. (2005). An overview of advanced materials for hydrogen storage. *Journal of Materials Processing Technology*, *162*, 169–177. https://doi.org/10.1016/j.jmatprotec.2005.02.027

Dornheim, M., Doppiu, S., Barkhordarian, G., Boesenberg, U., Klassen, T., Gutfleisch, O., & Bormann, R. (2007). Hydrogen storage in magnesium-based hydrides and hydride composites. *Scripta Materialia*, *56*, 841–846. https://doi.org/10.1016/j.scriptamat.2007.01.003

Erdogan, F.O., Celik, C., Turkmen, A.C., Sadak, A.E., & Cücü, E. (2023). Hydrogen storage behavior of zeolite/graphene, zeolite/multiwalled carbon nanotube and zeolite/green plum stones-based activated carbon composites. *Journal of Energy Storage*, 72, 108471. https://doi.org/10.1016/j.est.2023.108471

Gómez-Gualdrón, D.A., Moghadam, P.Z., Hupp, J.T., Farha, O.K., & Snurr, R.Q. (2016). Application of consistency criteria to calculate BET areas of micro and mesoporous metal-organic frameworks. *Journal of the American Chemical Society*, *138*, 215–224. https://doi.org/10.1021/jacs.5b10266

Gujral, H.S., Singh, G., Baskar, A.V., Guan, X., Geng, X., Kotkondawar, A.V., Rayalu, S., Kumar, P., Karakoti, A., & Vinu, A. (2022). Metal nitride-based nanostructures for electrochemical and photocatalytic hydrogen production. *Science and Technology of Advanced Materials*, *23*, 76–119. https://doi.org/10.1080/14686996.2022.2029686

Gygi, D., Bloch, E.D., Mason, J.A., Hudson, M.R., Gonzalez, M.I., Siegelman, R.L., Darwish, T.A., Queen, W.L., Brown, C.M., & Long, J.R. (2016). Hydrogen storage in the expanded pore metal-organic frameworks $M_2$ (Dobpdc) (M = Mg, Mn, Fe, Co, Ni, Zn). *Chemistry of Materials*, *28*, 1128–1138. https://doi.org/10.1021/acs.chemmater.5b04538

Holechek, J.L., Geli, H.M.E., Sawalhah, M.N., & Valdez, R. (2022). A global assessment: Can renewable energy replace fossil fuels by 2050? *Sustainability*, *14*, 4792. https://doi.org/10.3390/su14084792

Hwang, H.T., & Varma, A. (2014). Hydrogen storage for fuel cell vehicles. *Current Opinion in Chemical Engineering*, *5*, 42–48. https://doi.org/10.1016/j.coche.2014.04.004

Iosub, V., Matsunaga, T., Tange, K., & Ishikiriyama, M. (2009). Direct synthesis of $Mg(AlH_4)_2$ and $CaAlH_5$ crystalline compounds by ball milling and their potential as hydrogen storage materials. *International Journal of Hydrogen Energy*, *34*, 906–912. https://doi.org/10.1016/j.ijhydene.2008.11.013

Jain, I.P. (2009). Hydrogen the fuel for 21st century. *International Journal of Hydrogen Energy*, *34*, 7368–7378. https://doi.org/10.1016/j.ijhydene.2009.05.093

Jia, Y., Sun, C., Shen, S., Zou, J., Mao, S.S., & Yao, X. (2015). Combination of nanosizing and interfacial effect, future perspective for designing Mg-based nanomaterials for hydrogen storage. *Renewable and Sustainable Energy Reviews*, *44*, 289–303. https://doi.org/10.1016/j.rser.2014.12.032

Jiao, Y., Zheng, Y., Jaroniec, M., & Qiao, S.Z. (2015). Design of electrocatalysts for oxygen- and hydrogen-involving energy conversion reactions. *Chemical Society Reviews*, *44*, 2060–2086. https://doi.org/10.1039/C4CS00470A

Jorgensen, S.W. (2011). Hydrogen storage tanks for vehicles: Recent progress and current status. *Current Opinion in Solid State and Materials Science*, *15*, 39–43. https://doi.org/10.1016/j.cossms.2010.09.004

Kannan, K., & Tari, V. (Eds.). (2025). *Novel energy storage and conversion technologies for two-dimensional MXenes and MBenes*. https://www.amazon.com/dp/9798369395462

Kim, J., Shapiro, L., & Flynn, A. (2015). The clinical application of Mesenchymal stem cells and cardiac stem cells as a therapy for cardiovascular disease. *Pharmacology & Therapeutics*, *151*, 8–15. https://doi.org/10.1016/j.pharmthera.2015.02.003

Kumar, N., Singh, H., & Kumar, A. (2021). *Renewable energy and green technology* (1st ed., p. 11). CRC Press. https://doi.org/10.1201/9781003175926

Langer, F.M., Tzimas, E., Kaltschmitt, M., & Peteves, S. (2007). Techno-economic assessment of hydrogen production processes for the hydrogen economy for the short and medium term. *International Journal of Hydrogen Energy*, *32*, 3797–3810. https://doi.org/10.1016/j.ijhydene.2007.05.027

Lelieveld, J., Klingmüller, K., Pozzer, A., Burnett, R.T., Haines, A., & Ramanathan, V. (2019). Effects of fossil fuel and total anthropogenic emission removal on public health and climate. *Earth, Atmospheric & Planetary Sciences*, *116*, 7192–7197. https://doi.org/10.1073/pnas.1819989116

Lim, K.L., Kazemain, H., Yaakob, Z., & Daud, W.R.W. (2010). Solid-state materials and methods for hydrogen storage: A critical review. *Chemical Engineering & Technology*, *33*, 213–226. https://doi.org/10.1002/ceat.200900376

Mazloomi, K., & Gomes, C. (2012). Hydrogen as an energy carrier: Prospects and challenges. *Renewable and Sustainable Energy Reviews*, *16*, 3024–3033. https://doi.org/10.1016/j.rser.2012.02.028

Munir, A., Ahmad, A., Sadiq, M.T., Sarosh, A., Abbas, G., & Ali, A. (2021). Synthesis and characterization of carbon-based composites for hydrogen storage application. *Engineering Proceedings*, *12*(1), 52. https://doi.org/10.3390/engproc2021012052

Naeem, A., Ting, V.P., Hintermair, U., Tian, M., Telford, R., Halim, S., Nowell, H., Hołyńska, M., Teat, S.J., Scowen, I.J., & Nayak, S. (2016). Mixed-linker approach in designing porous zirconium-based metal-organic frameworks with high hydrogen storage capacity. *Chemical Communications*, *52*(50), 7826–7829. https://doi.org/10.1039/c6cc03787a

Nair, A.A.S., & Sundara, R. (2016). Palladium cobalt alloy catalyst nanoparticles facilitated enhanced hydrogen storage performance of graphitic carbon nitride. *The Journal of Physical Chemistry C*, *120*, 9612–9618. https://doi.org/10.1021/acs.jpcc.6b01850

Niaz, S., Manzoor, T., & Pandith, A.H. (2015). Hydrogen storage materials, methods and perspectives. *Renewable and Sustainable Energy Reviews*, *50*, 457–469. https://doi.org/10.1016/j.rser.2015.05.011

Ouyang, L.Z., Qin, F.X., & Zhu, M. (2006). The hydrogen storage behavior of $Mg_3La$ and $Mg_3LaNi_{0.1}$. *Scripta Materialia*, *55*(12), 1075–1078. https://doi.org/10.1016/j.scriptamat.2006.08.052

Prabhukhot, P.R., Wagh, M.M., & Gangal, A.C. (2016). A review on solid state hydrogen storage material. *Advances in Energy and Power*, *4*(2), 11–22. https://doi.org/10.13189/aep.2016.040202

Pudukudy, M., Yaakob, Z., Mohammad, M., Narayanan, B., & Sopian, K. (2014). Renewable hydrogen economy in Asia-opportunities and challenges: An overview. *Renewable and Sustainable Energy Reviews*, *30*, 743–757. https://doi.org/10.1016/j.rser.2013.11.015

Rauch, S., Polak-Kraśna, K., Tian, M., Holyfield, L.T., Mays, T.J., Bowen, C.R., & Burrows, A.D. (2017). Hydrogen storage in polymer-based processable microporous composites. *Journal of Materials Chemistry A*, *5*, 18752–18761. https://doi.org/10.1039/C7TA05232D

Reilly Jr, J.J., & Wiswall Jr, R.H. (1968). Reaction of hydrogen with alloys of magnesium and nickel and the formation of $Mg_2NiH_4$. *Inorganic Chemistry*, *7*(11), 2254–2256. https://doi.org/10.1021/ic50069a016

Ren, J., Musyoka, N.M., Langmi, H.W., Mathe, M., & Liao, S. (2017). Current research trends and perspectives on materials-based hydrogen storage solutions: A critical review. *International Journal of Hydrogen Energy*, *42*, 289–311. https://doi.org/10.1016/j.ijhydene.2016.11.195

Rochat, S., Polak-Kraśna, K., Tian, M., Holyfield, L.T., Mays, T.J., Bowen, C.R., & Burrows, A.D. (2017). Hydrogen storage in polymer-based processable microporous composites. *Journal of Materials Chemistry A*, *5*, 18752–18761.

Rosen, M.A., & Fayegh, S.K. (2016). The prospects for hydrogen as an energy carrier: An overview of hydrogen energy and hydrogen energy systems. *Energy, Ecology & Environment*, *1*, 10–29. https://doi.org/10.1007/s40974-016-0005-z

Rusman, N.A.A., & Dahari, M. (2016). A review on the current progress of metal hydrides material for solid-state hydrogen storage applications. *International Journal of Hydrogen Energy*, *41*, 12108–12126. https://doi.org/10.1016/j.ijhydene.2016.05.244

Sadhasivam, T., Kim, H.T., Jung, S., Roh, S.H., Park, J.H., & Jung, H.Y. (2017). Dimensional effects of nanostructured $Mg/MgH_2$ for hydrogen storage applications: A review. *Renewable and Sustainable Energy Reviews*, *72*, 523–534. https://doi.org/10.1016/j.rser.2017.01.107

Sapountzi, F.M., Gracia, J.M., Weststrate, C.J., Fredriksson, H.O.A., & Niemantsverdriet, J.W. (2017). Electrocatalysts for the generation of hydrogen, oxygen and synthesis gas. *Progress in Energy and Combustion Science*, *58*, 1–35. https://doi.org/10.1016/j.pecs.2016.09.001

Schlapbach, L. (2002). Hydrogen as a fuel and its storage for mobility and transport. *MRS Bulletin*, *27*, 675–679. https://doi.org/10.1557/mrs2002.220

Schlapbach, L., & Zuttel, A. (2001). Hydrogen storage materials for mobile applications. *Nature*, *414*, 353–358. https://doi.org/10.1038/35104634

Schneemann, A., White, J.L., Kang, S., Jeong, S., Wan, L.F., Cho, E.S., Heo, T.W., Prendergast, D., Urban, J.J., Wood, B.C., & Allendorf, M.D. (2018). Nanostructured metal hydrides for hydrogen storage. *Chemical Reviews*, *118*(22), 10775–10839. https://doi.org/10.1021/acs.chemrev.8b00313

Shahi, R.R., Tiwari, A.P., Shaz, M.A., & Srivastava, O.N. (2013). Studies on de/rehydrogenation characteristics of nanocrystalline $MgH_2$ co-catalyzed with Ti, Fe and Ni. *International Journal of Hydrogen Energy*, *38*, 2778–2784. https://doi.org/10.1016/j.ijhydene.2012.11.073

Shao, H., Xin, G., Zheng, J., Li, X., & Akiba, E. (2012). Nanotechnology in Mg-based materials for hydrogen storage. *Nano Energy*, *1*, 590–601. https://doi.org/10.1016/j.nanoen.2012.05.005

Sharma, A. (2020). Investigation on platinum loaded multi-walled carbon nanotubes for hydrogen storage applications. *International Journal of Hydrogen Energy*, *45*, 2967–2974. https://doi.org/10.1016/j.ijhydene.2019.11.093

Siddiqui, N., Faiyaz, T., & Tari, V. (2022). Ganga and Yamuna: Through the lens of National Green Tribunal. *Nature, Environment and Pollution Technology*, *21*(4), 1657–1664. https://doi.org/10.46488/NEPT.2022.v21i04.018

Srivastava, S., Shet, S.P., Priya, S.S., Sudhakar, K., & Tahir, M. (2022). Molecular simulation of copper-based metal-organic framework (Cu-MOF) for hydrogen adsorption. *International Journal of Hydrogen Energy*, *47*, 15820–15831. https://doi.org/10.1016/j.ijhydene.2022.03.089

Tari, V. (2015). Earth, man, and environment. In Pandurang Y. Patil (Ed.), *Definition principles and scope of environment* (pp. 1–20). DPS Publishing House.

Tari, V., Gupta, R., & Siddiqui, N. (2021). Impact of climate change on Upper Ganga Ramsar Site, UP, India. In *Monitoring and evaluating the ecological health of wetlands* (1st ed., pp. 92–105). IGI Global. https://doi.org/10.4018/978-1-7998-9498-8.ch006

Tari, V., Gupta, V., Joshi, A., Kurve, P., & Chandra, N. (2024a). Biodiversity loss due to forest diversion. In A. Rathore (Ed.), *Impact of societal development and infrastructure on biodiversity decline* (pp. 137–165). IGI Global. https://doi.org/10.4018/979-8-3693-6950-0.ch010

Tari, V., Siddiqui, N., Rathi, D., Siddiqui, N.N., & Wahyuni, D.K. (2024b). Need for an evolved groundwater justice in rural area Uttar Pradesh, India. *Nature, Environment and Pollution Technology*, *23*(1), 125–137. https://doi.org/10.46488/NEPT.2024.v23i01.009

Trahey, L., Brushett, F.R., Balsara, N.P., Ceder, G., Cheng, L., Chiang, Y.M., Hahn, N.T., Ingram, B.J., Minteer, S.D., Moore, J.S., & Mueller, K.T. (2020). Energy storage emerging: A perspective from the Joint Center for Energy Storage Research. *Proceedings of the National Academy of Sciences*, *117*(23), 12550–12557. https://doi.org/10.1073/pnas.1821672117

Vajo, J.J. (2011). Influence of nano-confinement on the thermodynamics and dehydrogenation kinetics of metal hydrides. *Current Opinion in Solid State & Materials Science*, *15*, 52–61. https://doi.org/10.1016/j.cossms.2010.11.001

Valenti, G. (2016). Hydrogen liquefaction and liquid hydrogen storage. In *Compendium of Hydrogen Energy* (pp. 27–51). Woodhead Publishing. https://doi.org/10.1016/B978-1-78242-362-1.00002-X

Viditha, V., Srilatha, K., & Himabindu, V. (2016). Hydrogen storage studies on Palladium-doped carbon materials (AC, CB, CNMs) @ metal–organic framework-5. *Environmental Science and Pollution Research*, *23*, 9355–9363. https://doi.org/10.1007/s11356-015-5194-z

Wani, S.A., Tari, V., & Mansoor, S.I.U. (2025). A comprehensive analysis of wastewater management challenges in India: Infrastructure and policy rerspectives. In Dhiman, S., & Mukherjee, G. (Eds.), *Resource Resurgence*. Springer Water. Springer, Cham. https://doi.org/10.1007/978-3-032-02750-4_11

Wei, L., & Mao, Y. (2016). Enhanced hydrogen storage performance of reduced graphene Oxide hybrids with Nickel or its metallic mixtures based on Spillover mechanism. *International Journal of Hydrogen Energy*, *41*, 11692–11699. https://doi.org/10.1016/j.ijhydene.2016.04.030

Xu, Y., Zhou, Y., Li, Y., & Ding, Z. (2024). Research progress and application prospects of solid state hydrogen storage technology. *Molecules*, *29(8)*, 1767. https://doi.org/10.3390/molecules29081767

York, R. (2012). Do alternative energy sources displace fossil fuels? *Nature Climate Change*, 2(6), 441–443. https://doi.org/10.1038/nclimate1451

Zhang, C., Kong, R., Wang, X., Xu, Y., Wang, F., Ren, W., Wang, Y., Su, F., & Jiang, J.X. (2017). Porous carbons derived from hypercrosslinked porous polymers for gas adsorption and energy storage. *Carbon*, *114*, 608–618. https://doi.org/10.1016/j.carbon.2016.12.064

Zhang, X., Lin, R.B., Wang, J., Wang, B., Liang, B., Yildirim, T., Zhang, J., Zhou, W., & Chen, B. (2020). Optimization of the pore structures of MOFs for record high hydrogen volumetric working capacity. *Advanced Materials*, *32*, 1907995. https://doi.org/10.1002/adma.201907995

Zhao, T., Ji, X., Jin, W., Yang, W., & Li, T. (2017). Hydrogen storage capacity of single-walled carbon nanotube prepared by a modified arc discharge. *Fullerenes, Nanotubes and Carbon Nanostructures*, *25*, 355–358. https://doi.org/10.1080/1536383X.2017.1305358

Zhou, C., & Szpunar, J.A. (2016). Hydrogen storage performance in Pd/Graphene nanocomposites. *ACS Applied Materials & Interfaces*, *8*, 25933–25940. https://doi.org/10.1021/acsami.6b07122

Zhou, H.C., Long, J.R., & Yaghi, O.M. (2012). Introduction to metal-organic frameworks. *Chemical Reviews*, *112*, 673–674. https://doi.org/10.1021/cr300014x

Zuttel, A. (2003). Materials for hydrogen storage. *Materials Today*, *6*, 24–33. https://doi.org/10.1016/S1369-7021(03)00922-2

# Index

Pages in *italics* refer to figures and pages in **bold** refer to tables.

For Product Safety Concerns and Information please contact our EU representative GPSR@taylorandfrancis.com
Taylor & Francis Verlag GmbH, Kaufingerstraße 24, 80331 München, Germany

www.ingramcontent.com/pod-product-compliance
Ingram Content Group UK Ltd.
Pitfield, Milton Keynes, MK11 3LW, UK
UKHW020848191026
476747UK00003B/40
*9781032954110*